METAL CATALYSED REACTIONS IN IONIC LIQUIDS

Catalysis by Metal Complexes

Volume 29

The titles published in this series are listed at the end of this volume.

METAL CATALYSED REACTIONS IN IONIC LIQUIDS

by

PAUL J. DYSON
Ecole polytechnique Fédérale de Lausanne (EPFL), Switzerland

and

TILMANN J. GELDBACH
Ecole polytechnique Fédérale de Lausanne (EPFL), Switzerland

Springer

A C.I.P. Catalogue record for this book is available from the Library of Congress.

ISBN-10 1-4020-3914-X (HB)
ISBN-13 978-1-4020-3914-0 (HB)
ISBN-10 1-4020-3915-8 (e-book)
ISBN-13 978-1-4020-3915-7 (e-book)

Published by Springer,
P.O. Box 17, 3300 AA Dordrecht, The Netherlands.

www.springeronline.com

Printed on acid-free paper

Printed in the Netherlands.

FOREWORD

Sceptics were predicting that the field of ionic liquids would slow down or even reverse – but such a situation could not be further from the truth. In fact, in the last year or so, more and more researchers and technologists in increasingly diverse areas have become interested in ionic liquids. While this is great for the subject, it presents somewhat of a nightmare to anyone foolish enough to contemplate writing a book on the subject, even a book restricted to metal catalysed reactions facilitated by ionic liquids.

At the time we started writing on this book, we were still optimistic that we could give a comprehensive overview on metal-catalysed reactions in ionic liquids. Yet, a literature search with the term "ionic liquid" gives some 3,100 references up to the end of 2004, of which more than 1,000 stem from 2004 alone. Thus, at some point we were forced to concede defeat and it is likely that some important contributions have been overlooked. This is probably particularly true for those papers, which are less easily accessible and for which no translated versions exist. We tried to include papers that appeared during the writing for as long as possible, but it could be that some of the most recent publications do not get the detailed description they should deserve.

Nevertheless, we hope to have succeeded in providing a useful overview of what has been done in the field of metal catalysed reactions in ionic liquids, covering most of the literature until early 2005. Not all reactions described on the following pages are strictly within the theme of this series – "Catalysis with Metal Complexes" – as reactions catalysed by simple metal salts are also described.

There are now several nomenclatures for ionic liquids in use and we hope that the system we settled on is sufficiently easy to comprehend.

Finally, we would like to thank those to help us spotting the most obvious errors in various chapters, notably Dr. Andre Ohlin, Dr. Zhaofu Fei, Adrian Chaplin and Prof. Paul Pregosin.

Lausanne, May 2005

TABLE OF CONTENTS

4 Hydroformylation

5 Oxidation Reactions

6 Carbon-Carbon Coupling Reactions

Chapter 1

INTRODUCTION

Changing the way chemists think and work

1.1 A Brief History of Biphasic Catalysis

Like the first paper dealing with ionic liquids – their history will be described later on – the origin of the concept of biphasic catalysis is not entirely clear. In the book edited by Cornils and Herrmann entitled 'Aqueous-Phase Organometallic Catalysis', Cornils and Kuntz credit the field to Manassen at the Weizmann Institute in Israel.[1] Due to the popularity of this book with those working in the field, many authors have subsequently cited the seminal work of Manassen, published in 1973, as representing the first paper on biphasic catalysis.[2] Yet, at the same time, Joó at the University of Debrecen in Hungary was also working on the concept of aqueous-organic biphasic catalysis.[3] In a recent publication, Cornils (one of the pioneers of the field), acknowledges the seminal contribution by Joó: "He [Joó] has always been recognised as the earliest pioneer in this field, but I was not previously aware that his first publications on the subject in 1973 (which were somewhat hidden away in Hungarian journals) appeared at the same time as those of Manassen's group, and that therefore the two developed the ideas independently."[4]

However, in 1972, one year before Manassen's and Joó's first publications, Parshall working at Dupont in Delaware reported that the hydrogenation and alkoxycarbonylation of alkenes could be catalysed by $PtCl_2$ when dissolved in tetraalkylammonium chloride/tin dichloride at temperatures of less than 100°C.[5] It was even noted that the product could be separated by decantation or distillation. Parshall recognised the separation problem associated with homogeneous catalysis and in this paper states, 'An approach that seems under utilised is the use of molten salts as stable, non-volatile solvents from which organic products are readily separated by distillation … or decantation'. An even earlier report by a group working for the U.S.

Industrial Chemicals Co. on biphasic catalysis can be found in a 1962 edition of the *Journal of the American Chemical Society* describing hydrogenations with $[Co(CN)_5H]^{3-}$ under aqueous-organic conditions.[6] Although we are not yet aware of any publication prior to that, it is highly likely that biphasic catalysis has a longer history than thought until recently. Yet, it was the systematic studies of Joó and Manassen that really laid the foundations to the field.

On an industrial scale, biphasic catalysis was first employed in the Shell Higher-Olefin Process (SHOP), which involves the use of two immiscible organic solvents, one containing the catalyst and substrates, the other being the formed product, which is immiscible with the reaction phase. The process consists of the nickel-catalysed oligomerisation of ethylene to give the majority of the worlds linear α-olefins, which are used in various applications, most importantly as detergents.[7] Subsequently, it did not take long for the first aqueous-organic biphasic processes to be implemented on an industrial scale (see below).

In the past decades, a number of alternatives to common organic solvents and water for use in biphasic catalysis have emerged. These include fluorous solvents, supercritical fluids and ionic liquids. The concept of fluorous biphasic catalysis originated from the work of Horváth and Rábai who were searching for a novel approach to selectively oxidise methane to methanol, and noticed that the solubility of methanol in fluorous solvents is low while the solubility of oxygen is extremely high.[8] When they realised that the miscibility of perfluoroalkanes, perfluorodialkyl ethers and perfluorotrialkylamines is low with many common organic solvents they demonstrated the utility of the technique, and elegantly showed that at low temperature the fluorous and organic solvents formed two phases, which on warming gave a single phase.[8,9] By careful design the reaction can be carried out homogeneously with the added advantage of a biphasic separation of catalyst and products after the reaction. This technique is particularly applicable to those reactions where non-polar compounds are converted to products with a higher polarity since the solubility of compounds in the fluorous phase decreases with increasing polarity. Apart from oxidation, many different reactions have been studied in fluorous phases including hydroformylation,[10] hydrogenation,[11] Diels-Alder,[12] C-C cross coupling reactions [13] and polymerisations,[14] to name but a few.

Supercritical fluids, another class of alternative reaction medium, have been used as solvents in catalysis for many years. Ipatiev reported the isomerisation of cyclohexane to methylcyclopentane under supercritical conditions using Al_2O_3 as the catalyst[15] and shortly afterwards communicated the oligomerisation of supercritical ethylene catalysed by $AlCl_3$.[16] However, it took another 60 years before supercritical fluids were used as solvents for homogeneous catalysts, which was reported by Kramer and Leder working at Exxon. In their patent, the isomerisation of hexane to methylpentane or dimethylbutane in a supercritical hexane/CO_2 mixture employing $AlBr_3$ as the catalyst is described ($AlCl_3$ is only poorly soluble under the conditions used).[17] To date, many homogeneously and heterogeneously catalysed reactions in supercritical fluids are known, and those employing gaseous substrates, such as hydrogenation and oxidation reactions, are particularly well characterised.[18] The use of supercritical CO_2 in industrial processes such as decaffeination of coffee[19] and polymerisation of ethylene[20] has been known for many years. Recently the first continuous flow supercritical fluid plant, with a capacity of up to 1,000 tons per year, has been developed by Thomas Swan & Co with the intention of conducting small-scale custom synthesis of high value products.[21]

Ionic liquids continue to receive increasing interest as alternative solvents for catalysis and the historical developments in the field are described in Section 1.4.

1.2 The Importance of Biphasic Catalysis

In a relatively short time biphasic catalysis, and notably aqueous-organic biphasic catalysis, has had an enormous impact on the way many chemists work. The study of alternative solvents is clearly of considerable academic interest, but it is perhaps the demand for cleaner chemical processes that has led to the explosion of research activity in this area. A major challenge for chemists today, when developing new products or more economic routes to existing products, is to do so with a consideration of minimising the impact that the synthesis of the product may have on the environment. Critical issues include the amount of materials and energy used in manufacture, the dispersion of chemicals in the environment, the use of resources and the durability and recyclability of the products. There are numerous approaches by which the above criteria can be achieved. These include the development of new synthetic pathways that require fewer steps and/or are more selective, alternative reaction conditions and improved solvents or methods for higher selectivity and energy minimisation. One of the most effective methods that influences all of the above criteria is to reduce or remove volatile organic compounds, i.e. solvents. Solvents are among the most ubiquitous classes of chemicals throughout society and the chemical industry because their applications are so broad and varied. In particular, solvents are used on a huge scale in manufacturing as a reaction medium for chemical transformations.

An ideal system would not require any solvent at all. However, solvents are extremely helpful for a number of reasons. They allow reactants to come together in a homogeneous fashion improving selectivity and yields. They also have an important role as a heat transfer medium, dissipating heat liberated by an exothermic reaction or supplying the thermal energy required to initiate an endothermic one. As such, elimination of solvents from many manufacturing processes is highly unlikely, especially for processes involving reactants and products that are solids under ambient conditions, and biphasic catalysis represents one of the best ways of improving the environmental efficiency of a chemical process. Apart from reducing the amount of volatile organic solvents, there are numerous other benefits obtained by using biphasic catalytic processes, of which preventing the loss of the catalyst, which tends to be both expensive and potentially toxic, is the most obvious.

The effectiveness of biphasic catalysis can be illustrated with reference to the synthesis of aldehydes *via* hydroformylation of alkenes in an aqueous-organic process pioneered by Kuntz. While the majority of aldehydes prepared from alkene hydroformylation were done so in organic solvents, in 1975 research showed that rhodium complexes bearing sulfonated phosphine ligands can be immobilised in water and are able to catalyse the hydroformylation of propene with virtually complete retention of rhodium in the aqueous phase.[22] Catalyst loss is always a major problem in the production of bulk chemicals of this nature and accordingly, Rhône-Poulenc filed a series of patent applications in order to protect the discovery. The hydroformylation reaction showed considerable promise and after a few years of development, Rhône-Poulenc joined forces with Ruhrchemie to develop a continuous biphasic hydroformylation process.[23] While many benefits were envisaged, virtually everything to do with the technology was unknown. But in less than two years a completely new process that had no precedent was developed, tested and, using a scale-up factor of 1:24,000, went into operation. The beauty of the process is that the substrates, propene, H_2 and CO, are gases at room temperature and dissolve in the water, whereas the product, which contains over 96% *n*-butyraldehyde, is a liquid under the process conditions and forms a second phase which is easily removed and is virtually free from rhodium contamination. During the production of the first 2,000,000 tonnes of *n*-butyraldehyde only 2 kg of rhodium were lost, which in the parts-per-billion range.

Ever since the plant went online, a great deal of fundamental research has been conducted, the process has been scaled-up further and new plants have come into operation. The economic and environmental benefits of the Ruhrchemie-Rhône-Poulenc process have been closely scrutinised. Overall the cost of the production is reduced, but it is the benefits to the environment that are the greatest.[24] These include the replacement of toxic organic solvents by water, the conservation of energy resources as the process operates under milder conditions compared to conventional methods, a higher selectivity towards the desired *n*-butyraldehyde product and the low loss of the toxic rhodium catalyst. Overall, the efficiency is extremely impressive and the environmental emissions are almost zero. Rhône-Poulenc is now also using aqueous-organic biphasic processes to conduct C-C coupling reactions for the small-scale production of vitamin precursors.[25]

1.3 The Alternatives to Aqueous-Organic Biphasic Processes

Not surprisingly, the most well developed biphasic system is that using water and organic solvents, despite the first industrial biphasic process involving only organic solvents. Obviously, water is the solvent of choice as it is abundant, cheap, non-flammable, non-toxic and has many other desirable properties such as being polar (and therefore relatively easy to separate from apolar compounds), high thermal conductivity, heat capacity and heat of evaporation. Nevertheless, alternative solvents to water for applications in biphasic catalysis are needed for several reasons:

- Many catalysts do not dissolve in water and thus their ligands must be modified with hydrophilic groups that can be expensive, especially for chiral ligands. Furthermore, some catalysts are unstable in water and decompose.
- The solubility of organic compounds in water is often poor. While this may be an advantage for product separation, lower reaction rates and yields are frequently observed. This problem may be overcome by increasing the stirrer rate or the temperature of the aqueous-organic biphasic system. Other protocols have also been developed to overcome solubility problems including the addition of phase transfer reagents, co-solvents, detergents and surfactants as well as the use of ultrasound and microwave dielectric heating.
- The high specific heat capacity of water means that distillation requires high energy, and reaction mixtures are difficult to heat or cool rapidly.[26]
- Other, more general problems faced when using water in catalysis include, for example, viscosity and pH effects.

Reaction media such as ionic liquids, fluorous solvents and supercritical fluids may offer a solution in avoiding some of the above mentioned problems. In addition to the physicochemical properties of solvents that are crucial when selecting a solvent for a particular task, within the context of producing cleaner chemical processes, other criteria are also important.

These include toxicity, flammability, explosivity, stratospheric ozone depletion, atmospheric ozone production and global warming potential.[27] Bearing all these criteria in mind, the alternative solvents mentioned above provide an excellent range of properties with considerable potential.

Various combinations of solvents that lead to biphasic and multiphasic systems have been evaluated in catalysis. For practical purposes, it is essential that the catalyst and product phase effectively separate after the reaction, and in this respect the most important solvent combinations used in multiphasic chemistry include aqueous-organic and fluorous-organic (see above), ionic liquid-organic,[28] ionic liquid-aqueous[29] and ionic liquid-supercritical (or dense phase) CO_2 (the CO_2 does not necessarily have to be supercritical).[30] Other biphasic systems have also been demonstrated such as ionic liquids incorporating fluorous groups which exhibit high miscibility with apolar compounds such as alkenes and resemble fluorous solvents in their phase behaviour with organic solvents[31] and phase vanishing systems.[32] Here an organic solvent containing one reagent with the same organic solvent containing another reagent is separated by a different solvent phase of higher density. The organic phase at the bottom diffuses through the central phase into the organic phase at the top. The reagent also slowly diffuses and reacts in a controlled manner with the substrate in the upper phase and only two phases remain once the reaction is complete. Triphasic processes employing three immiscible solvents have also been reported, for example, fluorous-organic-aqueous[33] and ionic liquid-water-organic systems.[34] Fluorous catalysis has even been reported without the need of a fluorous solvent[35] and ionic liquids may even be heterogenised on fixed-beds giving what has been termed supported ionic liquid catalysis.[36] In principle, however, there is no reason why not even more than three phases could be employed in a multiphasic process. One could envisage different phases supporting different catalysts, which transform a substrate to a product in a number of steps in a sequential fashion.

1.4 Key Developments in the Chemistry of Ionic Liquids

Ionic liquids are sometimes, especially in the older literature, also referred to as *molten salts*, *non-aqueous ionic liquids* or *room temperature molten salts*. While all of these names are entirely valid, their meaning has somewhat changed over the years. The term molten salt is now used less frequently in the field of ionic liquids and generally refers to salts with melting points greater than 100°C. The expression non-aqueous ionic liquid was originally coined to differentiate synthetic ionic liquids from water, since liquid water is partially composed of ions, although the dissociation constant for pure water is 10^{-14}. The last term, room temperature molten salts, is sometimes used to differentiate those that are truly low melting from molten salts with high melting points. Apart from ionic liquids, molten salts represent a huge field of study in their own right, based on inorganic salts and salt mixtures.[37] Applications for these range from organic waste processing by high temperature oxidation, through electrodeposition and separation, to the preparation of superconductors.

We intend to give an outline of the key historic events in the field based on the generally accepted definition of an ionic liquid. However, it is worth noting in advance that many ionic liquids have been around for a long time, but have just not been recognised as such. Davies describes such ionic liquids as 'incognito ionic liquids', and comments that a survey of the chemical catalogues reveals over 100 commercially available salts that melt below 150°C, which implies that many new ionic liquids could be easily generated by matching a desired cation from such a salt with a different anion, or vice versa.[38] Many phase transfer reagents that have been used for years in organic synthesis and catalysis are actually low melting salts and recently these types of compounds have been recognised as being ionic liquids and exploited as such.[39]

The work most often cited as describing the first ionic compound which is liquid at room temperature is that by Walden, published in 1914. The low melting salt, $[EtNH_3][NO_3]$, was prepared from the reaction of ethylamine with concentrated nitric acid,[40] and has a melting point of 12-14°C. However, in a review by Wilkes entitled, 'A short history of ionic liquids – from molten salts to neoteric solvents', he notes that some materials which would now be classed as ionic liquids were observed as far back as the mid 19th century.[41] These early studies on liquid salts did not lead to an explosion of interest in ionic liquids and it was not before the late 1940's that the next ionic liquids were discovered by Hurley and Wier at the Rice Institute in Texas. While looking for an inexpensive and facile method for aluminium electroplating they noted that by mixing powdered alkylpyridinium chlorides with $AlCl_3$ a reaction took place resulting in the formation of a liquid.[42] These ionic liquids incorporate organic cations, i.e. the type of cations used in the ionic liquids that now form the basis of modern synthetic applications, and chloroaluminate anions. While such anions are still being used in synthesis and catalysis, they have become less popular than other more inert anions. This is mainly due to their sensitivity towards air and moisture and the fact that extraction of certain organic products may result in the destruction of these particular ionic liquids. Studies involving low melting inorganic salts, typically eutectic mixtures based on $AlCl_3$ and NaCl, as solvents for organic reactions were also reported from time to time,[43] but room temperature ionic liquids were still not being used as solvents for synthesis.

Osteryoung, Wilkes, Hussey and Zaworotko working on electrochemical aspects of the chloroaluminates at the U.S. Air Force Academy were largely responsible for bringing ionic liquids to the attention of a wider scientific community.[44] They were studying chloroaluminates as solvents for transition metal complexes[45] and as reaction media for stoichiometric organic synthesis.[46] Chauvin and Osteryoung independently combined these two features, i.e. that ionic liquids could dissolve transition metal complexes and support organic chemistry. Chauvin showed that nickel complexes dissolved in acidic chloroaluminate ionic liquids represent an excellent system for the dimerisation of alkenes[47] while Osteryoung used Ziegler-Natta catalysts in acidic chloroaluminates to polymerise ethylene.[48]

It was Zaworotko who made the next leap forward, this being the synthesis of water-stable ionic liquids that contain tetrafluoroborate, hexafluorophosphate, nitrate, sulfate and acetate anions.[49] Wilkes comments in his introduction to the book *Ionic Liquids in Synthesis* that, 'This was such an obvious and useful idea that we marvelled that neither we nor others had tried to do it already.'[50] Although many years before an isolated example of an air and moisture stable room temperature ionic liquid, tetra-n-hexylammonium benzoate, was reported,[51] it is probably Zaworotko's discovery that can be cited as directly leading to the huge growth in the field. We shall not describe here all the breakthroughs that came in the application and development of ionic liquids as solvents for catalysis, as these will become apparent in the subsequent chapters. However, one person who stands out as having made a considerable contribution to the field, looking at both the fundamental properties and applications of ionic liquids is Seddon at the University of Belfast. He is perhaps the person who has done most to popularise ionic liquids resulting in such intensive research activity around the world. In addition, Grätzel and co-workers at the EPFL showed the scientific community how to make highly pure ionic liquids,[52] their interest being in the application of ionic liquids as electrolytes in solar cells.[53] Ionic liquids have since been utilised as in separation processes,[54] as extractants for heavy metals with potential applications in the nuclear processing industry,[55] as lubricants,[56] as matrices in MALDI mass spectrometry[57] and even as propellants for small satellites.[58] The first industrial process using ionic liquid technology in chemical synthesis has also been reported[59] and numerous others are expected to follow.

1.5 General Properties of Ionic Liquids that Lend Themselves to Catalysis

It is worth considering here the general properties ionic liquids possess which make them so attractive as solvents for conducting organic synthesis, and notably, transition metal catalysed reactions. There are already a large number of excellent reviews and edited volumes concerned (at least in part) with organic synthesis and catalysis in ionic liquids,[28,60-70] and from these collations, the following features have emerged:

- Ionic liquids do not evaporate as they have no detectable vapour pressure. Since reducing the emissions of volatile organic compounds is viewed as one of the most important ways of reducing pollution from the chemical industry (and other industries) it is perhaps this property of ionic liquids that makes them so attractive as potentially benign replacements to organic solvents. Volatile organics (reaction products) can thus be easily removed from the ionic liquid under vacuum, by distillation or using a carrier gas such as CO_2.
- Many metal catalysts (notably salts), organic compounds (especially polar compounds), gases and biocatalysts dissolve in ionic liquids, allowing homogeneously catalysed reactions to be performed.
- Ionic liquids are immiscible with many organic solvents and compounds, which lends themselves to biphasic or multiphasic catalytic reactions. Most are also immiscible with fluorous phases and some are immiscible with water. In the ideal biphasic process involving ionic liquids, a soluble polar substrate is converted to a less polar – and thus insoluble – product, which will then form a separate phase. Attaching fluorous groups to the ionic liquid cation can reverse the solubility properties.
- Ionic liquids have polarities comparable to lower alcohols, which are amongst the most widely used solvents in which to conduct homogeneously catalysed reactions. In contrast to lower alcohols, however, many ionic liquids are non-nucleophilic, which can have a pronounced effect on a catalysed reaction. For example, solvated catalytic intermediates are unlikely to form, and this feature could lead to different reaction mechanisms, and hence to different yields and selectivities. The non-nucleophilic environment presented by many ionic liquids is also less likely to deactivate a catalyst and can lead to increased turnover numbers, which is essential in biphasic processes where catalyst recycling and reuse is required.
- Ionic liquids have favourable thermal stabilities and operate over large temperature ranges. Many of the commonly used ionic liquids melt below room temperature and melting points as low as –80°C or below are not unusual. On the other hand they often only start to decompose above 300°C,

providing a large temperature range in which to conduct synthesis compared to molecular solvents.

• It is not always necessary to modify a catalyst for use in ionic liquids, and expensive ligands that modify the solubility properties of a catalyst are not necessarily required. For example, Wilkinson's catalyst and the highly active C-C coupling catalyst $Pd(PPh_3)_4$ have been used in ionic liquids, although quantification of catalyst loss was not reported, and this might be high. Many catalysts are salts and in general these are very well retained in ionic liquids, and where necessary, ligands that anchor a catalyst in an ionic liquid phase are available.

• Functional ionic liquids, often referred to as task-specific ionic liquids, can be prepared by incorporating functional groups on the organic cation, or using functional anions. These liquids have specific chemical properties and combined with a greater understanding of the factors controlling the physical properties of ionic liquids, designer ionic liquids are now being made. The ability to design an ionic liquid for a specific process is essential as there seems to be essentially no limit to the number of different ionic liquids that can be made.

• While very little toxicity data are available it would appear that many ionic liquids do not represent an immediate threat to health if handled properly. However, at present very little is known about their biodegradability and aquatic toxicity and it is envisaged that more data are needed prior to their use on a large scale. In addition, ionic liquids composed of fluoride containing anions such as $[BF_4]^-$ or $[PF_6]^-$ cannot be incinerated and therefore their use on a large scale might be problematic.

One of the chief concerns regarding the large-scale use of ionic liquids is their cost. However, ionic liquids can be made on a large scale – 100 tons is now perfectly feasible – with a cost around 10 € per litre, although some ionic liquids will remain considerable more expensive. Bearing in mind that the ionic liquid is probably going to be recycled and reused repeatedly this cost is relatively inexpensive.

Based on the need to design chemical products and processes that reduce or eliminate the use and generation of hazardous substances, combined with the properties outlined above, it is not difficult to understand why ionic liquids are under intensive study as solvents in which to conduct catalysed reactions. Academics, curious about the prospect of conducting chemistry in such a unique environment, together with those in the chemical industry who are keen to exploit their properties, have made considerable progress in a very short time, and in the remainder of this book we trace the nascent studies of reactions catalysed by metals in ionic liquids to a field that we predict will soon have an considerable impact on the production of chemical products.

1.6 References

[1] Cornils, B.; Kuntz, E. G. In *Aqueous-Phase Organometallic Catalysis, Concepts and Applications*, Cornils, B., Herrmann, W. A., Eds.; Wiley-VCH: Weinheim, 1998, p 271.
[2] Manassen, J. In *Catalysis: Progress in Research*, Bassolo, F., Burwell, R. L., Eds.; Plenum Press: London, 1973.
[3] Joó, F.; Beck, M. T. *Magy. Kém. Folyóirat* **1973**, *79*, 189.
[4] Cornils, B. *Angew. Chem. Int. Ed.* **2003**, *42*, 2704-2705.
[5] Parshall, G. W. *J. Am. Chem. Soc.* **1972**, *94*, 8716-8719.
[6] Kwiatek, J.; Mador, I. L.; Seyler, J. K. *J. Am. Chem. Soc.* **1962**, *84*, 304-305.
[7] Turner, A. H. *J. Am. Oil Chem. Soc.* **1983**, *60*, 202.
[8] Horváth, I. T. *Acc. Chem. Res.* **1998**, *31*, 641-650.
[9] Horváth, I. T.; Rábai, J. *Science* **1994**, *266*, 72-75.
[10] Horváth, I. T.; Kiss, G.; Cook, R. A.; Bond, J. E.; Stevens, P. A.; Rábai, J. *J. Am. Chem. Soc.*, **1998**, *120*, 3133-3134.
[11] Rutherford, D.; Juliette, J. J. J.; Rocaboy, C., Horváth I. T.; Gladysz, J. A. *Catal. Today* **1998**, *42*, 381-388.
[12] Myers, K. E.; Kumar, K. *J. Am. Chem. Soc.* **2000**, *122*, 12025-12026.
[13] (a) Moreno-Manas, M.; Pleixats R.; Villarroya, S. *Organometallics* **2001**, *20*, 4524-4528; (b) Schneider, S.; Bannwarth, W. *Angew. Chem. Int. Ed.* **2000**, *39*, 4142-4125; (c) Moineau, J.; Pozzi, G.; Quici, S.; Sinou, D. *Tetrahedron. Lett.* **1999**, *40*, 7683-7686; (d) Schneider, S.; Bannwarth, W. *Helv. Chim. Acta* **2001**, *84*, 735-742.
[14] Haddleton, D. M.; Jackson, S. G.; Bon, S. A. F. *J. Am. Chem. Soc.* **2000**, *122*, 1542-1543.
[15] Ipatiev, V.; Dovgelevich, N. *J. Russ. Phys. Chem. Soc.* **1911**, *43*, 1431.
[16] Ipatiev, V.; Rutala, O. *Ber. Dtsch. Chem. Ges.* **1913**, *46*, 1748.
[17] Kramer, G. M.; Leder, F. US Patent 3,880,945, **1975**.
[18] (a) Baiker, A. *Chem. Rev.* **1999**, *99*, 453-474: (b) Jessop, P. G.; Ikariya, T.; Noyori, R. *Chem. Rev.* **1999**, *99*, 475-493.
[19] Demirbas, A. *Energy Conserv. Manag.* **2001**, *42*, 279-294.
[20] Kendall, J. L.; Canelas, D. A.; Young, J. L.; DeSimone, J. M. *Chem. Rev.* **1999**, *99*, 543-563.
[21] (a) Institute of Applied Catalysis, *iAc News*, **2002**, *9*, 1; (b) Press release PA66/02, University of Nottingham, 11 July **2002**; (c) Gray, W. K.; Smail, F. R.; Hitzler, M. G.; Ross, S. K.; Poliakoff, M. *J. Am. Chem. Soc.* **1999**, *121*, 10711-10718.
[22] Kuntz, E. French Patent 2,314,910, **1975**.
[23] Cornils, B. *J. Mol. Catal. A* **1999**, *143*, 1-10.
[24] Cornils, B.; Wiebus E. In *Aqueous-Phase Organometallic Catalysis*, Cornils, B., Herrmann, W. A., Eds. Wiley-VCH: Weinheim, **1998**, pp 259-268.
[25] An interesting review describing the status and directions of industrial aqueous biphasic catalysis is available: Cornils, B. *Org. Proc. Res. Devel.* **1998**, *2*, 121-127.

[26] Lancaster, M. *Green Chemistry: An Introductory Text*, Royal Society of Chemistry: Cambridge, **2002**.
[27] Anastas, P. T. In *Clean Solvents, Alternative Media for Chemical Reactions and Processing*, ACS Symposium Series 819, **2002**, pp 1-10.
[28] For a review see: Dupont, J.; de Souza, R. F.; Suarez, P. A. Z. *Chem. Rev.* **2002**, *102*, 3667-3692.
[29] Dyson, P. J.; Ellis, D. J.; Welton, T. *Can. J. Chem.* **2001**, *79*, 705-708.
[30] Leitner, W. *Acc. Chem. Res.* **2002**, *35*, 746-756.
[31] van den Broeke, J.; Winter, F.; Deelman, B. J.; van Koten, G. *Org. Lett.* **2002**, *4*, 3851-3854.
[32] Ryu, I.; Matsubara, H.; Yasuda, S.; Nakamura, H.; Curran, D. P. *J. Am. Chem. Soc.* **2002**, *124*, 12946-12947.
[33] Quici, S.; Cavazzini, M.; Ceragioli, S.; Montanari, F.; Pozzi, G. *Tetrahedron Lett.* **1999**, *40*, 3647-3650.
[34] Carmichael, A. J.; Earle, M. J.; Holbrey, J. D.; McCormac, P. B.; Seddon, K. R. *Org. Lett.* **1999**, *1*, 997-1000.
[35] Wende, M.; Gladysz, J. A. *J. Am. Chem. Soc.* **2003**, *125*, 5861-5872.
[36] Mehnert, C. P.; Mozeleski, E. J.; Cook, R. A. *Chem. Commun.* **2002**, 3010-3011.
[37] For example see: *Advances in molten salts, from structural aspects to waste processing*, Gaune-Escard, M., Ed.; Begell House: New York, **1999**.
[38] Davis Jr., J. H.; Fox, P. A. *Chem. Commun.* **2003**, 1209-1210.
[39] Bradaric, C. J.; Downard, A.; Kennedy, C.; Robinson, A. J.; Zhou, Y. *Green Chem.* **2003**, *5*, 143-152.
[40] Walden, P. *Bull. Acad. Imper. Sci. (St. Petersburg)* **1914**, 1800.
[41] Wilkes, J. S. *Green Chem.* **2002**, *4*, 73-80.
[42] Hurley, F. H.; Wier, T. P. *J. Electrochem. Soc.*, **1951**, *98*, 207-212.
[43] Pagni, R. M. In: *Advances in Molten Salt Chemistry*, Mamantov, G., Mamantov, C. B., Braunstein, J., Eds.; Vol. 6, Elservier: New York, **1987**, pp 221-346.
[44] (a) Robinson, J.; R. A.; Osteryoung, *J. Am. Chem. Soc.* **1979**, *101*, 323-327. (b) Wilkes, J. S.; Levisky, J. A.; Wilson, R. A.; Hussey, C. L. *Inorg. Chem.* **1982**, *21*, 1263-1264.
[45] Appleby, D.; Hussey, C. L.; Seddon, K. R.; Turp, J. E. *Nature* **1986**, *323*, 614-616.
[46] Boon, J. A.; Levisky, J. A.; Pflug, J. L.; Wilkes, J. S. *J. Org. Chem.* **1986**, *51*, 480-483.
[47] Chauvin, Y.; Gilbert, B.; Guibard, I. *J. Chem. Soc., Chem. Commun.* **1990**, 1715-1716.
[48] Carlin, R. T.; Osteryoung, R. A. *J. Mol. Catal.* **1990**, *63*, 125-129.
[49] Wilkes, J. S.; Zaworotko, M. J. *J. Chem. Soc., Chem. Commun.* **1992**, 965-966.
[50] Wilkes, J. S. In *Ionic Liquids in Synthesis*, Wasserscheid, P., Welton, T., Eds.; Wiley-VCH: Weinheim, **2002**, pp 1-6.
[51] Swain, C. G.; Ohno, A.; Roe, D. K.; Brown, R.; Maugh II, T. *J. Am. Chem. Soc.* **1967**, *89*, 2648-2649.
[52] Bonhôte, P.; Dias, A.-P.; Papageorgiou, N.; Kalyanasundaram, K.; Grätzel, M. *Inorg. Chem.* **1996**, *35*, 1168-1178.
[53] (a) Wang, P.; Zakeeruddin, S. M.; Exnar, I.; Grätzel, M. *Chem. Commun.* **2002**, 2972-2973. (b) Wang, P.; Zakeeruddin, S. M.; Comte, P.; Exnar, I.; Grätzel, M. *J. Am. Chem. Soc.* **2003**, *125*, 1166-1167.
[54] (a) Visser, A. E.; Swatloski, R. P.; Reichert, W. M.; Mayton, R.; Sheff, R.; Wierzbicki, A.; Davis, J. H.; Rogers, R. D. *Environ. Sci. Technol.* **2002**, *36*, 2523-2529. (b) Bates, E. D.; Mayton, R. D.; Ntai, I.; Davis Jr. J. H. *J. Am. Chem. Soc.* **2002**, *124*, 926-927.
[55] Visser, A. E.; Swatloski, R. P.; Reichert, W. M.; Mayton, R.; Sheff, S.; Wierzbicki, A.; Davis, J. H.; Rogers, R. D. *Chem. Commun.* **2001**, 135-136.

[56] (a) Liu, W. M.; Chen, Y. X.; Yu, L. G. *Chem. Commun.* **2001**, 2244-2245. (b) Li, W. M.; Ye, C. F.; Gong, Q. Y.; Wang H. Z.; Wang, P. *Tribology Lett.* **2002**, *13*, 81-85.
[57] (a) Armstrong, D. W.; Zhang, L. K.; He L. F.; Gross, M. L. *Anal. Chem.* **2001**, *73*, 3679-3686. (b) Carda-Broch, S.; Berthod, A.; Armstrong, D. W. *Rapid Commun. Mass Spec.* **2003**, *17*, 553-560.
[58] Gamero-Castano, M.; Hruby, V. *J. Propuls. Power*, **2001**, *17*, 977-987.
[59] News of the week in *Chem. Eng. News* **2003** March 31, 9.
[60] Seddon, K. R. *J. Chem. Technol. Biotechnol.* **1997**, *68*, 351.
[61] (a) Welton, T. *Chem. Rev.* **1999**, *99*, 2071. (b) Welton, T. *Coord. Chem. Rev.* **2004**, *248*, 2459-2477.
[62] Olivier, H. *J. Mol. Catal. A* **1999**, *146*, 285-289.
[63] Wasserscheid, P.; Keim, W. *Angew. Chem. Int. Ed.* **2000**, *39*, 3773-3789.
[64] Gordon, C. M. *App. Catal. A* **2001**, *222*, 101-117.
[65] Sheldon, R. *Chem. Commun.* **2001**, 2399-2407.
[66] Zhao, D.; Wu, M.; Kou, Y.; Min, E. *Catal. Today* **2002**, *74*, 157-189.
[67] Olivier-Bourbigou, H.; Magna, L. *J. Mol. Catal. A* **2002**, *182-183*, 419-437.
[68] *Ionic Liquids, Industrial Applications for Green Chemistry*, Rogers, R. D., Seddon, K. R. Eds.; ACS Symposium Series 818, **2002**.
[69] *Ionic Liquids in Synthesis*, Wasserscheid, P., Welton, T., Eds.; Wiley-VCH: Weinheim, **2002**.
[70] Jain, N.; Kumar, A.; Chauhan, S.; Chauhan, S. M. S. *Tetrahedron*, **2005**, *61*, 1015-1060.

Chapter 2

IONIC LIQUIDS – PROPERTIES AND PREPARATION

2.1 Introduction

As described in the introductory chapter, biphasic catalysis has been around for a long time, but despite a few notable successes such as the Shell Higher Olefin Process (SHOP) and the Rhône-Poulenc-Ruhrchemie hydroformylation process, very few biphasic processes have made it into the industrial arena. The limitations of the solvents used so far in biphasic (or multiphasic) catalysis appear to be overcome by ionic liquids, and even if the perfect ionic liquid is not yet available, then there seems to be almost no limit to the number of new ionic liquids that can be made. It has been estimated that up to 10^{18} different ionic liquids may exist[1,2] and with such a vast number to choose from it is essential that understanding increases in order to allow accurate predictions of their properties and functions, opening up the possibility of designer solvents.

The fact that ionic liquids are solvents composed entirely of ions creates a new set of challenges to chemists and chemical engineers who are not used to handling such materials in the capacity of a solvent. At present, the most popular solvents are those which are volatile, since solvents are added to facilitate reactions, but ultimately they must be removed from the product. Accordingly, alternative techniques are required for the isolation of reaction products from ionic liquids, as well as for their disposal. Furthermore, if catalysed reactions are to be understood in ionic liquids, then new procedures may be required to delineate reaction mechanisms and identify catalytic intermediates. With respect to this latter goal, some simple solutions have already been devised. For example, by using labelled substrates as opposed to preparing deuterated ionic liquids, reactions have been monitored *in situ* by NMR spectroscopy.[3] Alternatively, various mass spectrometric methods can be used to analyse catalysts present in extremely low concentrations in ionic liquids.[4-6] And as ionic liquids do not evaporate, nanoparticles formed during a reaction can be studied directly by TEM, not necessitating separation of the particles from the solvent.[7]

2.2 Composition and Physical Properties of Ionic Liquids

The generation of low melting salts requires the use of organic cations with inorganic or organic anions. Many organic cations, when matched with an appropriate anion, give rise to ionic liquids and some of the principal cations are illustrated in Figure 2.1.

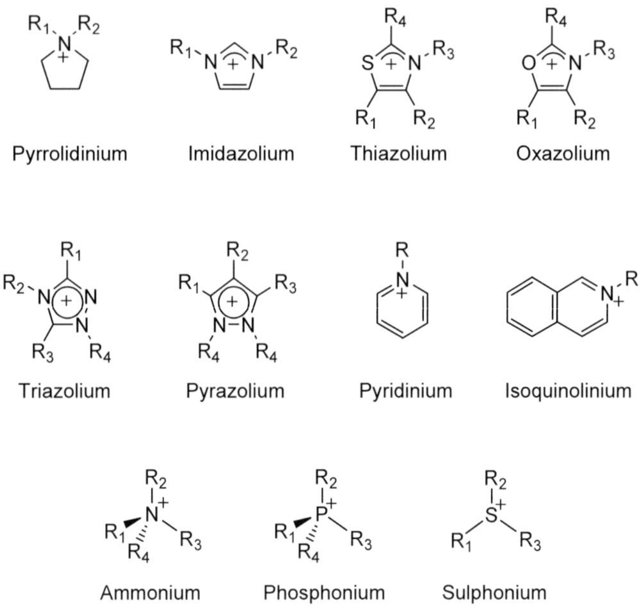

Figure 2.1: Examples of organic cations used to provide ionic liquids

The prediction of the melting point of a given anion-cation combination remains difficult and the very low melting points of some ionic liquids are not yet entirely understood. Nevertheless, some trends appear to have emerged. A fairly general prerequisite is that the charge should be delocalised over several atoms or buried inside a pocket so that it cannot be approached too closely. Additionally, melting points tend to be lower when the cation and/or anion is asymmetrical. The phenomenon of low melting salts is not restricted to singly charged species and even multiply charged cations can give rise to ionic liquids. For example, the phosphazene-based salt shown in Figure 2.2 has an overall charge of +6, albeit distributed over six pyridinium groups, and yet it is a liquid at room temperature.[8]

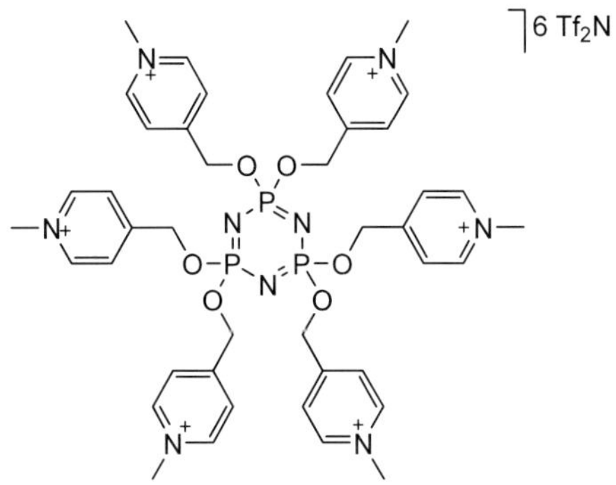

Figure 2.2: A phosphazene centred hexacationic salt that is liquid at room temperature

At present the most frequently encountered type of cation is based on the 1,3-disubstituted imidazolium moiety, but it is likely that other cations will gain importance in the future. The popularity of imidazolium-based ionic liquids is owed, at least in part, to the fact that simple (and minor) modifications lead to significant changes in their physical properties. For example, the melting points of the 1-alkyl-3-methylimidazolium salts can be tuned by varying the length of the alkyl substituent. The melting point first decreases on going from C_1 to $C_{6\text{-}8}$, but then increases again with longer alkyl groups until beyond a certain length a liquid crystalline phase is reached. This is illustrated with the example of some 1-alkyl-3-methylimidazolium tetrafluoroborate salts, which are listed in Table 2.1.

Table 2.1: Melting points of 1-alkyl-3-methylimidazolium tetrafluoroborate salts

Ionic liquid	Melting point [°C]	Ref
$[C_1C_1im][BF_4]$	10	[9]
$[C_2C_1im][BF_4]$	6	[10]
$[C_4C_1im][BF_4]$	–81	[11]
$[C_6C_1im][BF_4]$	–82	[10]
$[C_8C_1im][BF_4]$	–78	[10]
$[C_{10}C_1im][BF_4]$	–4	[10]

It is not only the melting point that changes with the alkyl chain length, but also a number of other physicochemical properties[12] including miscibility with other solvents. While, for example, $[C_1C_1im][BF_4]$ is completely miscible with water at room temperature $[C_8C_1im][BF_4]$ only becomes miscible at ca. 80°C. This property has been exploited in thermomorphic catalysis in which the reaction is conducted at temperatures exceeding 80°C under homogeneous conditions. After the reaction the product extraction is conducted at low temperature with the catalyst retained in the ionic liquid phase and the product in the aqueous phase.[13]

In addition to the structure of the cation, the type of anion present also has a pronounced influence on the physical properties of the ionic liquid. The melting points of some $[C_4C_1im]^+$ based ionic liquids with different anions are listed in Table 2.2. Also, the miscibility depends markedly on the anion present. For example, exchanging $[BF_4]^-$ with $[PF_6]^-$ in the 1-alkyl-3-methylimidazolium liquids leads to hydrophobic liquids that only dissolve very low concentrations of water. Thus, it can be relatively easy to modify the physical properties of an ionic liquid to match the requirements of a particular process.

Table 2.2: Melting points of 1-butyl-3-methylimidazolium salts

Ionic liquid	Melting point [°C]	Ref
$[C_4C_1im]Cl$	41	[11]
$[C_4C_1im][PF_6]$	–64	[11]
$[C_4C_1im][BF_4]$	–81	[11]
$[C_4C_1im][Tf_2N]$	–4	[14]
$[C_4C_1im][OTf]$	16	[14]

Chlorometallate ionic liquids are worth mentioning in this context as the ratio of cation to anion not only has a marked influence on the physical properties of the resulting solvent, but also on the chemical behaviour. These ionic liquids are made from the direct reaction of the solid [cation]Cl and metal chloride like $AlCl_3$, $SnCl_2$ or CuCl, which together collapse to form a liquid.[14,15] The identity of the anion present depends upon the stoichiometry (mole fraction) of the reactants used. When the mole fraction of [cation]Cl:$AlCl_3$ is 0.5 the $[AlCl_4]^-$ anion is essentially the only species present. If the mole fraction of $AlCl_3$ employed is greater than 0.5 then multinuclear species such as $[Al_2Cl_7]^-$ and $[Al_3Cl_{10}]^-$ are formed and the ionic liquid is Lewis acidic, whereas below 0.5 a (Lewis) basic ionic liquid is formed. For this reason chloroaluminate ionic liquids are often written as [cation]Cl-$AlCl_3$, rather than [cation]$AlCl_4$.

These features can be exploited in synthesis and catalysis. Acidic ionic liquids have proved to be particularly useful as they strip away halide ligands attached to catalysts thereby generating coordinatively unsaturated species that are more reactive towards the substrates. Ionic liquids based on chlorocuprates are prepared in a similar manner from the direct reaction of the [cation]Cl and CuCl.[16] Like the chloroaluminate ionic liquids they also contain a complex mixture of anionic species including $[CuCl_2]^-$, $[Cu_2Cl_3]^-$ and $[Cu_3Cl_4]^-$. The mole fraction between an ammonium halide and e.g. $AlCl_3$ not only affects the chemical properties but also exerts a strong effect on the melting point of the mixture, as shown in Figure 2.3 for the $[C_2C_1im]$-$AlCl_3$ system.[17]

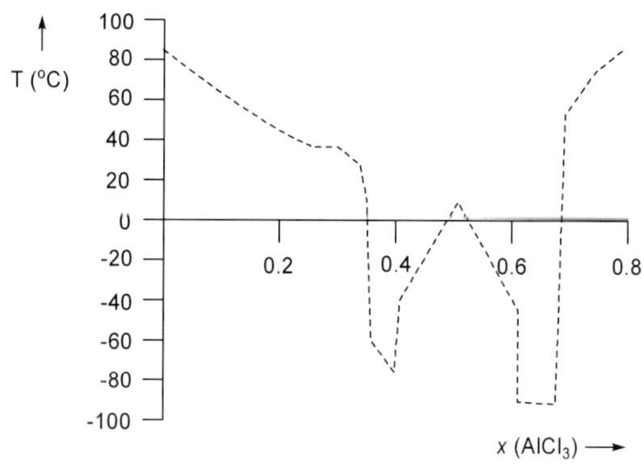

Figure 2.3: Phase diagram for $[C_2C_1im]$-$AlCl_3$

Although ionic liquids do not boil it is useful to know the temperature range under which they can operate. For many ionic liquids, the onset of decomposition takes place at temperatures exceeding 250°C and often much higher.[18] However, 1,3-dialkylimidazolium phosphates were found to slowly decompose below 200°C[19] and when the liquid is composed of less common cations or anions, the thermal stability may be considerably lower. Nevertheless, since the overwhelming majority of catalysed reactions take place at much lower temperatures, thermally induced solvent decomposition tends not to be a problem, although chemically induced decomposition can be a real concern! For example, thermal decomposition of $[C_4C_1im][BF_4]$ takes place a much lower temperature if nucleophiles are present.[20]

From the perspective of a chemical engineer, who eventually will scale up any ionic liquid based process developed in the laboratory, whether a solvent melts at –30°C or –80°C makes very little difference. The viscosity of the solvent, on the other hand, is much more important, which for ionic liquids is usually high. Table 2.3 lists the viscosity data of a series of ionic liquids together with some molecular solvents for comparison purposes. Considerably more extensive series of physical properties of ionic liquids can be found elsewhere,[21,22] the aim here is to illustrate general trends and principles.

Table 2.3: Viscosity of some common solvents and ionic liquids

Ionic liquid	Viscosity [cP]	Temp. [°C]	Ref.
H_2O	0.890	25	[23]
CH_3OH	0.544	25	[23]
CH_2Cl_2	0.413	25	[23]
DMSO	1.987	25	[23]
Ethylene glycol	16.1	25	[23]
$[(C_6H_{13})(C_2H_5)_3N][Tf_2N]$	167	25	[24]
$[C_4py][BF_4]$	103	25	[25]
$[C_4py][Tf_2N]$	57	25	[25]
$[C_2C_1im][BF_4]$	32	25	[25]
$[C_4C_1im][BF_4]$	154	20	[26]
$[C_4C_1im][PF_6]$	371	20	[26]
$[C_4C_1im][Tf_2N]$	52	20	[14]
$[C_4C_1im][CF_3CO_2]$	73	20	[14]
$[C_6C_1im][BF_4]$	314	20	[26]
$[(CC{=}C)C_1im][BF_4]$	6110	20	[27]
$[(CC{=}C)C_1im][BF_3CHCH_3CH_2CN]$	25	20	[27]
$[C_2C_1im][BF_4][BF_3CF_3]$	26	25	[28]
$[C_4C_1im][BF_4][BF_3CF_3]$	49	25	[28]

It is clear from the table that the viscosity of ionic liquids is very high compared to that of molecular solvents, although it decreases markedly with increasing temperature. For example, $[C_4C_1im][BF_4]$ has a viscosity of 154 cP at 20°C, which decreases to 40 cP at 50°C and 12 cP at 90°C. Many of the commonly employed anions are highly symmetric almost spherical molecules and using less symmetric anions may be another way to markedly lower the viscosity. Examples of such anions comprise the fluoroborate anions $[BF_3CHCH_3CH_2CN]^-$[29] and $[BF_3CF_3]^-$,[28] shown in Figure 2.4. At present, the long-term stability of these anions remains unknown, but it is likely that under the reactions conditions not untypical of many catalysed reactions, they might decompose. Nonetheless these examples provide clues as to how ionic liquids with low viscosities may be generated and the design of low ionic liquids with low viscosity may prove to be essential for the progression from the laboratory to a chemical plant.

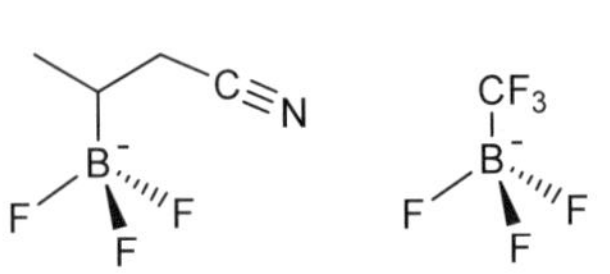

Figure 2.4: Anions that provide low viscosity ionic liquids; (left) $[BF_3CHCH_3CH_2CN]^-$ and (right) $[BF_3CF_3]^-$

Both the melting point and viscosity of an ionic liquid are highly dependent on their purity and values in the literature differ accordingly. In the 1-alkyl-3-methylimidazolium tetrafluoroborate and hexafluorophosphate ionic liquids, which are the ones most widely used in catalysis thus far, halide impurities are usually present to varying degrees, due to the metathesis route commonly used for their preparation (see Section 2.3). The effect of different chloride concentrations present on the viscosity of $[C_4C_1im][BF_4]$ is shown in Figure 2.5.[16]

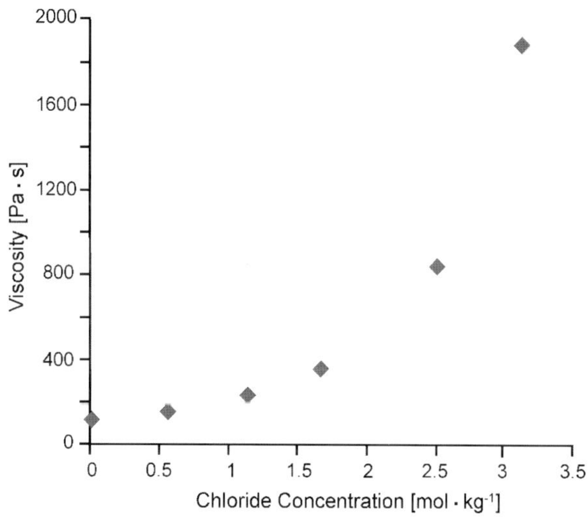

Figure 2.5: Graph of chloride concentration versus viscosity for $[C_4C_1im][BF_4]$

The densities of ionic liquids have a propensity to be higher than those of water or most organic solvents and tend to lie in the range 1.2-2.0 g/cm^3, although ionic liquids with higher densities can be prepared.[30] A high density may be desirable to induce fast phase separation in e.g. biphasic catalysis, but could, on the other hand, result in less efficient mixing of the two phases during the reaction.

The heat capacities for a series of 1-alkyl-3-methylimidazolium ionic liquids have been determined and found to be comparable to other organic solvents, but lower than those with strong hydrogen bonding interactions such as water and acids.[31] It has been found that the heat capacity of ionic liquids increases linearly with temperature. Surface tension has also been measured for a series of 1-alkyl-3-methylimidazolium ionic liquids, showing that it varies considerably, but can be very low.[32] Essentially, the smaller the cation, i.e. the shorter the alkyl group, the lower the surface tension, but further studies are required on a larger range of ionic liquids before a more detailed analysis can be made.

Physical properties such as melting, boiling (or decomposition) temperatures, viscosity and density of a solvent are essential when choosing a solvent for a particular application. Furthermore, it is also important to have some assessment of the solvent polarity as this attests the solvation capability for a solute as well as knowing about other properties such as donor and acceptor numbers and other solvent scales and parameters.[33] Solvent polarity is not easily quantified since it depends on the interplay of electrostatic, inductive, dispersion, charge-transfer and hydrogen bonding forces,[34] but despite the problems quantifying solvent polarity, numerous methods have been devised, which are based on various physical and chemical properties.

The polarity of some 1-alkyl-3-methylimidazolium ionic liquids has been probed using the solvatochomic dye Nile Red and their polarity was found to be comparable to that of lower alcohols as compared in Table 2.4.[35] These data put an end to the well cited myth that ionic liquids exhibit incredibly high polarities greater than that of any organic solvent. Ionic liquid polarity has also been measured by determining the ^{14}N hyperfine coupling constant of the nitroxide radical using EPR spectroscopy and these data agree well with the values obtained using the solvatochromic dye.[36]

Table 2.4: E_T^N values for some common molecular solvents and ionic liquids

Ionic liquid	E_T^N
H_2O	201.7
MeOH	217.7
EtOH	218.2
CH_3Cl	222.5
MeCN	225.0
n-hexane	247.0
$[C_4C_1im][BF_4]$	217.2
$[C_4C_1im][PF_6]$	218.5
$[C_4C_1im][Tf_2N]$	218.0
$[C_6C_1im][BF_4]$	216.8
$[C_6C_1im][PF_6]$	216.8

As such, polar organic solvents like dichloromethane and propan-2-one are miscible with ionic liquids, solvents of low polarity show partial miscibility and non-polar solvents such as hexane are immiscible. The polarity of ionic liquids can be modified by incorporation of various groups onto the cation and, for example, perfluorinated chains have the effect of considerably lowering polarity such that the resulting ionic liquid is miscible with organic solvents of low polarity and immiscible with polar organic solvents.

In general, ionic complexes tend to dissolve well in ionic liquids whereas neutral complexes are less soluble, but often dissolve on heating. For example, Wilkinson's catalyst, $Rh(PPh_3)_3Cl$, dissolves in $[C_4C_1im][BF_4]$ after warming, but it is usually poorly retained in the ionic liquid on product extraction as it also dissolves in organic solvents. Accordingly, charged ligands are required to improve its immobilisation in ionic liquids and many examples of ligand design are cited throughout this book, in particular in Section 4.4.

The solubility of reactive gases such as H_2, CO and O_2 is also very important in many catalysed reactions such as hydrogenation, carbonylation, hydroformylation and oxidation. The general trend for the solubility of relevant gases is as follows:[37-40]

$$CO_2 >>> C_2H_4 \approx CH_4 > O_2 > CO > H_2$$

It is interesting to note that ionic liquids dissolve very high amounts of CO_2 (considerably higher than in organic solvents)[41-43] and therefore could find uses in C_1 chemistry.[44-46] Carbon dioxide is already being successfully used to extract organic products from ionic liquids and continuous flow processes are particularly interesting.[47-49] While the solubility of neat ionic liquids in supercritical CO_2 is very low, a considerable amount may be dissolved in the presence of a co-solvent or large amounts of substrate.[50] The solubility of ethylene and methane in ionic liquids is essentially the same as that observed in organic solvents. Hydrogen and CO tend to be less soluble in ionic liquids compared to organic solvents. In general, it would appear that gases with large dipole or quadrupole moments and those which can form specific (non-covalent) interactions with ionic liquids have the highest solubilities.[27] However, it is worth noting that by careful design of an ionic liquid it should be possible to increases (or decrease) the solubility of a particular gas. Since the solubility of CO was determined in an extensive series of ionic liquids some general trends could be appreciated. For the 1-butyl-3-methylimidazolium cation the solubility of CO increases according to the series $[BF_4]^- < [PF_6]^- < [SbF_6]^- < [CF_3CO_2]^- < [Tf_2N]^-$. This coincides with an increase in the size of the anion, and a decrease in π^* interactions. From the series of $[Tf_2N]^-$ ionic liquids additional trends are apparent; the solubility increases with the chain length of the alkyl substituent in the pyridinium and imidazolium liquids (the solubility also increases with chain length for 1-methyl-3-alkylimidazolium tetrafluoroborate salts). The Kamlet–Taft parameters, α, β, and π^*, have been evaluated for a number of ionic liquids.[51] In general, π^* is higher for ionic liquids (at least those studied) and varies with both anion and cation: the α and β values are generally moderate; the former depends mainly on the cation whereas the latter depends mainly on the anion. It has also been shown that, at least for CO_2 and ethylene, regular solution theory can be used to model gas solubilties in ionic liquids at low pressures.[52] Gas solubility data relevant to specific catalysed reactions are presented in the appropriate chapters.

The donor[53] and acceptor number[54] of a solvent, corresponding to the Lewis basicity and acidity of the solvent, respectively, are useful parameters widely studied for molecular solvents, but thus far largely ignored by the ionic liquid community. The acceptor numbers for a limited range of ionic liquids containing diphenylcyclopropenone has been determined by Raman spectroscopy.[55] More generally, the Lewis acidity and basicity of an ionic liquid is determined by the nature of the anion with strongly basic anions such as $[NO_3]^-$ at one extreme and acidic anions such as $[Al_2Cl_7]^-$ at the opposite end. Anions such as $[BF_4]^-$, $[PF_6]^-$ and $[Tf_2N]^-$ are considered as neutral, or only weakly coordinating, although recent studies have shown that caution must be applied to such generalisations.[56] In addition, with the introduction of functional groups onto ionic liquid cations, many systems now exist where the cation could provide the determining role.

Although the treatise above has focused only on those solvent properties, which are most relevant for applications in catalysis, many other properties could have been discussed, and a more detailed account that discusses these parameters in the context of stoichiometric organic reactivity can be found elsewhere.[57] It is worth noting that any generalisation regarding ionic liquid properties should be taken with caution since ionic liquids can be designed to exhibit specific properties. Also, in order to fully rationalise reaction mechanisms in a solvent an understanding of the microscopic dynamics, i.e. liquid structure and solvation dynamics, are necessary, and in the case of ionic liquids such fundamental studies are only beginning to be studied.

2.3 Synthesis of Ionic Liquids

The number of available ionic liquids continues to grow at an ever increasing rate, however only few are used by the wider community. At the same time, it has become fashionable to label charged compounds that not long ago would have been (appropriately) described as sticky oil, as ionic liquids, and a more considerate use of the term would sometimes be desirable.

The majority of the ionic liquids in use are prepared via salt metathesis reactions and one of the greatest challenges in the field of ionic liquids concerns their synthesis in high purity. Until ionic liquids of well-defined purity are commercially available at acceptable prices, the reliability of the data reported will remain somewhat ambiguous and catalyst performance in these new media must be compared with some caution. At present, many research groups working in the field prepare their solvents themselves, inevitably leading to variations in quality and purity data are seldom provided.

Throughout this book the reader will find examples where small amounts of (mainly halide or water) impurities can exert a huge effect on the outcome of a catalysed reaction. Also examples, where results could not be reproduced and different conclusions are reached, are not uncommon. Having given this warning it is also clear that the field is moving at an incredible pace and new synthetic routes and purification procedures are being developed all the time, and in general the quality of ionic liquids used presently is superior to those employed just a few years ago. Such changes are not untypical of relatively new research areas, especially one which has captured the imagination of such a large body of researchers worldwide.

Ideally, ionic liquids should be colourless (at least in the absence of functional groups), clear, odourless and free flowing liquids. While coloured impurities may subjectively appear more disturbing, they usually cause less problems then the less obvious halide contaminants. Especially hydrophilic ionic liquids comprised of e.g. $[BF_4]^-$ or $[OTf]^-$ are likely to contain some residual halide. If such ionic liquids are prepared via the commonly employed metathesis route, careful washing of a solution in, for example, dichloromethane with water is essential until addition of $AgNO_3$ to the water-phase does not lead to a positive reaction for silver halide. While this does not mean that there is no halide left in the ionic liquid, it could at least be regarded as a minimum standard that ought to be fulfilled. While halide impurities are easier to extract from hydrophobic ionic liquids, those based on $[PF_6]^-$, and to a lesser degree also those based on $[BF_4]^-$, are susceptible towards anion degradation and formation of HF. It is increasingly common to check the results for possible interference from the various impurities.

Numerous examples report on the beneficial effects when catalysis is performed in an ionic liquid, be it increased rate, selectivity, recycling potential or even the accessibility of products that were not available otherwise. Are all these results to be questioned due to insufficiently purified ionic liquids? Certainly not, rather the contrary is true. While ionic liquids of poor quality are likely to afford inferior results, it is only in few cases that the contaminant can be held responsible for any observed catalytic activity. An example are acid-catalysed reactions where HF from degradation of the anion could act as the catalyst.

Thus, while the absolute value of reported turnover frequencies and selectivities are likely to be subject to change, it appears that with increasing solvent quality results will change for the better, not the worse. Publications, which report significant improvements in ionic liquids relative to conventional solvents, can in most cases be taken for the real thing. Whether it makes sense to run a given reaction in an ionic liquid both from an economical as well as an ecological point of view is a different question and has to be decided in each individual case.

The most widely used synthetic approach used to prepare ionic liquids is summarised in Scheme 2.1. The scheme depicts the preparation of imidazolium-based ionic liquids although the method is widely applicable to other types of cations, notably pyridinium systems.

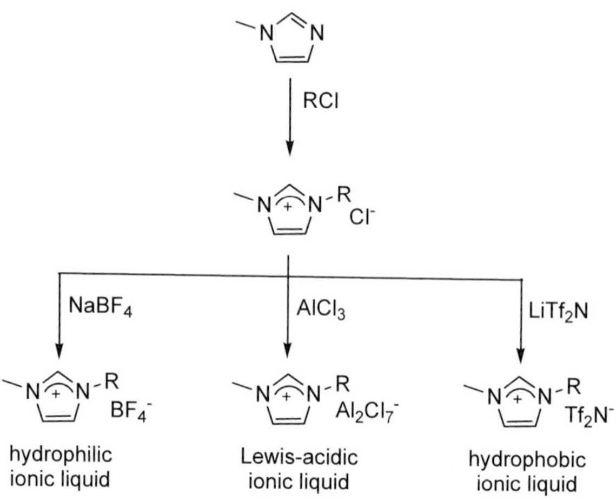

Scheme 2.1: Standard route to 1-alkyl-3-methylimidazolium-based ionic liquids

In the first step an alkylhalide (a chloride is shown in the scheme, but bromides and iodides are also employed), reacts with methylimidazole to form the imidazolium halide.[17,58] In many cases solvent is not required during the reaction, although a solvent may be required for separation of the minor excess of one of the starting materials and if the alkylhalide is composed of a long alkyl chain. Unfortunately the reaction does not always go to completion which means that considerable efforts are required to purify the salt product from the unreacted starting materials. The melting points of the halide salts tend to be high and replacement with alternative anions such as $[BF_4]^-$ or $[Tf_2N]^-$ via metathesis must be undertaken. Recrystallisation of the halide salt (e.g. from acetonitrile) prior to anion metathesis is advisable as removal of organic impurities at a later stage is considerably more difficult. The salt by-product formed during the anion exchange reaction is removed by filtration and/or washing with water in a biphasic purification. However, this step is problematic as quantitative removal of e.g. NaCl is difficult and residual halide impurities can have a significant effect on the resulting physical and chemical properties. To alleviate this problem an alternative synthesis of $[C_4C_1im][BF_4]$ has been developed as outlined in Equation 1. By using tetrafluoroboric acid, HCl is produced in place of NaCl, which can be removed more effectively by washing and under vacuum since it is volatile

$$[C_4C_1im]Cl + HBF_4 \longrightarrow [C_4C_1im][BF_4] + HCl\,(g) \qquad (1)$$

A similar method can be used for the preparation of hexafluorophosphate liquids, but the $[PF_6]^-$ anion is prone to decomposition with the release of HF and care must be taken to avoid this problem.

Due to the problems associated with the metathesis route a number of alternative synthetic procedures have been reported that give rise to halide-free ionic liquids together with analytical methods for determining low halide contamination levels[59,60] in ionic liquids. Two halide-free routes are illustrated in Scheme 2.2. In the first route alkylimidazoles are methylated with trimethyloxonium tetrafluoroborate such that the only by-product is dimethyl ether, which is highly volatile and very easily removed.[61] The method is most effective when the alkyl chain is short, but as long as good stoichiometries are used it can be applied to a wide range of alkyl imidazole precursors. The manufacture of ionic liquids on a large scale using this method would not be unreasonable to perform, but the cost of the trimethyloxonium tetrafluoroborate precursor could prove to be inhibitive. Alternatively, reaction of alkyl imidazoles with dimethyl carbonate affords a zwitterionic salt (hailde-free) which on subsequent treatment with HBF_4 under carefully controlled conditions affords tetrafluoroborate based ionic liquids in highly pure form.[62]

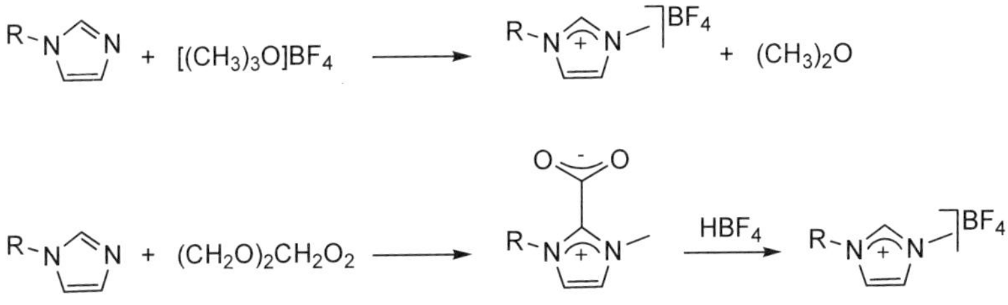

Scheme 2.2: Routes to 'halide-free' imidazolium salts

A relatively inexpensive route to halide-free tetrafluoroborate ionic liquids has been reported, shown in Scheme 2.3. In this one-pot reaction between glyoxal, methylamine, *n*-butylamine, formaldehyde, and tetrafluoroboric acid a mixture of dialkylimidazolium tetrafluoroborate salts is produced.[63] The mixture can be used without further purification being necessary.

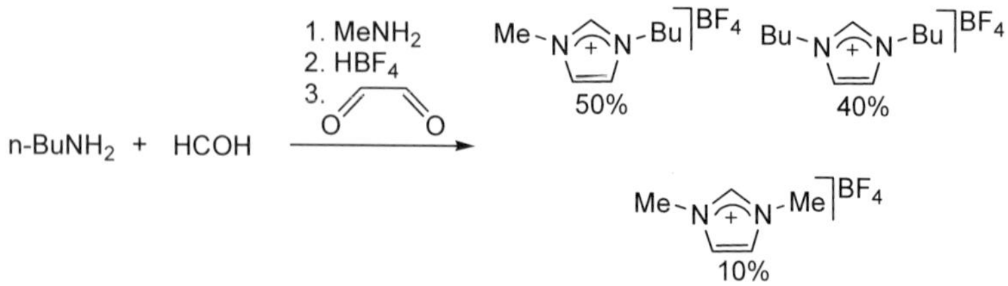

Scheme 2.3: One-pot (inexpensive) synthesis of 1-alkyl-3-methylimidazolium tetrafluoroborate ionic liquids

An increasing number of ionic liquids are no longer based on tetrafluoroborate or hexafluorophosphate anions, although these remain the most widely studied as solvents for catalysis. Alkylation with appropriate precursors that simultaneously generate the anion represents the most effective method. Alkylation of the 1-alkylimidazole derivatives are carried out in trichloroethane, a solvent chosen for its stability toward strongly alkylating agents, its moderately high boiling point, and the insolubility of the imidazolium salts in this medium.[14]

Alternative activation sources have also been applied to the synthesis of ionic liquids, notably, the use of ultrasound[64] and microwave dielectric heating, which not only reduces the reaction time, but can also increase the conversion.[65,66]

There has been some scepticism whether it will be possible to produce ionic liquids on a large (multi-tonne) scale and in the advent of such a product, then it would be intolerably expensive. However, a number of recent papers have hopefully dispelled such perceptions and a number of companies are ready to prepare ionic liquids on very large scales should there be demand. It is likely that the ionic liquids prepared on a large scale at low cost will not be based on the imidazolium cation, although even this cannot be ruled out, but instead will contain phosphonium[67] or N-octylpicolinium cations.[68] Lonza, for example, has the facility to prepare N-octylpicolinium-based ionic liquids with chloroaluminate or tetrafluoroborate anions on a scale of 100 tonnes. It is worth noting, however, that large quantities of ionic liquids may not be essential for their application in biphasic catalysis.

2.4 Functionalised Ionic Liquids

Ionic liquids with functional groups attached to the cation are largely based on the imidazolium unit and are often prepared using the metathesis route. Commencing with 1-methylimidazole and an appropriately functionalised alkylhalide the quaternisation method tends to be very effective and electron withdrawing groups attached to the alkylhalide enhance the electrophility and thus shorten the reaction time. This method is suitable for the synthesis of nearly all ionic liquids as long as they are stable towards base. Because the basicity of imidazole is relatively strong in some cases, elimination of hydrogen halide or a Hoffmann-type elimination can occur.[69]

A wide variety of functional groups have been introduced directly onto the imidazolium moiety including alcohols,[70,71] carboxylic acids,[62,72] amine and amide,[73] phosphine,[74] urea and thioureas,[75] thiols,[76,77] alkynes,[78,79] alkenes,[29] dienes,[80] fluorous chains[4,81,82] and chiral moieties.[83] Figure 2.6 shows a series of functionalised imidazolium cations that in combination with appropriate anions give rise to low melting salts. The figure is not comprehensive and it is also worth pointing out that not all the functionalised systems have been prepared with catalytic applications in mind. For example, the thiol functionalised ionic liquids were prepared in order to stabilise gold nanoparticles, although here potential catalytic applications can be envisaged. Additionally, the $-P(=O)(OEt)_2$ derivatised system was inspired by potential tribological properties.

The use of functionalised ionic liquids in catalysis is still very much in its infancy. The presence of a functional group may serve any of the following purposes in order to improve the catalytic reaction: (i) assist in the activation of the catalyst; (ii) generate a novel catalytic species; (iii) improve the stability of the catalyst; (iv) optimise immobilisation and recyclability; (v) facilitate product isolation and (vi) influence the selectivity of the reaction. In addition, the functionalised ionic liquid has to meet certain requirements to be of use in synthesis/catalysis. Most important is their inertness with respect to the reaction in question. Also, the right balance between catalyst stabilisation and activation has to be found and this might require careful tuning of the reaction conditions.

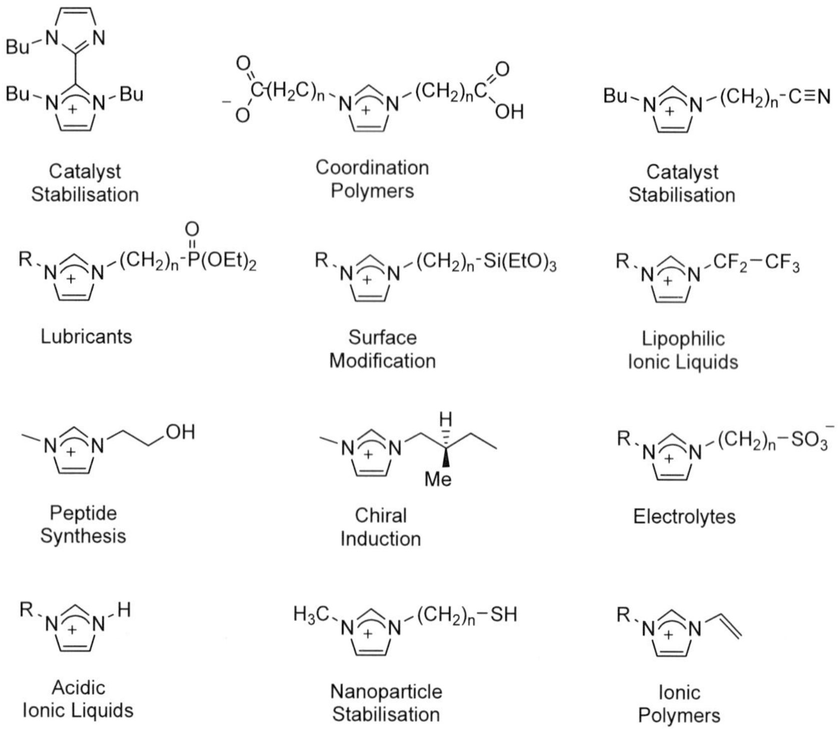

Figure 2.6: Examples for functionalised imidazolium salts

The conversion of one functionalised salt to another is possible and may sometimes provide the only route to certain derivatives. For example, a carboxylic ester can be converted to a carboxylic acid, affording Brønsted-acidic ionic liquids, as shown in Scheme 2.4.[72] As mentioned above, halide contamination in ionic liquids prepared by the traditional salt metathesis route can lead to marked differences in their properties. The strategy used to prepare the carboxylic acid functionalised ionic liquids efficiently eliminates halide impurities by progressing through a zwitterionic intermediate, which undergoes protonation with acids such as HBF_4 and HOTf to afford highly pure, halide-free, ionic liquids.

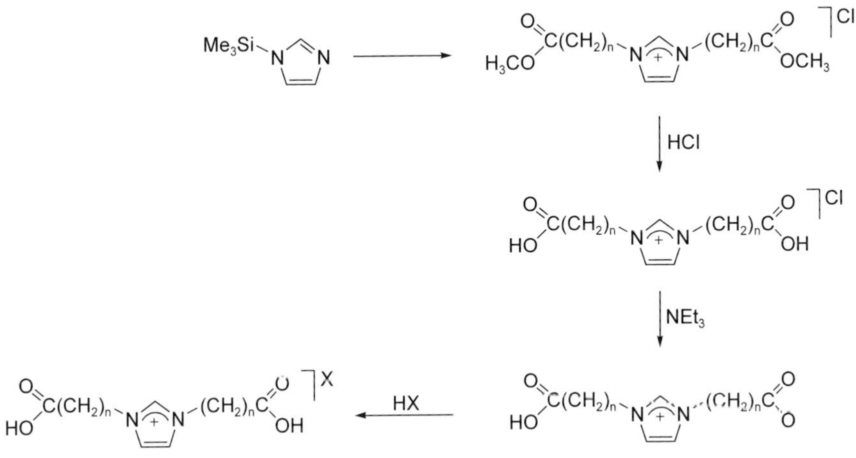

Scheme 2.4: Preparation of carboxylic acid-functionalised ionic liquids via a zwitterionic intermediate

Sulfonic acid functionalised ionic liquids may also be prepared via a zwitterionic intermediate from a Michael-type addition, as shown in Scheme 2.5.[84,85] In the first step 1-methylimidazole reacts with the sulfonic acid precusor 1,3-propane sultone to form a zwitterionic intermediate. Protonation with Brønsted acids affords ionic liquids with high purity that have proven to be highly efficient reaction media in, for example, esterification reactions.[86] Task-specific ionic liquids may also be prepared using semi-combinatorial methods.[87]

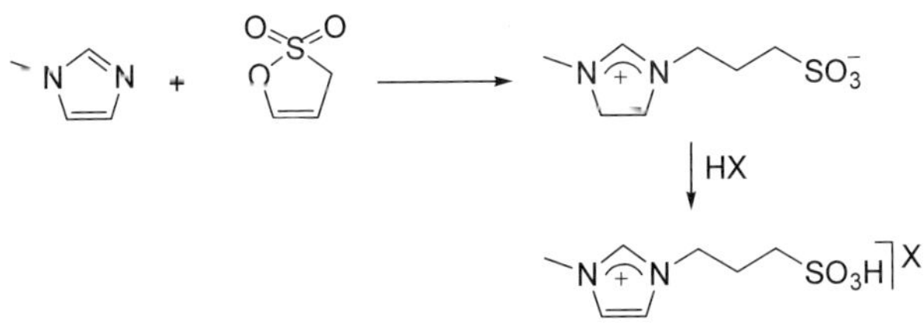

Scheme 2.5: Preparation of sulfonic-acid functionalised ionic liquids

It is also relatively easy to functionalise imidazolium cations at the 2-position.[88] For example, the phosphine derivatised salts shown in Figure 2.7 illustrate such a substitution pattern and they are easily prepare by virtue of the acidity of the 2-proton.[74] Substitution of the 2-proton tends to yield relatively high melting salts instead of ionic liquids. Despite this limitation the imidazolium-phosphine compounds are good ligands for catalysis improving the immobilisation potential of complexes dissolved in ionic liquids.

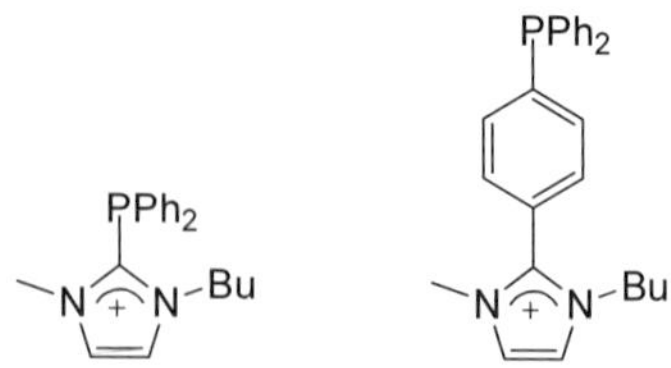

Figure 2.7: Phosphine-functionalised imidazolium salts

Functionalised ionic liquids based on cations other than imidazolium have also been developed. For example, pyridinium cations functionalised with pentafluorosulfanyl[89] or alkyl-nitrile groups[90] have been prepared as cheaper alternatives to their imidazolium-based counterparts (see Figure 2.8). The latter have been evaluated in palladium catalysed C-C cross coupling reactions and improved catalyst retention and stability were observed in the nitrile-functionalised ionic liquid compared to the simple alkyl-analogue. Consequently, the nitrile-functionalised ionic liquid solution can be reused repeatedly without significant decrease in activity (see Chapter 6 for further information).

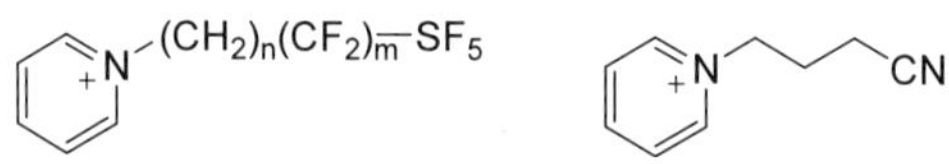

Figure 2.8: Functionalised pyridinium-based ionic liquids.

Although much less developed than the functionalisation of ionic liquid cations, some efforts have been made in the development of ionic liquids containing functionalised anions. Many 'off the shelf' anions have been combined with 1,3-dialkylimidazolium cations to afford ionic liquids with properties that could be beneficial in certain catalysed reactions and some of these are illustrated in Figure 2.9. Anions such as carboranes[91] and metal carbonylates[92] may be regarded as interesting curiosities, but can nevertheless be exploited in organic synthesis and catalysis (see Chapters 3 and 9, respectively).

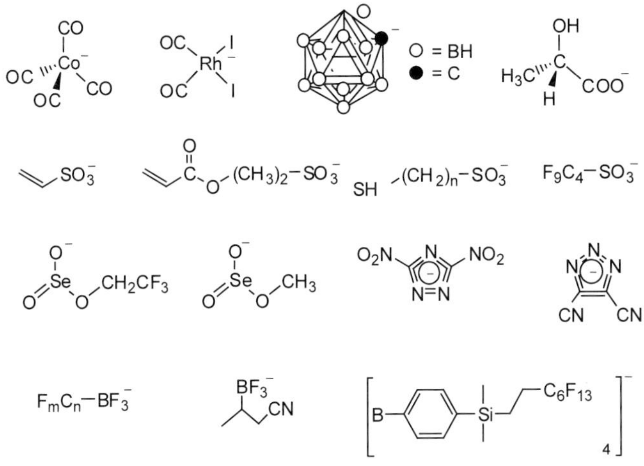

Figure 2.9: Selection of functional anions employed in ionic liquids

The Monsanto catalyst for carbonylation of methanol is anionic, *viz.* $[Rh(CO)_2(I)_2]^-$, and has been combined with the $[C_4C_1im]^+$ cation to afford a room temperature ionic liquid which might have some benefits in catalysis, although this has yet to be shown.[4] Aminoacid anions could prove useful in catalysis[93-95] and a series of imidazolium-based ionic liquids with natural amino acid anions has been prepared.[96] Perfluoroated anions may lead to ionic liquids that dissolve apolar solvents or substrates and can also be used to immobilised catalysts bearing perfuorated ligands. As and example, an ionic liquid containing a highly lipophilic anion, *viz.* $[B\{C_6H_4(SiMe_2CH_2CH_2C_6F_{13})\text{-}p\}_4]^-$, was employed in hydrosilylation of olefins with a perfluorinated rhodium catalyst.[97]

Not all the anions shown in Figure 2.9 have been used in ionic liquids that have been evaluated in catalysis, however, all could prove useful in the future. The functionalised fluoroborate anions, $[BF_3R]^-$ (R = CF_3, C_3F_7, C_4F_9 and CH_3CHCH_2CN), combined with imidazolium cations afford low viscosity ionic liquids, the latter has also been shown to be stable in hydrogenation reactions.[28] The selenium-based anion $[SeO_2(OCH_3)]^-$, prepared from the reaction of SeO_2 with 0.5 equivalents of K_2CO_3, also affords low viscosity ionic liquids in combination with imidazolium cations and they have been used as reaction media for oxidative carbonylation of aromatic amines.[98]

Combining functionalised cations with functionalised anions may lead to highly task-specific ionic liquids, but very little has been undertaken in this area in the catalysis domain. However, a number of so-called ‘dual-functionalised’ ionic liquids have been reported[27] and it has been shown that combining cations with donor groups with anions with varying donor properties can have a marked influence on the nature of the catalyst and the outcome of the reaction.[99]

Chiral ionic liquids could become an important class of solvents,[94,100,101] which may even be relatively cheap when prepared from the natural chiral pool with anions such as lactate. However, as yet, their ability to induce enantiomeric excess in catalysed reactions has been only met with limited success.[102-104] With increasing evidence to suggest that carbene ligands derived from ionic liquid imidazolium cations are involved in certain catalysed reactions it is likely that chiral ionic liquids that act as both solvent and ligand could prove highly useful in stereoselective reactions. Some examples of chiral cations that may lead to ionic liquids are shown in Figure 2.10.

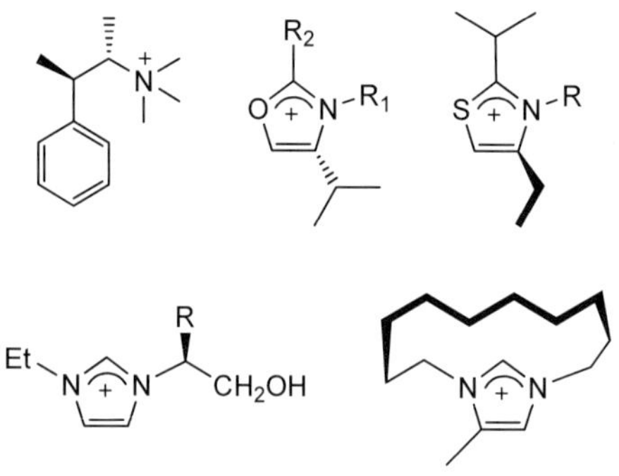

Figure 2.10: Examples of chiral ionic liquid cations

Under Mitsunobu conditions, see Scheme 2.6, it is possible to prepare imidazoles with chiral side chains that can subsequently be converted into ionic liquids.[105] One of the features of many of the chiral ionic liquids reported to date, that could limit their application in catalysis, is their high viscosity. However, by combining chiral imidazolium cations with anions described above that help to reduce viscosity, this problem could potentially be overcome.

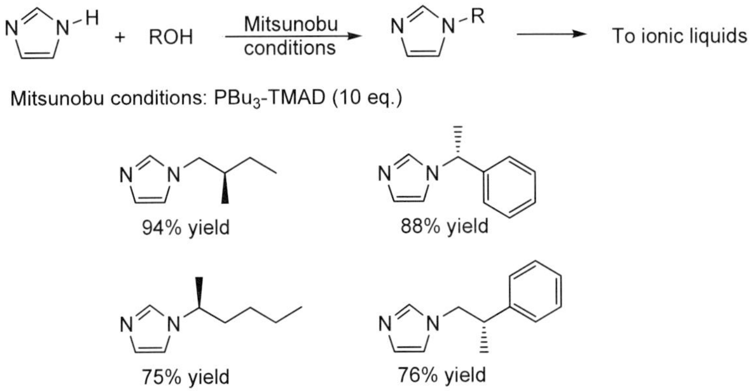

Scheme 2.6: Preparation of chiral imidazoles under Mitsunobu conditions

An area in which functionalised ionic liquids are already playing an important role in catalysis is heterogenisation on solid supports. The general concept involves the immobilisation of imidazolium and other cationic fragments onto solid supports using appropriate functional groups attached to the cation. An ionic catalyst then resides within the ionic matrix and several examples of such supported ionic liquid phase catalysts are provided in the subsequent chapters of this book. The concept is illustrated in Figure 2.11.

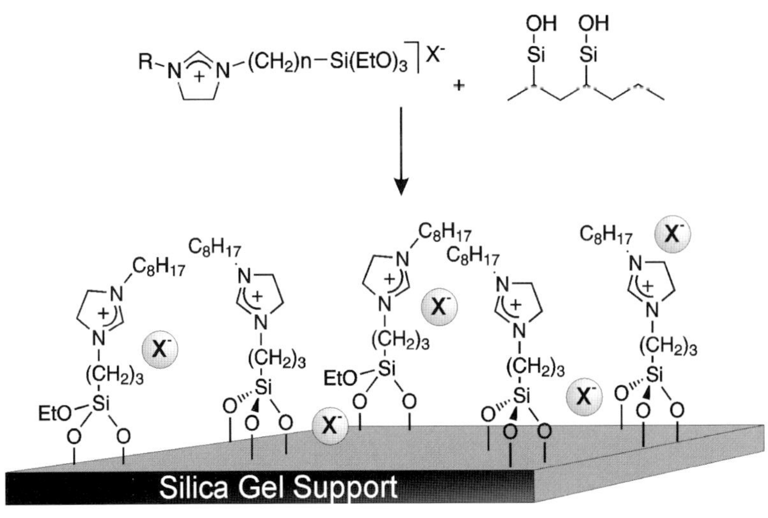

Figure 2.11: Surface modification using a functionalised ionic liquid

2.5 Toxicological and Environmental Concerns

The toxicological and environmental properties of ionic liquids are slowly emerging, but are becoming increasingly needed, as ionic liquids are beginning to be employed in larger quantities. While ionic liquids are classed as non-toxic and 'green' this is not necessarily the case, and such generalisation must be taken with extreme caution. It would appear that many ionic liquids are more toxic than commonly used organic solvents.

A strategy for the risk-assessment of ionic liquids has been developed,[106] and the aquatic biodegradation of both imidazolium cations and a range of anions have been investigated.[107] It was found that 1-butyl-3-methylimidazolium based ionic liquids with a range of different anions (those currently widely used in catalysis) are essentially poorly biodegradable. However, with a modified cation, viz. the 3-methyl-1-(propoxymethylcarbonyl)-imidazolium cation, the analogous series of ionic liquids were considerable more prone to biodegredation, indicating again how designer ionic liquids can overcome specific problems.[108] Likewise, ionic liquids containing the octylsulfate anion, which is cheap, widely available and has a well-documented toxicology, have been developed and are probably amongst the greenest available.[109] Despite being prepared via the convensional metathesis route they are readily isolated in halide free form and are now prepared on the ton scale.

Ionic liquids have been shown to exhibit both antimicrobial and antifungal activities.[110] In general, if the alkyl groups are short then they are not active, but with C_8 chains and above activities become comparable to effective reagents. The implication here is that if an ionic liquid can enter microbial or fungal cells then it is likely that they could also enter mamalian cells, exhibiting toxicity or potentially being mutagenic. In fact, preliminary assays using human cells have been reported.[111,112] However, the great advantage of ionic liquids is that they are non-volatile and therefore cannot be ingested by inhalation. So as long as they are handled properly and fully removed from organic products, their expose to humans should approach zero and therefore not present a serious health risk.

2.6 References

[1] Gorman, J. *Science News Online* **2001**, *160*, 156.
[2] Holbrey, J. D.; Seddon, K. R. *Clean Prod. Process.* **1999**, *1*, 223-236.
[3] Owens, G. S.; Durazo, A.; Abu-Omar, M. M. *Chem. Eur. J.* **2002**, *8*, 3053.
[4] Dyson, P. J.; McIndoe J. S.; Zhao, D. *Chem. Commun.* **2003**, 508-509.
[5] Dyson, P. J.; McIndoe, J. S. *Inorg. Chim. Acta* **2003**, *354*, 68-74.
[6] Dyson, P. J.; Khalaila, I.; Luettgen, S.; McIndoe, J. S.; Zhao, D. *Chem. Commun.* **2004**, 2204-2205.
[7] Gelesky, M. A.; Umpierre, A. P.; Machado, G.; Correia, R. R. B.; Magno, W. C.; Morais, J.; Ebeling, G.; Dupont, J. *J. Am. Chem. Soc.* **2005**, *127*, 4588-4589.
[8] Omotowa, B. A.; Phillips, B. S.; Zabinski, J. S.; Shreeve, J. M. *Inorg. Chem.* **2004**, *43*, 5466-5471.
[9] Ellis, D. J., PhD-Thesis University of York, **2002**.
[10] Holbrey, J. D.; Seddon, K. R. *J. Chem. Soc., Dalton Trans.* **1999**, 2133-2139.
[11] Huddleston, J. G.; Visser, A. E.; Reichert, W. M.; Willauer, H. D.; Broker, G. A.; Rogers, R. D. *Green Chem.* **2001**, *3*, 156-164.
[12] Tokuda, H.; Hayamizu, K.; Ishii, K.; Bin Hasan Susan, A.; Watanabe, M. *J. Chem. Phys. B* **2005**, *109*, 6103-6110.
[13] Dyson, P. J.; Ellis, D. J.; Welton, T. *Can. J. Chem.* **2001**, *79*, 705-708.
[14] Bonhôte, P.; Dias, A.-P.; Papageorgiou, N.; Kalyanasundaram, K.; Grätzel, M. *Inorg. Chem.* **1996**, *35*, 1168-1178.
[15] Øye, H.A. ; Jagtoyen, M.; Oksefjell, T.; Wilkes, J. S. *Mater. Sci. Forum*, **1991**, *73-75*, 183.
[16] Chauvin, Y.; Olivier-Bourbigou, H. *Chemtech* **1995**, *25*, 26-30.
[17] Hitchcock, P. B.; Seddon, K. R.; Welton, T. *J. Chem. Soc., Dalton Trans.* **1993**, 2639-2643.
[18] Fredlake, C. P.; Crosthwaite, J. M.; Hert, D. G.; Aki, S. N. V. K.; Brennecke, J. F. *J. Chem. Eng. Data* **2004**, *49*, 954-964.
[19] Kosmulski, M.; Gustafsson, J.; Rosenholm, J. B. *Thermochim. Acta* **2004**, *412*, 47-53.
[20] Glenn, A. G.; Jones, P. B. *Tetrahedron Lett.* **2004**, *45*, 6967-6969.
[21] Hagiwara, R.; Ito, Y. *J. Fluorine Chem.* **2000**, *105*, 221-227.
[22] Wasserscheid, P.; Welton, T., Eds. *Ionic Liquids in Synthesis*, Wiley-VCH: Weinheim, **2002**.
[23] Lide, D. R. Ed.,*CRC Handbook of Chemistry and Physics*, CRC Press: Boca Raton, **1992**.
[24] Sun, J.; Forsyth, M.; MacFarlane, D. R. *J. Phys. Chem. B* **1998**, *102*, 8858-8864.
[25] Noda, A.; Hayamizu, K.; Watanabe, M. *J. Phys. Chem. B* **2001**, *105*, 4603-4610.
[26] Seddon, K. R.; Stark, A.; Torres, M. J. *Pure Appl. Chem.* **2000**, *72*, 2275-2287.
[27] Zhao, D.; Fei, Z.; Ohlin, C. A.; Laurenczy G.; Dyson, P. J. *Chem. Commun.* **2004**, 2500-2501.
[28] Zhou, Z.-B. ; Matsumoto, H.; Tatsumi, K. *Chem. Lett.* **2004**, *33*, 680-681.
[29] Zhao, D.; Fei, Z.; Geldbach, T. J.; Scopelliti, R.; Laurenczy, G.; Dyson, P. J. *Helv. Chim. Acta*, **2005**, *88*, 665-675.
[30] Ye, C.F.; Shreeve, J.M. *J. Org. Chem.* **2004**, *69*, 6511-6513.
[31] Holbrey, J. D.; Reichert, W. M.; Reddy, R. G.; Rogers, R. D. In *Ionic Liquids as Green Solvents: Progress and Prospects*, ACS Symposium Series 856, **2003**, 121-133.
[32] Law G.; Watson, P. R. *Langmuir*, **2001**, *17*, 6138-6141.

[33] Reichardt, C. *Solvents and Solvent Effects in Organic Chemistry*, 3rd Ed., Wiley-VCH: Weinheim, **2002**.
[34] Reichardt, C. *Nachr. Chem. Tech. Lab.*, **1997**, *45*, 759-763.
[35] Carmichael, A. J.; Seddon, K. R. *J. Phys. Org. Chem.* **2000**, *13*, 591-595.
[36] Kawai, A.; Hidemori, T.; Shibuya, K. *Chem. Lett.* **2004**, *33*, 1464-1465.
[37] Anthony, J. L.; Maginn, E. J.; Brennecke, J. F. *J. Phys. Chem. B* **2001**, *105*, 10942-10949.
[38] Anthony, J.L.; Anderson, J.L.; Maginn, E.J.; Brennecke, J.F. *J. Phys. Chem. B* **2005**, *109*, 6366-6374.
[39] Husson-Bor, P.; Majer, V.; Costa Gomes, M. F. *J. Chem. Eng. Data* **2003**, *48*, 480-485.
[40] Camper, D.; Becker, C.; Koval, C.; Noble, R. *Ind. Eng. Chen. Res.* **2005**, *44*, 1928-1933.
[41] Blanchard, L. A.; Gu, Z.; Brennecke, J. F. *J. Phys. Chem. B.* **2001**, *105*, 2437-2444.
[42] Pérez-Salado Kamps, A.; Tuma, D.; Xia, J.; Maurer, G. *J. Chem. Eng. Data* **2003**, *48*, 746-749.
[43] Baltus, R. E.; Culbertson, B. H.; Dai, S.; Luo, H.; DePaoli, D. W. *J. Phys. Chem. B* **2004**, *108*, 721-727.
[44] Alvaro, M.; Baleizao, C.; Das, D.; Carbonell, E.; Gracía, H. *J. Catal.* **2004**, *228*, 254-258.
[45] Kawanami, H.; Sasaki, A.; Matsui, K. Ikushima, Y. *Chem. Commun.* **2003**, 896-867.
[46] Kim, Y.-J.; Cheong, M. *Bull. Korean Chem. Soc.* **2002**, *23*, 1027-1028.
[47] Sellin, M. F.; Webb, P. B.; Cole-Hamilton, D. J. *Chen. Commun.* **2001**, 781-782.
[48] Dzyuba, S. V.; Bartsch, R. A. *Angew. Chem. Int. Ed.* **2003**, *42*, 148-150.
[49] Reetz, M. T.; Wiesenhöfer, W.; Franciò, G. Leitner, W. *Adv. Synth. Catal.* **2003**, *345*, 1221-1228.
[50] Wu, W.; Zhang, J.; Han, B.; Chen, J.; Liu, Z.; Jiang, T.; He, J.; Li, W. *Chem. Commun.* **2003**, 1412-1413.
[51] Crowhurst, L.; Mawdsley, P. R.; Perez-Arlandis, J. M.; Salter P. A.; Welton, T. *Phys. Chem. Chem. Phys.* **2003**, *5*, 2790-2794.
[52] Camper, D.; Scovazzo, P.; Koval, C.; Noble, R. *Ind. Eng. Chem. Res.* **2004**, *43*, 3049-3054.
[53] Fawcett, W. R. *J. Phys. Chem.* **1993,** *97*, 9540-9546.
[54] Mayer, U.; Gutmann V.; Gerger, W. *Monatsh. Chem.* **1975**, *106*, 1235-1257.
[55] Kimura, Y.; Fukuda, M.; Fujisawa, T.; Terazima, M. *Chem. Lett.* **2005**, *34*, 338-339.
[56] Williams, D. B.; Stoll, M. E.; Scott, B. L.; Costa, D. A.; Oldham, W. J. *Chem. Commun.* **2005**, 1438-1440.
[57] Chiappe, C.; Pieraccini, D. *J. Phys. Org. Chem.* **2005**, *18*, 275-297.
[58] Suarez, P. A. Z.; Dullius, J. E. L.; Einloft, S.; de Souza, R. F.; Dupont, J. *Polyhedron* **1996**, *15*, 1217-1219.
[59] Villagran, C.; Deetlefs, M.; Pitner, W.R.; Hardacre, C. *Anal. Chem.*, **2004**, *76*, 2118-2123.
[60] Villagran, C.; Banks, C. E.; Hardacre, C.; Compton, R. G. *Anal. Chem.*, **2004**, *76*, 1998-2003.
[61] Olivier, H.; Favre, F. United States Patent 6,245,918, **2001**.
[62] Holbrey, J. D.; Reichert, W. M.; Tkatchenko, I.; Bouajila, E.; Walter, O.; Tommasi, I.; Rogers, R. D. *Chem. Commun.* **2003**, 28-29.
[63] de Souza, R. F.; Rech, V.; Dupont, J. *Adv. Synth. Catal.* **2002**, *344*, 153-155.
[64] Leveque, J. M.; Luche, J. L.; Petrier, C.; Roux, R.; Bonrath, W. *Green Chem.* **2002**, *4*, 357-360.

[65] Fraga-Dubreuil, J.; Famelart, M. H.; Bazureau, J. P. *Org. Proc. Res. & Dev.*, **2002**, *6*, 374-378.
[66] Singh, V.; Kaur, S.; Sapehiya, V.; Singh, J.; Kad, G. L. *Catal. Commun.* **2005**, *6*, 57-60.
[67] Bradaric, C. J.; Downard, A.; Kennedy, C.; Pobertson, A. J.; Zhou, Y. In *Ionic Liquids as Green Solvents: Progress and Prospects*, ACS Symposium Series 856, **2003**, 41-56.
[68] Dyson, P. J.; Geldbach, T. J.; Moro, F.; Taeschler, C.; Zhao, D. In *Ionic Liquids III : Fundamentals, Progress, Challenges, and Opportunities*, ACS Symposium Series 901, **2005**.
[69] Horváth, A. *Synthesis*, **1995**, 1183.
[70] Branco, L. C.; Rosa, J. N.; Moura Ramos, J. J.; Alfons, C. A. M. *Chem. Eur. J.* **2002**, *8*, 3671-3677.
[71] Fraga-Dubreuil, J.; Bazureau, J. P. *Tetrahedron Lett.* **2001**, *42*, 6097-6100.
[72] Fei, Z.; Zhao, D.; Geldbach, T. J.; Scopelliti, R.; Dyson, P. J. *Chem. Eur. J.* **2004**, *10*, 4886-4893.
[73] Lee, K.-M.; Lee, Y.-T.-Y.; Lin, J. B. *J. Mat. Chem.* **2003**, *13*, 1079.
[74] Brauer, D. J.; Kottsieper, K.; Liek, C.; Stelzer, O.; Waffenschmidt, H.; Wasserscheid, P. *J. Organomet. Chem.* **2001**, *630*, 177-184.
[75] Visser, A. E.; Swatloski, R. P.; Reichert, W. M.; Mayton, R.; Sheff, S.; Wierzbicki, A.; Davis, J. H.; Rogers, R. D. *Chem. Commun.* **2001**, 135-136.
[76] Itoh, H.; Naka, K.; Chujo, Y. *J. Am. Chem. Soc.* **2004**, *126*, 3026-3027.
[77] Kim, K. S.; Demberelnyamba, D.; Lee, H. *Langmuir* **2004**, *20*, 556-560.
[78] Schottenberger, H.; Wurst, K.; Horvath, U. E. I.; Cronje, S.; Lukasser, J.; Polin, J.; McKenzie, J. M.; Raubenheimer, H. G. *Dalton Trans.* **2003**, 4275-4281.
[79] Fei, Z.; Zhao, D.; Scopelliti, R.; Dyson, P. J. *Organometallics*, **2004**, *23*, 1622-1628.
[80] Geldbach, T. J.; Dyson, P. J. *J. Am. Chem. Soc.* **2004**, *126*, 8114-8115.
[81] Merrigan, T. L.; Bates, E. D.; Dorman, S. C.; Davis, J. H. *Chem. Commun.* **2000**, 2051-2052.
[82] Singh, R. P.; Manandhar, S.; Shreeve, J. M. *Tetrahedron Lett.* **2002**, *43*, 9497-9499.
[83] Bao, W. L.; Wang, Z. M.; Li, Y. X. *J. Org. Chem.* **2003**, *68*, 591-593.
[84] Mori, S.; Ida, K.; Ue, M. US Patent 4 892 944, **1990**.
[85] Cole, A. C.; Jensen, J. L.; Ntai, I.; Tran, K. L. T.; Weaver, K. J.; Forbes, D. C.; Davis, J. H. *J. Am. Chem. Soc.* **2002**, *124*, 5962-5963.
[86] Gui, J. L.; Cong, X. H.; Liu, D.; Zhang, X. T.; Hu, Z. D.; Sun, Z. L. *Catal. Commun.* **2004**, *5*, 473-477.
[87] Wasserscheid, P.; Drießen-Hölscher, B.; van Hal, R.; Steffens, C.; Zimmermann, J. *Chem. Commun.* **2003**, 2038-2039.
[88] Handy, S. T.; Okello, M. *J. Org. Chem.* **2005**, *70*, 1915-1918.
[89] Singh, R. P.; Winter, R. W.; Gard, G. L.; Gao, Y.; Shreeve, J. M. *Inorg. Chem.* **2003**, *42*, 6142-6146.
[90] Zhao, D.; Fei, Z.; Geldbach, T. J.; Scopelliti, R.; Dyson, P. J. *J. Am. Chem. Soc.*, **2004**, *126*, 15876-15882.
[91] Larsen, A. S.; Holbrey, J. D.; Tham, F. S.; Reed, C. A. *J. Am. Chem. Soc.* **2000**, *122*, 7264-7272
[92] Brown, R. J. C.; Dyson, P. J.; Ellis, D. J.; Welton, T. *Chem. Commun.* **2001**, 1862-1863.
[93] Earle, M. J.; McCormac, P. B.; Seddon, K. R. *Green Chem.* **1999**, 23-25.
[94] Wasserscheid, P.; Bösmann, A.; Bolm, C. *Chem. Commun.* **2002**, 200-201.
[95] Bicak, N. *J. Mol. Liquids*, **2005**, *116*, 15-18.

[96] Fukumoto, K.; Yoshizawa, M.; Ohno, H. *J. Am. Chem. Soc.* **2005**, *127*, 2398-2399.
[97] van den Broeke, J.; Winter, F.; Deelman, B.-J.; van Koten, D. *Org. Lett.* **2002**, *4*, 3851-3854.
[98] Kim, H. S.; Kim, Y. J.; Lee, H.; Park, K. Y.; Lee, C.; Chin, C. S. *Angew. Chem. Int. Ed.* **2002**, *41*, 4300-4003.
[99] Chiappe, C.; Pieraccini, D.; Zhao, D.; Fei, Z.; Dyson, P. J. *Chem. Commun*, in press.
[100] Levillain, J.; Dubant, G.; Abrunhosa, I.; Gulea, M.; Gaumont, A.-C. *Chem. Commun.* **2003**, 2914-2915.
[101] Jodry, J. J.; Mikami, K. *Tetrahedron Lett.* **2004**, *45*, 4429-4431.
[102] Ding, J.; Desikan, V.; Han, X.; Xiao, T. L.; Ding, R.; Jenks, W. S.; Armstrong, D. W. *Org. Lett.* **2005**, *7*, 335-337.
[103] Pégot, B.; Vo-Thanh, G.; Gori, D.; Loupy, A. *Tetrahedron Lett.* **2004**, *45*, 6425-6428.
[104] Kiss, L.; Kurtan, T.; Antus, S.; Brunner, H. *Arkivoc* **2003**, 69-76.
[105] Jin Kim, E.; Ko, S. Y.; Dziadulewicz, E. K. *Tetrahedron Lett.* **2005**, *46*, 631-633.
[106] Jastorff, B.; Störmann, R.; Ranke, J.; Mölter, K.; Stock, F.; Oberheitmann, B.; Hoffmann, W.; Hoffmann, J.; Nüchter, M.; Ondruschka, B.; Filser, J. *Green Chem.* **2003**, *5*, 136-142.
[107] Garcia, M. T.; Gathergood, N.; Scammells, P. J. *Green Chem.* **2005**, *7*, 9-14.
[108] Gathergood, N.; Garcia, M. T.; Scammells, P. J. *Green Chem.* **2004**, *6*, 166-175.
[109] Wasserscheid, P.; van Hal, R.; Bösmann, A. *Green Chem.* **2002**, *4*, 400-404.
[110] Pernak, J.; Sobaszkiewicz, K.; Mirska, I. *Green Chem.* **2003**, *5*, 52-56.
[111] Ranke, J.; Mölter, K.; Stock, F.; Bottin-Weber, U.; Poczobutt, J.; Hoffmann, J.; Ondruschka, B.; Filser, J.; Jastorff, B. *Ecotox. Environ. Safety*, **2004**, *58*, 396-404.
[112] Docherty, K. M.; Kulpa, C. F. *Green Chem.* **2005**, *7*, 185-189.

Chapter 3

HYDROGENATION

The catalytic reduction of unsaturated bonds in the presence of a hydrogen source, usually molecular hydrogen, is one of the most important chemical processes for the synthesis of fine and bulk chemicals. Various innovative hydrogenation catalysts have been developed during the last century. Of particular importance are homogeneous catalysts, which provide high activity, selectivity and, if required, enantioselectivity, which is difficult to achieve with heterogeneous catalysts. Of all catalytic transformations performed in the presence of ionic liquids, hydrogenation is probably the best investigated. However, as this is a very diverse field, little is yet understood in terms of how activity and selectivity can be optimised.

3.1 Introduction

One of the earliest reports on hydrogenation, in, what can possibly be considered as ionic liquid, dates back to 1982, when Buchanan *et al.* reported on the transfer-hydrogenation of polycyclic aromatic hydrocarbons in molten $SbCl_3$ (mp. 73°C) which acts as both reaction medium and catalyst, see Scheme 3.1.[1]

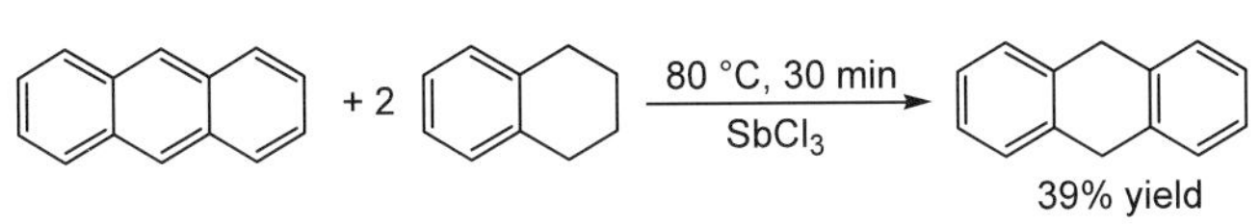

Scheme 3.1: Transfer-hydrogenation of anthracene with tetrahydronaphthalene as hydrogen source in molten $SbCl_3$

Although not a catalytic reaction, the reduction of arenes with electropositive metals like lithium, zinc or aluminium in chloroaluminate ionic liquids with gaseous HCl as the hydrogen source is worth mentioning here.[2] At ambient temperature, 9,10-dimethylanthracene was reduced almost quantitatively within 30 minutes, see Scheme 3.2. Unlike other reactions with dissolved metals as reducing agent, products containing conjugated double bonds were not observed.

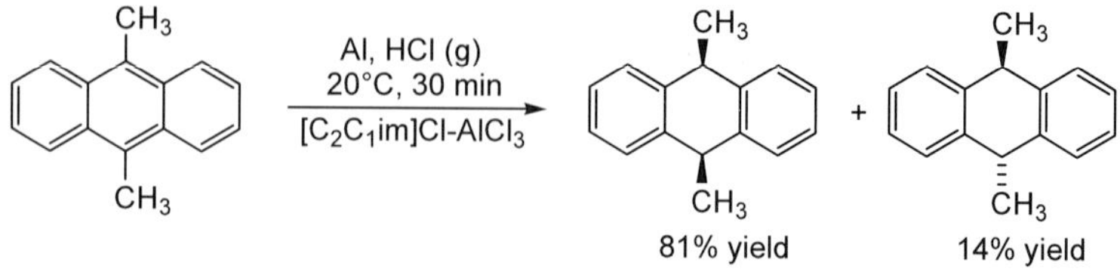

Scheme 3.2: Reduction of dimethylanthracene with dissolved aluminium as reducing agent and HCl as the hydrogen source

Chauvin and co-workers were the first to use room-temperature ionic liquids as media for hydrogenation reactions, employing $[Rh(nbd)(PPh_3)_2][PF_6]$ as catalyst for the reduction of 1-pentene.[3,4] Relative to acetone, hydrogenation rates were almost five times higher in $[C_4C_1im][SbF_6]$. Inferior rates were observed in $[C_4C_1im][PF_6]$, ascribed to the lower solubility of the substrate in this solvent, whereas $[C_4C_1im][BF_4]$ gave only poor results for which chloride impurities in the solvent were believed to be responsible. Catalyst immobilisation was good in all cases with less than 0.02% of rhodium detected in the organic phase and catalyst reuse was possible. This early work certainly stands out in the field as it triggered the development of biphasic ionic liquid catalysis. Furthermore, in the same paper, the authors also described the partial hydrogenation of cyclohexadiene to cyclohexene with 98% selectivity, showing the benefit of different substrate and product solubility in the ionic media.

Since these nascent studies, a wide range of transition metal catalysts, mainly based on rhodium and ruthenium, but also palladium, platinum and other metals, have been used together with ionic liquids to hydrogenate various substrates with a range of functional groups. These include alkenes, alkynes, arenes, carbonyls, carboxylates, nitrates and polymers. In general, the results obtained so far are very promising. Many of the catalysts can be recovered simply by phase separation and then used for further catalytic runs. Importantly, higher catalyst activities and chemo-, regio- and enantioselectivities are frequently observed in ionic liquids relative to conventional molecular solvents.

Immobilisation of the catalyst in an ionic liquid does not appear to affect the reaction mechanism relative to a homogeneous system.[5] However, if charged intermediates play a role in the catalytic cycle, these are likely to be stabilised, which might have an impact on the rate and selectivity of a given reaction. Like in other biphasic transformations, special protocols for the separation of the products, such as supporting ionic liquids on solid supports or using supercritical (or dense phase) carbon dioxide for product extraction, are attracting growing attention and these will be discussed in more detail below.

3.2 Hydrogen Solubility in Ionic Liquids

There was, and still is, some ambiguity concerning the solubility of hydrogen gas in ionic liquids and its effect on activity and selectivity of hydrogenation reactions. Initially, it was believed that H_2 is considerably more soluble in ionic liquids than in conventional solvents and in some of the early papers this hypothesis was used to explain, at least in part, higher turnover frequencies relative to conventional homogeneous systems.[6] Various methods have been applied to determine hydrogen solubility,[7-9] but it is only recently that reliable data have become available. High pressure NMR spectroscopy was successfully used to determine the H_2 solubility in a number of different ionic liquids, which are listed in Table 3.1.[10]

Table 3.1: Hydrogen concentration in common solvents and selected ionic liquids at 298 K and atmospheric pressure[a)]

Solvent	H_2 [mM]	Ionic liquid	H_2 [mM]
Water	0.81	$[C_4C_1im][BF_4]$	0.86
Methanol	3.75	$[C_4C_1im][PF_6]$	0.73
Ethanol	2.98	$[C_4C_1im][SbF_6]$	0.93
Benzene	4.47	$[C_4C_1im][OTf]$	0.97
Toluene	3.50	$[C_4C_1im][Tf_2N]$	0.77
Cyclohexane	3.63	$[C_4C_1im][CF_3COO]$	0.98
		$[C_6C_1im][BF_4]$	0.79
		$[C_8C_1im][BF_4]$	0.62

a) Calculated from the solubility at 10.1 MPa, assuming that it changes linearly with partial pressure.

This method offers the advantage that only dissolved hydrogen is observed and that it is significantly more sensitive than, for example, mass-flow control. With the latter approach large volumes of ionic liquid are necessary in order to generate reliable data. The solubility of hydrogen in the majority of the ionic liquids was found to be of the same order as in water, and lower than in many common organic solvents. Only in the somewhat exotic ionic liquid $[(C_6)_3(C_{14})P][PF_3(C_2F_5)_3]$ was a significantly higher hydrogen concentration, 1.84 mM, found. The hydrogen solubility increases with temperature, but this effect is rather small. However, by using mixtures of ionic liquid and *sc*CO_2, the hydrogen concentration can be increased significantly, as is shown in Figure 3.1.[11]

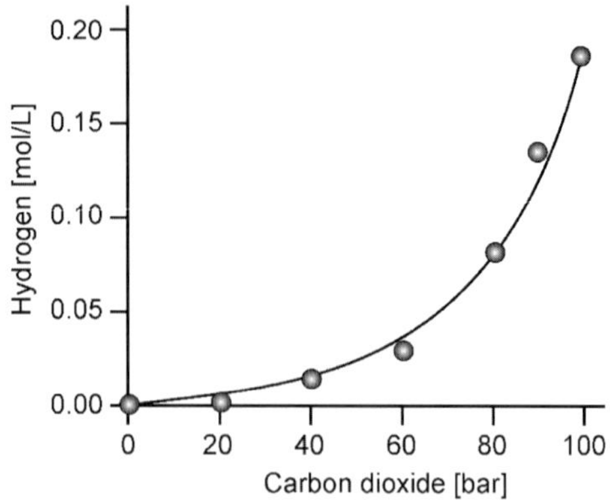

Figure 3.1: Solubility of hydrogen in $[C_2C_1im][Tf_2N]$ at $p(H_2)$ = 30 bar as a function of added CO_2 pressure

It is likely that at least some of the hydrogen that is transported to the catalyst is dissolved in the substrate rather than in the ionic liquid. Therefore it is difficult to decide whether mass-transfer of the gas into the ionic liquid is the limiting factor. From the data available it appears, however, that substrate solubility in the ionic liquid is of greater importance with respect to the reaction rate than hydrogen solubility.

An elegant study on the effect of different hydrogen solubility was undertaken by Jessop *et al.*, who investigated the enantioselective hydrogenation of tiglic and atropic acid.[12,13] Whereas tiglic acid requires a low H_2 concentration for good *ee*, the opposite applies to atropic acid.[14] The selectivity indeed showed a dependence on the ionic liquid employed. In agreement with the above solubility data, for tiglic acid a higher *ee* was obtained in $[C_4C_1im][PF_6]$ than in $[C_4C_1im][BF_4]$, 93 *vs.* 88%; for atropic acid, $[C_4C_1im][BF_4]$ gave better selectivity than $[C_4C_1im][PF_6]$, 37 *vs.* 22%.

3.3 Ionic Liquids in Hydrogenation Reactions

Imidazolium-based ionic liquids, mostly the $[BF_4]^-$ and $[PF_6]^-$ salts, are commonly employed for hydrogenation reactions and it has been observed, that anions like triflate and acetate, which are reasonably good ligands, can deactivate the catalyst.[15]

In the acidic ionic liquid $[C_2C_1im]Cl\text{-}AlCl_3$, stabilisation of Wilkinson's catalyst $RhCl(PPh_3)_3$ was observed.[16] When the complex is dissolved, a phosphine ligand dissociates and the vacant coordination site is presumably occupied by $[AlCl_4]^-$. Upon addition of hydrogen, the remaining chloride ligand is substituted for a hydride, see Scheme 3.3. Despite the stabilising effect provided by the anion, chloroaluminate ionic liquids are rarely used in hydrogenation due to their tendency to polymerise olefins.[17]

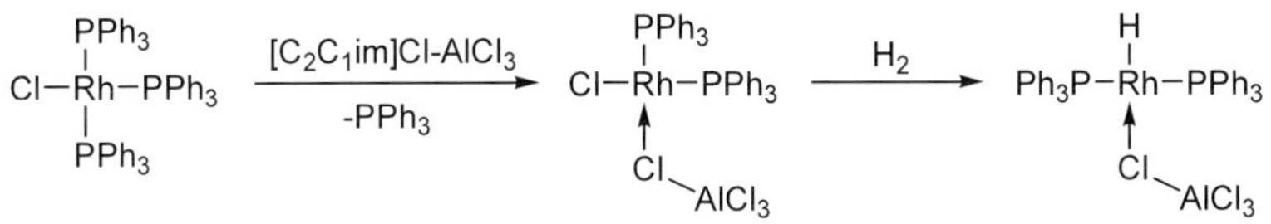

Scheme 3.3: Proposed stabilisation of $RhCl(PPh_3)_3$ in a chloroaluminate ionic liquid

A similar process probably applies when $RuCl_2(PPh_3)_3$ is dissolved in $[C_4C_1im]Cl\text{-}ZnCl_2$, and such a solution was used in the biphasic hydrogenation of 1-hexene.[18] Polymerisation was not observed, with the catalyst activity (TOF = 2,640 $mol{\cdot}mol^{-1}{\cdot}h^{-1}$) being comparable to that in $[C_4C_1im][BF_4]$. However, product separation was more facile using the chlorozincate ionic liquid as it solidifies at room temperature, allowing the product phase to be poured off.

The solubility of substrate and product in the ionic liquid may not only affect the rate of the reaction, but it can also have a significant influence on the selectivity of a given reaction. Usually, hydrogenation reactions afford products that are less polar than the starting material. If the substrate contains several functional groups that can be reduced, selectivity of the reaction may be tuned by varying the ionic liquid. An example of solvent-induced selectivity involves the reduction of dienes to monoenes. Under homogeneous conditions an alkene like butadiene can easily re-associate with the catalyst once it has been reduced to butene to undergo further hydrogenation. In ionic liquid biphasic catalysis, the solubility of the monoene is often significantly lower and the partially reduced product can be obtained with high selectivity.[19-22]

3.4 Homogeneous Catalysts

3.4.1 Monomolecular Catalysts

It is mostly complexes of ruthenium and rhodium that have been used to conduct hydrogenation reactions in ionic liquids and little attention has so far been paid to modifying the employed catalysts to improve their performance in the ionic environment. The majority of the catalysts used are identical to those employed in conventional homogeneous catalysis conducted in molecular solvents like, for example, $RhCl(PPh_3)_3$ and $RuCl_2(PPh_3)_3$.

Accordingly, many papers serve the purpose to demonstrate that a specific substrate can be reduced in the presence of an ionic liquid. It has been shown indeed that these established complexes tend to work rather well in ionic liquids and the rates of reaction often even exceed those obtained in homogeneous (single-phase) systems. In many cases, the actual catalyst is formed *in situ* by dissociation of a ligand, a step that could proceed faster in a highly ionic environment if the ligand is charged, although one that could be impeded by halide-contaminated ionic liquids. Yet, this effect cannot be generalised and it was shown that chloride dissociation from a metal centre can be thermodynamically disfavoured in ionic liquids.[23] Design of pre-catalysts bearing better leaving groups may be required to allow the generation of a vacant coordination site. It is assumed that after dissociation of an anionic ligand the resulting cationic species is stabilised in the ionic liquid, leading to increased lifetime and ultimately a higher recycling potential. In comparison to biphasic aqueous hydrogenation reactions, substrate solubility is usually significantly higher in ionic liquids, resulting in increased reaction rates.

In a number of publications, propanol is employed as a co-solvent, apparently serving a two-fold purpose: To decrease the viscosity of the ionic liquid and to assist in the catalyst-forming step. With common ruthenium catalyst precursors of the type $Ru(OAc)_2(PP)$ (PP = chelate diphosphine), a reaction sequence as shown in Scheme 3.4 has been proposed to account for catalyst activation.[24]

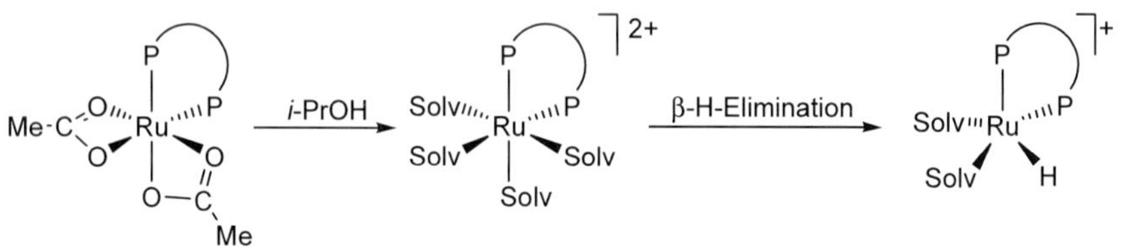

Scheme 3.4: Catalyst activation via β-hydrogen elimination from 2-propanol

The presence of polar co-solvents such as simple alcohols in biphasic systems could lead to increased catalyst leaching, even if the active complex is charged, and ligand-modification may be necessary in order to minimise loss of the catalyst.

Ideally, biphasic catalysis is performed in such a way that mass-transfer from one phase to the other does not restrict the rate of the reaction. An elegant solution to overcome this potential limitation is reversible two phase-single phase reaction conditions. An example of a temperature-controlled reversible ionic liquid-water partitioning system has been demonstrated for the hydrogenation of 2-butyne-1,4-diol, see Figure 3.2.[25]

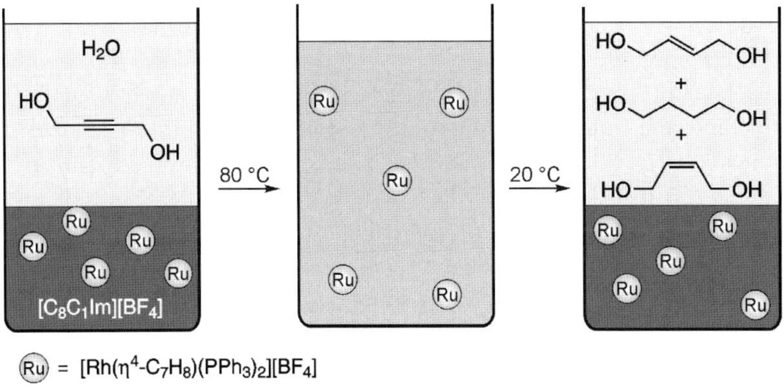

Figure 3.2: Hydrogenation under reversible two phase-single phase reaction conditions

An ionic liquid with a relatively long alkyl chain, $[C_8C_1im][BF_4]$, was selected as it is forms two phases with water at low temperature, but above 80°C gives rise to a homogeneous system. Using the hydrophobic and ionic catalyst $[Rh(nbd)(PPh_3)_2][BF_4]$, 2-butyne-1,4-diol was hydrogenated at 60 bar, 80°C to give a mixture of the corresponding alkene and alkane. After the reaction, the system was cooled to room temperature and two phases formed enabling facile separation of the products without catalyst contamination. The ionic liquid-catalyst system could be reused several times without depreciation in activity. Some characteristics of this particular system are not yet optimal in that, for example, the chosen ionic liquid contains *ca.* 12% water at ambient temperature and therefore not all of the formed product is obtained by decantation. Although 2-butyne-1,4-diol is a relatively easy substrate to hydrogenate and, for example, has been hydrogenated in supercritical CO_2 with the stainless steal reactor serving as the catalyst,[26] the reaction in $[C_8C_1im][BF_4]$ illustrates the potential of eliminating organic solvents completely from reactions involving water soluble substrates.

There are many examples of hydrogenation reactions in ionic liquids and only a selection will be shown in greater detail, however a summary of hydrogenation reactions performed in ionic liquids is provided in Table 3.2.

Table 3.2: Overview of hydrogenation catalysis in ionic liquids

Ionic liquid	Catalyst	Substrate	Comments	Ref
Homogeneous Ruthenium Catalysts				
$[C_4C_1im][BF_4]$	$RuCl_2(PPh_3)_3$	Olefins	25-40 bar, 30-60 °C; <3 % catalyst leaching; products isolated by decantation; catalyst solution remains active for at least four cycles.	[21]
$[C_4C_1im]Cl\text{-}ZnCl_2$	$RuCl_2(PPh_3)_3$	1-Hexene	25 bar, 60 °C; no apparent deactivation after the fifth cycle; product separation facile as ionic liquid solidifies at room temperature.	[18]
$[C_4C_1im][PF_6]$	$RuCl_2(TPPMS)_3(DMSO)$	1-Hexene	15-40 bar, 60-120 °C; product isolated by decantation; pressure and temperature effects studied.	[27]
$[C_4C_1im][PF_6]$	$RuCl_2(DPPE)_2$	$CO_2 + NH(n\text{-}C_3H_7)_2$	55 bar, 80 °C; *N,N*-dipropylformamide as product; good selectivity; product extracted with *sc*-CO_2.	[28]
$[C_8C_1im][BF_4]$	$[Ru(nbd)(PPh_3)_2]^+$	2-Butyne-1,4-diol	60 bar, 80 °C; product isolated by decantation; reversible ionic liquid-water two phase-single phase system.	[25]
$[C_xC_yim][BF_4]$	$RuHCl(CO)(PCy_3)_2$	Acrylonitrile butadiene rubber	10-80 bar, 100-160 °C; toluene as co-solvent; product isolated by decantation; catalyst can be reused several times without change in activity and selectivity.	[6] [29]
$[C_4C_1im][BF_4]$	$[RuCl_2(BINAP)]_2$	2-Arylacrylic acids	25-100 bar, 20 °C; *i*PrOH as co-solvent; catalyst active for at least four cycles; product isolated by decantation.	[30]
$[C_4C_1im][PF_6]$	$[RuCl_2(tol\text{-}BINAP)]_2$	Methyl-3-hydroxy-2-methylenebutanoate	*i*PrOH as co-solvent; influence of H_2-pressure studied.	[7]
$[C_4C_1im][PF_6]$ large variety	$Ru(OAc)_2(tol\text{-}BINAP)$	Tiglic acid Atropic acid	5-100 bar, 20 °C; *i*PrOH as co-solvent, H_2O as co-solvent; product extracted with *sc*CO_2; catalyst solution can be reused at least four times; studies of different H_2-pressure, co-solvents and additives on enantioselectivity; studies on the *ee* by changing ionic liquid, strong effect of the anion; effect of CO_2-expanded ionic liquids.	[12] [13]

Ionic liquid	Catalyst	Substrate	Comments	Ref
$[C_4C_1im][PF_6]$ $[C_4C_1im][BF_4]$ $[C_3C_1C_1im][Tf_2N]$	$RuCl_2$(4,4'-subst.-BINAP)$(DMF)_2$	*β*-Keto esters	105 bar, 25°C; methanol as co-solvent; in $[C_3C_1C_1im][Tf_2N]$ activity is unchanged for four cycles.	[31] [32] [33]
$[C_4C_1im][PF_6]$ $[C_4C_1im][BF_4]$ $[C_3C_1C_1im][Tf_2N]$	$RuCl_2$(4,4'-subst.-BINAP)(diamine)	Aromatic ketones	50 bar, *i*PrOH as co-solvent; presence of acidic proton on the imidazolium deteriorates catalytic activity; catalyst activity decreases markedly after the third run.	[33]
large variety	$RuBr_2$(disubst.-BINAP)	Ethyl acetoacetate	40 bar, 50 °C; *ee* dependent on the nature of the ionic liquid anion; selectivity increases in the second cycle; small amounts of water lead to a drastic decrease in selectivity.	[34]
$[C_4C_1im][PF_6]$ $[C_4C_1im][OTf]$	$[RuCp^*(\eta^4\text{-sorbic acid})]^+$	Sorbic acid	16 bar, 20-60 °C; MTBE as co-solvent; higher rates than in ethylene glycol; no catalytic activity in $[C_4C_1m][OTf]$ due to coordination of the triflate; catalyst reused three times; detailed kinetic studies; product isolated by decantation.	[15]
$[C_{1\text{-}8}C_1im][BF_4]$	$RuCl_2$(*p*-cymene)(PTA)	Arenes	60 bar, 90 °C; chloride-free ionic liquids are important to achieve high TOF; catalyst solution can be reused for at least five runs.	[35]
$[C_4C_1im][BF_4]$	$[RuCl(p\text{-cymene})(TRIPHOS)]^+$	Arenes	60 bar, 90 °C; catalyst now believed to be heterogeneous.	[36]
		Homogeneous Rhodium Catalysts		
$[C_4C_1im][PF_6]$	$RhCl(PPh_3)_3$	Olefins	48 bar, 50 °C; *sc*CO_2 as co-solvent, product extracted with *sc*CO_2.	[28]
$[C_4C_1im][BF_4]$	$RhCl(TPPTS)_3$	Polymer	90-100 °C, 3 bar; toluene as co-solvent; H_2O as co-solvent; product isolated by decantation and additional extraction.	[37]
$[C_4C_1im][BF_4]$ $[C_4C_1im][PF_6]$ $[C_4C_1im]Cl\text{-}AlCl_3$	$RhCl(PPh_3)_3$ $[Rh(cod)_2]^+$	Cyclohexene	25 °C, 10 bar; product isolated via decantation, < 2% catalyst leaching; $[C_4C_1im]Cl\text{-}AlCl_3$ leads to polymerisation of the substrate.	[17]

Ionic liquid	Catalyst	Substrate	Comments	Ref
$[C_4C_1im]$[various]	$[Rh(nbd)(PPh_3)_2]^+$	Olefins	30 °C, 1 bar; strong influence of the nature of the anion due to different substrate solubility; reuse of catalyst solution possible; leaching <0.02 %.	[3]
$[C_4C_1im][SbF_6]$	$[Rh(cod)(DIOP)]^+$	α-Acetamido cinnamic acid	30 °C, 1 bar; *i*PrOH as co-solvent.	[3]
$[C_4C_1im][PF_6]$	$[Rh(cod)(Et\text{-}DuPHOS)]^+$	α-Acetamido cinnamic acid	5-100 bar, 20°C; *i*PrOH as co-solvent; *ee*-value dependent on hydrogen concentration in the ionic liquid; significant decrease in activity after the fourth cycle due to leaching; influence of H_2-pressure studied.	[7]
$[C_4C_1im][PF_6]$ $[C_8C_1im][BF_4]$ $[C_4py][CB_{11}H_{12}]$	Rhodocarborane + (*R*)-BINAP	Acetophenone Benzylformate	12 bar, 50°C; enantioselectivity in ionic liquid much higher than in THF; product isolated by distillation.	[38]
$[C_4C_1im][PF_6]$	$[Rh(cod)(Me\text{-}DuPHOS)]^+$	Enamides	2 bar, 25 °C; *i*PrOH as co-solvent; catalyst recycling possible.	[39]
$[C_4C_1im]$[various] $[C_8C_1im][BF_4]$	$Rh(ndb)_2$ + chiral bidentate phosphines	Enamides	1 bar, 20°C; methanol, *i*PrOH or water as co-solvent.	[40]
$[C_4C_1im][SbF_6]$	Rh(cod) with ionic phosphine ligand	Enamides	1 bar, 20°C; activity drops after the third cycle; enantioselectivity remains basically unchanged.	[41]
		Other Metals		
$[C_4C_1im]BF_4$	$K_3[Co(CN)_5]$	Butadiene	25 bar, 25-60 °C; selectivity dependent on reaction temperature; no recycling possible.	[21]
$[C_4C_1im][BF_4]$ $[C_4C_1im][PF_6]$ $[C_4C_1im][OTf]$	$Co(acac)_2$ $Co(acac)_3$ $Ni(acac)_2$ $Fe(acac)_3$	Butadiene	15-45 bar, 25-70 °C; investigation on catalyst concentration; $[OTf]^-$ as anion significantly changes the selectivity in favour of butane; recycling is possible; TOFs are rather low.	[20]
$[C_4C_1im][BF_4]$	$[Ir(H)_2(NCCH_3)_3(P\textit{i}Pr_3)]^+$	Alkynes	Dimerisation/hydrogenation, depending on the substrate; toluene as co-solvent; product isolated by distillation.	[5]

Ionic liquid	Catalyst	Substrate	Comments	Ref
various	Ir-XYLIPHOS	Trimethylindoline	40 bar, 50°C; no significant ionic liquid anion effect; lipophilic ionic liquids give higher TOFs and *ee*'s.	[42]
$[C_nC_1im][various]$ $[C_4C_1py][Tf_2N]$	$[IrCl(cod)]_2$ + PN-ligand	Imines	30-100 bar, 0-40°C; *sc*CO_2 as co-solvent; product extracted with *sc*CO_2.	[11]
$[C_4C_1im][PF_6]$ $[C_4C_1im][BF_4]$	$Pd(acac)_2$	Dienes	5-25 bar, 50 °C; product isolated by decantation; selectivity to monoene product similar in conventional solvents; less than 0.1% leaching; catalyst solution active for more than 15 runs.	[19]

The hydrogenation of sorbic acid to afford *cis*-3-hexanoic acid proceeds with up to 93% selectivity in $[C_4C_1im][PF_6]$ with $[RuCp^*(\eta^4\text{-sorbic acid})][OTf]$, **1**, as catalyst, as shown in Scheme 3.5.[15] Increasing the hydrogen pressure led to a drastic decrease in selectivity whereas the effect of temperature was found to be less pronounced.

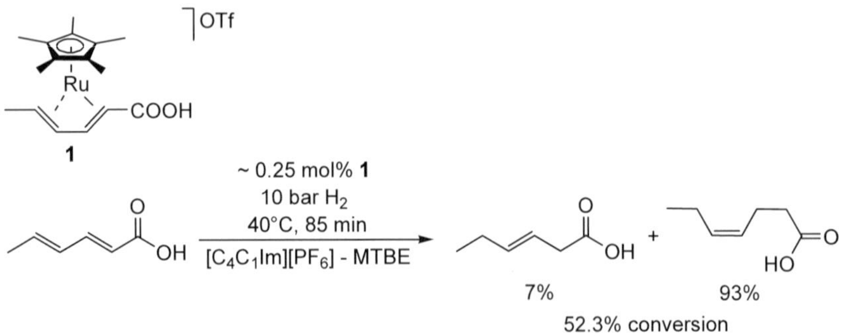

Scheme 3.5: Reduction of sorbic acid with $[RuCp^*(\eta^4\text{-sorbic acid})][OTf]$ as catalyst

Only very few examples of homogeneous catalysts based on other metals have been reported, namely iridium, cobalt and palladium. With palladium(II) acetylacetonate, highly selective (up to 100%) hydrogenation of dienes to monoenes was achieved. The high selectivity was attributed to the formation of a thermodynamically favoured η^3-allyl Pd-intermediate and also to the higher solubility of the dienes in the ionic phase compared to the monoenes.[19] The catalytic system can be recycled up to fifteen times following a simple phase separation.

The cobalt salt $K_3[Co(CN)_5]$ was evaluated for the selective hydrogenation of dienes in $[C_4C_1im][BF_4]$ and at low temperature gave 100% selectivity for the reduction of butadiene to 1-butene.[21] The catalytic system was, however, deactivated after one run and it is proposed that strong hydrogen bonding interactions between $[C_4C_1im]^+$ cations and $[Co(CN)_5]^{3-}$, to form a species of the formula $[C_4C_1im]_3[Co(CN)_5]$, prevent the coordination of the diene substrate to the metal centre.

Of particular importance are enantioselective hydrogenation reactions. Here, the usually expensive chiral ligands make recyclable systems particularly attractive. Enantioselective hydrogenation in an ionic liquid-propanol mixture was first attempted with $[Rh(cod)(DIOP)][PF_6]$, **2**, in $[C_4C_1im][SbF_6]$ for the reduction of α-acetamidocinnamic acid to give (*S*)-phenylalanine, see Scheme 3.6.[3] The selectivity was only moderate, 64% *ee*, but catalyst recycling was possible with less than 0.02% rhodium detected in the organic phase.

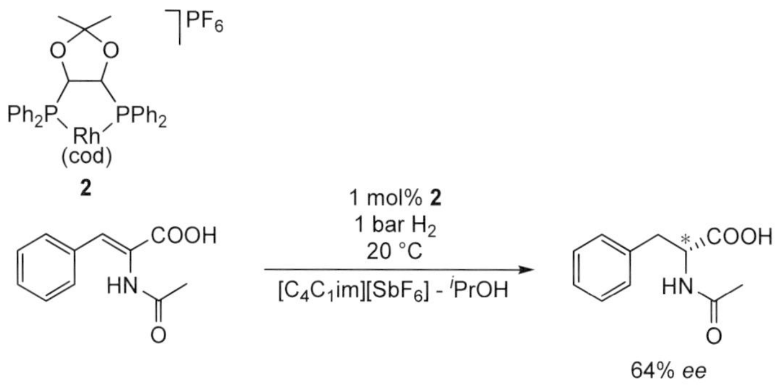

Scheme 3.6: Enantioselective reduction of α-acetamidocinnamic acid

Chiral rhodium-DuPHOS complexes are highly efficient catalyst for the enantioselective hydrogenation of enamides. One drawback of these catalysts is that they are easily oxidised and inert conditions are required for optimal results. The methyl- and ethyl substituted Rh-DuPHOS compounds, **3a** and **3b**, have been successfully applied in the reduction of α-acetamidocinnamic acids in $[C_4C_1im][PF_6]$, Scheme 3.7.[7,39] While activities and selectivities are slightly lower compared to the homogeneous reaction in 2-propanol, the ionic liquid-immobilised catalyst is less prone to oxidation and recycling is feasible at least three times.

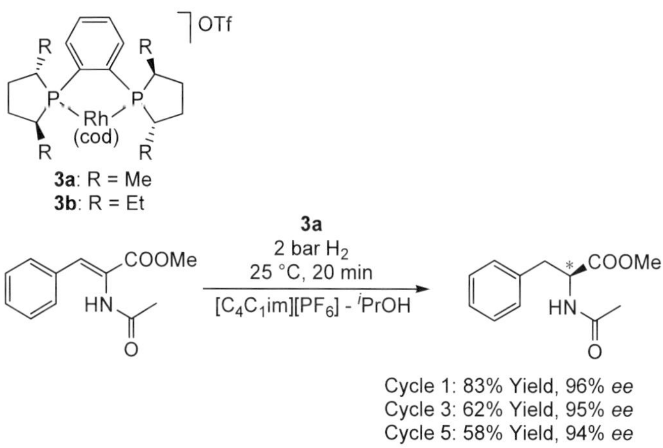

Scheme 3.7: Asymmetric reduction of α-acetamidocinnamate

Numerous other bidentate phosphine ligands have also been tested in this reaction.[40] Interestingly, addition of water to the employed ionic liquids $[C_8C_1im][BF_4]$ and $[C_4C_1im][Tf_2N]$ led to increased *ee*s and/or reaction rates compared to common organic solvents when ferrocene-based di-phosphines such as **4** and **5**, shown in Figure 3.3, were used.

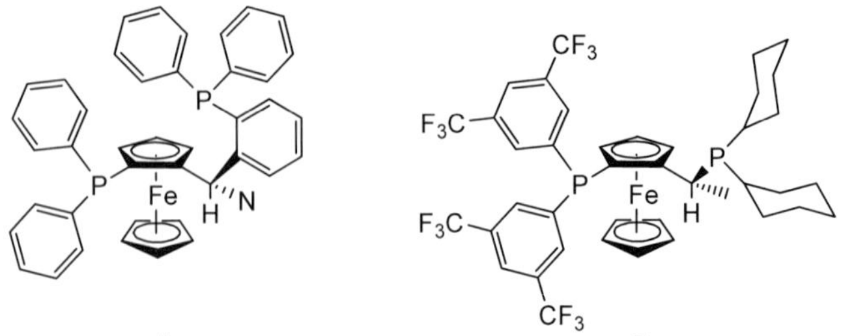

4 **5**

Figure 3.3: TANIAPHOS and JOSIPHOS-type ligands used in the asymmetric reduction of enamides

Modification of a phosphine ligand with imidazolium substituents affords the highly charged rhodium catalyst **6**, illustrated in Scheme 3.8, which was successfully used for the enantioselective reduction of *N*-acetyl-phenylethenamine in $[C_4C_1im][SbF_6]$/2-propanol.[41] Use of this charged phosphine ligand reduced catalyst leaching to below ICP-OES detection limits, relative to 2% catalyst loss with the underivatised ligand, and this is mirrored in improved retention of catalytic activity.

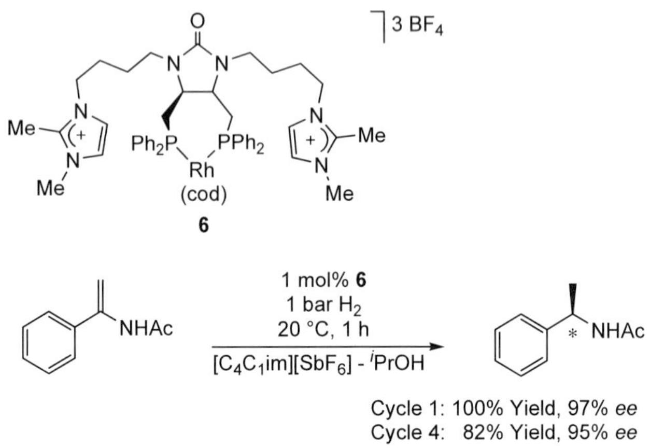

Scheme 3.8: Asymmetric reduction of *N*-acetylphenylethenamine

The enantioselective hydrogenation of imines was also undertaken with iridium complexes bearing bidentate phosphine[42] or phosphine-oxazoline[11] ligands. With the latter ligands the presence of CO_2 as co-solvent was found to lead to an increase in catalytic activity in the reduction of *N*-(phenylethylidene)aniline when $[C_2C_1im][Tf_2N]$ was used. With other ionic liquids, rate enhancement was also observed, though to a lesser extent. The positive effect was attributed to both reduced viscosity and increased hydrogen solubility in the solvent mixture. Besides the structure of the ligand, the selectivity of the reaction depended on the coordinating ability of the ionic liquid anion and highest *ee*s (78%) were found with $[C_2C_1im][BARF]$, while $[C_4C_1im][BF_4]$ gave only 30% *ee*. Using *sc*CO_2 to extract the product gave a highly stable catalytic system with no significant decrease in activity after seven runs.

Instead of charged moieties, the presence of polar functional groups may be sufficient to enhance catalyst immobilisation in ionic liquids. The ruthenium complex **7a**, containing a BINAP ligand with phosphonic acid substituents in the 4,4'-position, has been employed in the asymmetric hydrogenation of β-keto esters, see Scheme 3.9.[31,32] Bulky or electron-donating substituents in this position were found to markedly increase the enantioselectivity relative to unsubstituted BINAP. In $[C_4C_1C_1im][Tf_2N]$, deterioration in activity and selectivity occurs only after the fourth cycle, whereas in $[C_4C_1im][BF_4]$ and $[C_4C_1im][PF_6]$, deactivation takes place much more rapidly. For both **7a** and **7b** no appreciable leaching occurred during product extraction and it is believed that loss of catalytic activity is caused from decomposition of the active hydride species.

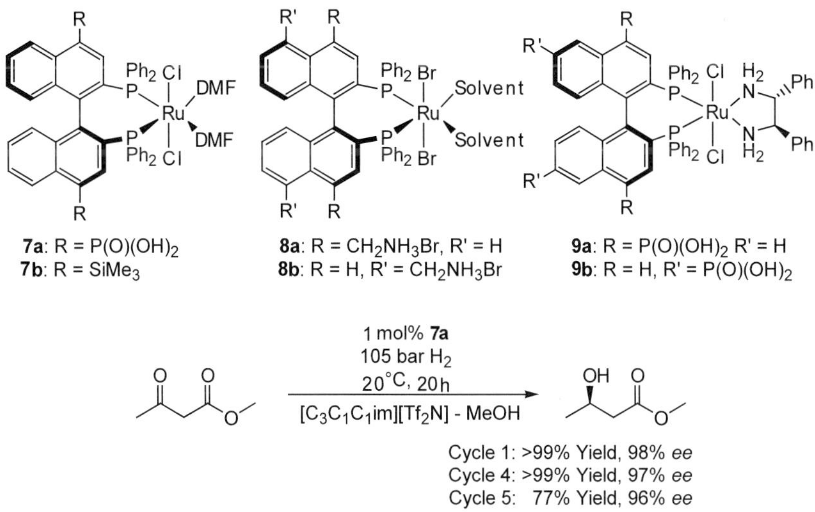

Scheme 3.9: Asymmetric hydrogenation of *β*-keto esters

When **8a** and **8b** are employed as catalysts for the hydrogenation of ethyl acetoacetate, the enantioselectivity shows a pronounced dependence on the nature of the ionic liquid anion with $[PF_6]^-$ giving the lowest selectivity, as shown in Table 3.3.[34] An increase in selectivity was observed after recycling of the catalyst solution, but this observation is not yet understood. The presence of small amounts of water were found to be detrimental to the *ee*. Unfortunately, recycling of the catalyst was restricted to two runs and it is therefore not clear whether the presence of the charged ammonium groups on the BINAP ligand offer any advantage relative to the polar, uncharged groups.

Table 3.3: Effect of the ionic liquid on the enantioselectivity in the reduction of ethyl acetoacetate with catalyst **8a** (50°C, 40 bar H_2)

Ionic Liquid	*ee* [%]
$[C_4C_1im][PF_6]$	75
$[C_4C_1im][BF_4]$	86
$[C_4C_1im][Tf_2N]$	78
$[C_4py][Tf_2N]$	85

Similar complexes, **9**, with DPEN as ligand were tested in the hydrogenation of aromatic ketones.[33] With **9a**, excellent enantioselectivity was achieved while **9b** gave much poorer results. The presence of an acidic proton on the imidazolium led to inferior results, presumably due to catalyst deactivation via formation of carbene complexes. Accordingly, best activities were obtained with $[C_3C_1C_1im][Tf_2N]$. After three catalytic runs, activity decreased markedly, however, the selectivity remained high.

A somewhat more unusual system consists of the rhodacarborane catalyst precursor [*closo*-1,3-{μ-(η^2-3-CH_2=$CHCH_2CH_2$)}-3-H-3-PPh_3-3,1,2-$RhC_2B_9H_{10}$], **10,** illustrated in Scheme 3.10, which contains what is termed a suicide alkene ligand,. Together with the chiral co-ligand (*R*)-BINAP, it has been used to catalyse the reduction of ketones to alcohols under hydrogenation conditions.[38] The catalyst was immobilised in either $[C_8C_1im][BF_4]$, $[C_4C_1im][PF_6]$ or an ionic liquid containing a 1-carbadodecaborate anion, $[C_4py][CB_{11}H_{12}]$. The latter ionic liquid showed a beneficial effect, which was attributed to a catalytically active species in which the carborane anion of the ionic liquid is bound to the metal centre. In the hydrogenation of acetophenone and ethyl benzoylformate to their corresponding alcohols, quantitative conversion and high *ee* (>97%) were obtained in the ionic liquids, while only moderate conversion and enantioselectivity was observed in THF. The catalyst could be recycled at least six times without deactivation or loss of selectivity following the removal of the reaction product under high vacuum.

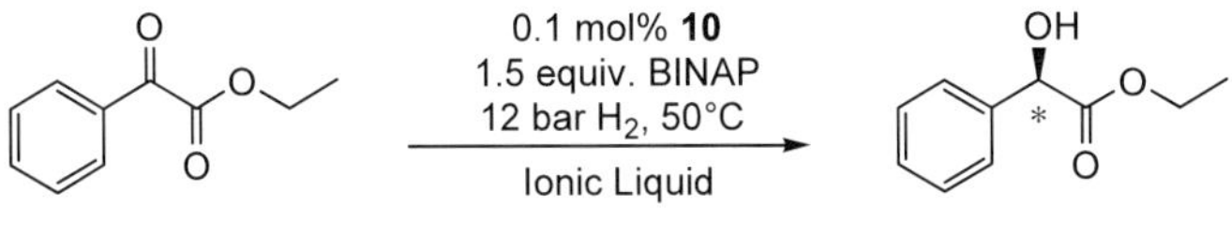

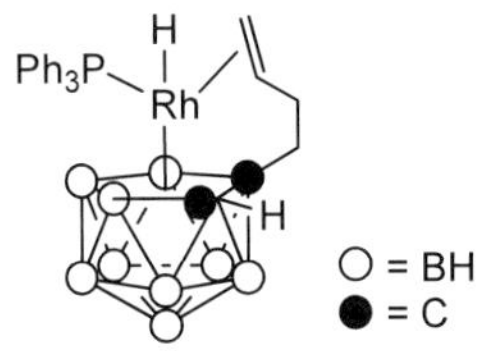

10

Solvent	TOF [h⁻¹]	*ee* [%]
THF	107	85.7
$[C_4C_1Im][BF_4]$	213	98.2
$[C_8C_1Im][PF_6]$	201	99.3
$[C_4Py][CB_{11}H_{12}]$	306	99.5

Scheme 3.10: Asymmetric hydrogenation of ethyl benzoylformate with a rhodacarborane catalyst

3.4.2 Cluster-Catalysed Hydrogenation Reactions

Although the true nature of the catalyst is not always clear when mononuclear complexes are used as pre-catalysts in hydrogenation reactions, with cluster compounds employed as the pre-catalyst, it is frequently found that the active species is colloidal and review articles are available on how to distinguish between homogeneous and heterogeneous regimes.[43,44] A summary of hydrogenation reactions with metal-clusters in ionic liquids is provided in Table 3.4.

Table 3.4: Cluster-catalysed hydrogenation reactions in ionic liquids

Ionic liquid	Cluster	Comments	Ref
$[C_4C_1im][BF_4]$	$[Ru_4H_4(\eta^6\text{-}C_6H_6)_4]^{2+}$	60 bar, 90°C; arenes as substrate; similar performance as in water; catalyst solution could be reused several times, product isolated by distillation or decantation.	[35] [45]
$[C_4C_1im][BF_4]$	$[Ru_6C(CO)_{16}]^{2-}$	50 bar, 100°C; styrene and other olefins as substrate; active species probably colloidal; with styrene similar activity to conventional solvents; cyclohexadiene is selectively reduced to cyclohexene.	[22]
$[C_4C_1im]$[various] $[C_4C_1C_1im]$[various] $[C_8C_1py][BF_4]$	$[Os_4H_3(CO)_{12}]^-$ $[WOs_3H(CO)_{12}]^-$	50 bar, 100°C; styrene as substrate; significantly higher activity in ionic liquid than in octane or methanol; cluster is probably the active species.	[22]

The performance of four different anionic carbonyl clusters in the hydrogenation of styrene, *viz* $[HFe_3(CO)_{11}]^-$, $[HWOs_3(CO)_{14}]^-$, $[H_3Os_4(CO)_{12}]^-$ and

$[Ru_6C(CO)_{16}]^{2-}$ has been investigated in octane, methanol and $[C_4C_1im][BF_4]$.[22] The activities of both the ruthenium and the iron cluster are essentially independent of the nature of the solvent – the latter being almost inactive – whereas performance of the osmium-containing cluster is considerably superior in the ionic liquid, but much lower than for the ruthenium compound. While $[Ru_6C(CO)_{16}]^{2-}$ decomposes under the reaction conditions to form colloidal ruthenium, $[H_3Os_4(CO)_{12}]^-$ appears to remain intact and the superior turnover frequencies in the ionic liquid relative to the molecular solvents are likely to be due to better stabilisation of the charged cluster catalyst in the ionic liquid.

Arene-hydrogenation was accomplished with the tetranuclear complex $[Ru_4H_4(\eta^6\text{-}C_6H_6)_4][BF_4]_2$ in $[C_4C_1im][BF_4]$[35,45] and this cluster has also been used to evaluate the effect of hydrogen concentration on catalytic activity.[10] The chloride-salt of the same cluster has already been tested in aqueous biphasic catalysis,[46] and turnover frequencies in ionic liquid biphasic catalysis are of the same order of magnitude. The main advantage relative to the aqueous process lies in the ease of product separation – especially if different arenes are to be reduced with the same batch of ionic liquid; in aqueous catalysis arene traces remain in the water and they are difficult to remove. Under the conditions applied, 60 bar at 90°C, it cannot be excluded that at least part of the catalytic activity is due to formation of nanoparticles, although experiments have not been performed to test this hypothesis.

3.4.3 Transfer-Hydrogenation

Examples of aqueous biphasic transfer-hydrogenation are rare,[47-49] and likewise, only few reports on transfer-hydrogenation in ionic liquids are available. Ionic liquid biphasic transfer-hydrogenation certainly suffers from the fact that highly polar hydrogen sources such as 2-propanol or formic acid/triethylamine azeotrope are required and only very few ionic liquids will form a stable biphasic system with these reagents. Furthermore, strong bases are often employed to generate the active catalyst *in situ*, which further limits the choice of ionic liquid. As such, when imidazolium-based ionic liquids are used it is necessary to replace the proton at the 2-position with, for example, an alkyl group, otherwise deprotonation and the subsequent formation of carbene complexes is highly likely to take place. An overview of transfer-hydrogenation in ionic liquids is provided in Table 3.5.

Table 3.5: Transfer-hydrogenation reactions in ionic liquids

Ionic Liquid	Catalyst	Comment	Ref
$[C_4C_1C_1im][PF_6]$	Chiral Ru catalyst with imidazolium tag on the η^6-arene	35-40°C; acetophenone as substrate; *i*PrOH/KOH or formic acid/NEt_3 azeotrope as hydrogen source; use of general precursor for biphasic catalysis; excellent *ee*; catalyst reuse possible four times.	[50]
$[C_4C_1im][BF_4]$ $[C_4C_1im][PF_6]$	Chiral Ru catalyst with imidazolium tag on the diamine ligand	20°C; arylketones as substrate; formic acid/NEt_3 azeotrope as hydrogen source; catalyst recycled four times; *ee* > 90%.	[51]
$[C_2R_3P][OTs]$	$[Rh_2(OAc)_4]$ + ligand	120-150°C; acetophenone as substrate; *i*PrOH/KOH as hydrogen source; rather low activity.	[52]
$[C_4C_1im][PF_6]$	Pd/C	150°C; functionalised arenes as substrate; ammonium formate as hydrogen source; microwave-accelerated catalysis; reduction of C=C, C≡C and NO_2.	[53]

Activated carbon containing 10% palladium has been used in the microwave-assisted transfer hydrogenation of various organic substrates with nitro, alkene and alkyne groups at high temperature, ca. 150°C.[53] These functional groups were reduced selectively in substrates also containing C=O or C≡N bonds. The catalyst was immobilised in $[C_4C_1im][PF_6]$ and either ammonium formate or triethylammonium was used as hydrogen source, see Scheme 3.11. The products were isolated in high purity after extensive extraction of the ionic liquid phase with methyl-*tert*-butyl ether and the catalyst could be reused, but its activity decreased notably with each run.

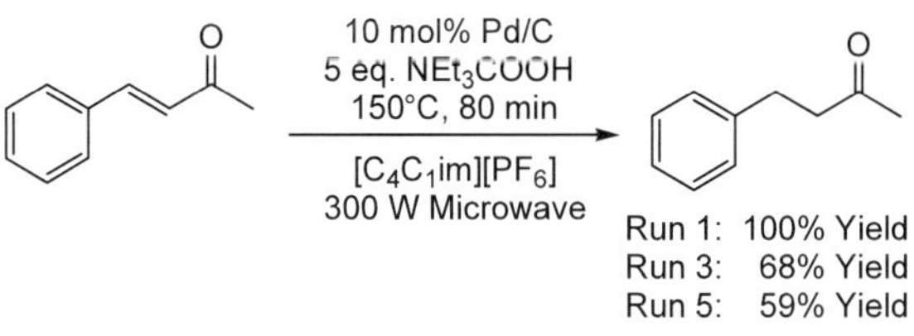

Scheme 3.11: Microwave-assisted transfer-hydrogenation of cinnamaldehyde with palladium on activated charcoal

A variety of phosphonium tosylate ionic liquids have been used for the enantioselective transfer-hydrogenation of acetophenone with a rhodium catalyst using 2-propanol as the hydrogen source.[52] The active catalyst is formed *in situ* from $Rh_2(OAc)_4$ and an excess of chiral ligand, either (*1S,2R*)-*cis*-aminoindanol or (-)-DIOP with only the latter ligand affording good selectivities. Reactions were performed at relatively high temperatures, typically 120-150°C, but conversion after 16 hours was only poor to moderate.

Due to the high melting point of the employed tosylates (between 68°C and 152°C), product separation is facile at the end of the reaction.

Complexes of the type [Ru(η^6-arene)(L)Cl]$^+$ (L = diamine or aminoalcohol) are widely used catalysts for enantioselective transfer-hydrogenation reactions.[54] In conjunction with highly polar co-solvents/reagents, modification of the catalyst is required to ensure sufficient retention in the ionic liquid. Attaching an imidazolium tag to the η^6-bound arene of the common catalyst precursor [Ru(η^6-arene)Cl_2]$_2$ results in a versatile starting material for biphasic catalysis, **11**, which allows the use of conventional, easily accessible ligands, as shown in Scheme 3.12.[50] The utility of such modified dimeric complexes was demonstrated in the highly enantioselective transfer-hydrogenation of acetophenone and related compounds.

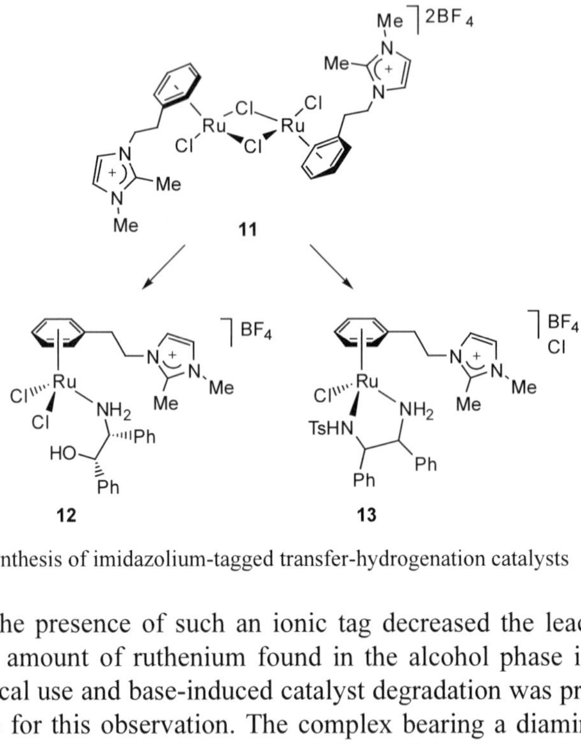

Scheme 3.12: Synthesis of imidazolium-tagged transfer-hydrogenation catalysts

Although the presence of such an ionic tag decreased the leaching significantly, the amount of ruthenium found in the alcohol phase is still too high for practical use and base-induced catalyst degradation was proposed to be responsible for this observation. The complex bearing a diamine-ligand, **13**, proved to be more stable than that containing the aminoalcohol, **12**. An ionic liquid solution containing **12** affords only 5% yield upon recycling, whereas 66% and 57% yield were obtained with **13** in the second and third run, respectively.

Despite the presence of the imidazolium tag, best results were obtained with the conventional catalyst **14** in combination with formic acid/triethyl amine azeotrope as hydrogen source and only after the fifth cycle was a marked decrease in activity observed (see Scheme 3.13). As this reagent forms a homogeneous phase together with $[C_4C_1C_1im][PF_6]$, a product extraction step with hexane or diethyl ether was also necessary. The remaining solution can be recharged with ketone and formic acid and reused. However, this gradually leads to increasing volume of the reaction solution and thus decreased turnover frequency due to dilution. Accordingly, the solution containing the catalyst was be washed with water after product extraction and dried *in vacuo* prior to the next catalytic cycle. This procedure could not be applied when complex **13** was used as the presence of the tag also increases the hydrophilicity of the catalyst.

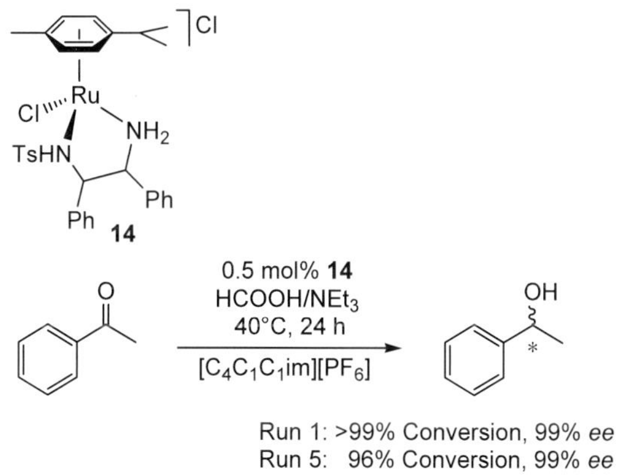

Scheme 3.13: Transfer-hydrogenation of acetophenone with $HCOOH/NEt_3$ as hydrogen source

Improved catalyst recycling was also achieved in the reduction of acetophenone with formic acid/triethyl amine azeotrope with the η^6-benzene analogue of **14** bearing an imidazolium moiety on the diamine ligand instead of on the coordinated arene (see Figure 3.4).[51]. In $[C_4C_1im][PF_6]$ the stability of the catalyst was superior in repeated runs when a mixture of $[RuCl_2(\eta^6\text{-benzene})]_2$ and **15** was employed rather than the neutral complex $RuCl(\eta^6\text{-benzene})(TsDPEN)$.

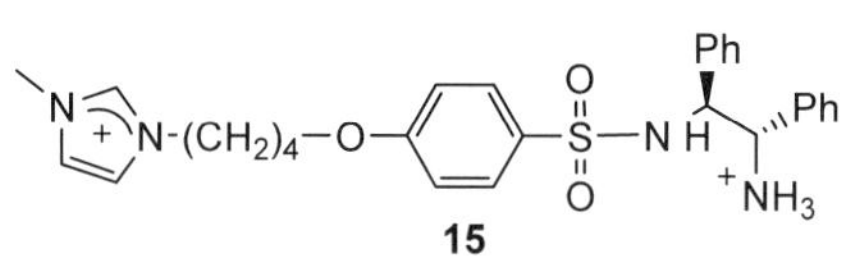

Figure 3.4: A chiral DPEN-ligand bearing an imidazolium tag

3.5 Nanoparticle Catalysts

Soluble transition metal nanoparticles (or nanoclusters) have attracted increasing attention in recent years and their preparation, structural characterisation and application as catalysts in alternative solvents are currently under intense scrutiny.[55-59] Modern nanoparticle catalysts are defined by their controlled composition, size, surface-stabilising molecules and solubility in an appropriate solvent. Due to their intrinsic charge and steric bulk, ionic liquids can create an electrostatic and steric stabilisation of nanoparticles similar to polyoxo-anions and tetraalkylammonium salts.[55,60-62] The interaction between ionic liquids and nanoparticles is demonstrated by the removal of gold nanoparticles from water with $[C_4C_1im][PF_6]$.[63] Accordingly, ionic liquids are finding increased use as media for the controlled preparation of nanoparticles which are often obtained within a fairly narrow size distribution. Functional groups on the ionic liquid like thiols[64,65] or nitriles[66] can further influence the size, solubility properties and stability of nanoparticles. A summary of nanoparticle-catalysed hydrogenation reaction is given in Table 3.6.

Nanoparticles of ruthenium, rhodium and iridium have been prepared in ionic liquids and tested in biphasic hydrogenation by Dupont and co-workers. Ruthenium nanoparticles of *ca.* 2.5 ± 0.4 nm were obtained *in situ* in $[C_4C_1im][PF_6]$ in the reduction of 1-hexene with RuO_2.[67] Depending on the anion of the ionic liquid, the reaction rate varies between 147 and 943 $mol{\cdot}mol^{-1}{\cdot}h^{-1}$, which compares with 3,300 $mol{\cdot}mol^{-1}{\cdot}h^{-1}$ in neat substrate. Whereas the particles formed in neat 1-hexene quickly lose their activity, reaction rates remain relatively stable when the catalyst is immobilised in the ionic liquid and reuse is possible more than 15 times. Ruthenium particles have also been applied to the partial hydrogenation of benzene.[68] A solution of Ru(cod)(cot) in $[C_4C_1im][PF_6]$ was treated with hydrogen (4 bar) at 75°C and the formed colloid isolated by centrifugation. A maximum of 39% selectivity for cyclohexene was observed, however at only 1% benzene conversion, and selectivity decreases to 11% at 15% conversion.

Although ionic liquids have a beneficial effect on stabilising the particles, the addition of further stabilisers may be necessary. Highly recyclable palladium particles of 2-5 nm size were prepared from the reduction of $Pd(OAc)_2$ with H_2 in $[C_4C_1im][PF_6]$ in the presence of 25 mol% phenanthroline.[69] Even after ten repeated hydrogenation runs of cyclohexene no deactivation was observed. In the absence of phenanthroline, black palladium particles precipitate and activity rapidly decreases.

Table 3.6: Overview of nanoparticle-catalysed hydrogenation reactions in ionic liquids

Ionic liquid	Metal	Substrate	Comments	Ref
$[C_4C_1im]$[various]	Ru	1-Hexene	4 bar, 75 °C; mean particle size 2.5 nm; precursor RuO_2; catalyst active for at least 17 runs.	[67]
$[C_4C_1im]$[various]	Ru	Olefins Arenes	4 bar, 75 °C; mean particle size 2.6 nm; precursor Ru(cod)(cot); low selectivity for the partial reduction of benzene to cyclohexene.	[68]
$[C_4C_1im][PF_6]$	Rh Ir	Olefins Arenes	4 bar, 75 °C; mean particle size 2.1 nm (Ir) and 2.5 nm (Rh); precursor $RhCl_3{\cdot}3H_2O$ and $[IrCl(cod)]_2$; water-free conditions are important; Rh-particles conglomerate; with Ir significant loss of activity after first cycle; product isolated by distillation or decantation.	[70]
$[C_4C_1im][PF_6]$	Ir	Olefins	4 bar, 75 °C; mean particle size 2.5 nm; precursor $[IrCl(cod)]_2$; active for at least seven cycles.	[71]
$[C_4C_1im][PF_6]$	Ir	Ketones	4 bar, 75 °C; precursor $[IrCl(cod)]_2$; active for at least 15 cycles, products isolated by decantation.	[72]
$[C_4C_1im][PF_6]$	Pd	Olefins	1 bar, 20-60 °C; particle size 2-5 nm; phenanthroline-protected nanoparticles; no deactivation observed after 10 cycles, products isolated by decantation.	[69]
large variety	Pd/C	Cinnamaldehyde Citral	3-40 bar, 30-60 °C; reaction significantly slower than in common solvents but much higher selectivity; activity decreases significantly after first cycle but then remains stable; product extracted with diethyl ether.	[73]
$[C_4C_1im][PF_6]$	Pd, Pt, Rh	Olefins Benzene	1-4 bar, 40-75 °C; PVP-stabilised particles, size *ca.* 3 nm for all metals; PCl_2, H_2PtCl_6 and $RhCl_3$ as precursor.	[74]
$[C_4C_1Im][PF_6]$	Pt	Olefins Arenes	4 bar, 75 °C; mean particle size 2-3 nm; precursor $Pt_2(dba)_3$; catalyst solution active for at least seven cycles without significant decrease in TOF, products isolated by distillation or decantation.	[75]

Rhodium, palladium and platinum nanoparticles, stabilised with poly (*N*-vinyl-2-pyrrolidone) (PVP), have also been evaluated in the hydrogenation of olefins and benzene.[74] After five catalytic runs, no aggregation and change in particle size was observed.

The selective reduction of either the C=C or C=O bond of *α,β*-unsaturated aldehydes is a very important reaction.[76] The reduction of the C=C bond in cinnamaldehyde and citral has been catalysed by 10% palladium on activated carbon in a range of ionic liquids.[73] Compared with organic solvents, the reaction rates are markedly lower in the ionic liquids. Much of the metal surface is inside the pores of the carbon support and depending on the size of the ionic liquid, diffusion into these pores will be limited and pore-blockage can occur. Much higher selectivity towards the product of C=C bond reduction was achieved in ionic liquids compared to organic solvents. By tuning the reaction conditions such as temperature and H_2 pressure, and using the appropriate ionic liquid, 100% conversion and 100% selectivity may be achieved. Of the 12 different ionic liquids screened, 100% conversion and selectivity could be obtained for the reduction of cinnamaldehyde to hydrocinnamaldehyde in $[C_4C_1im][BF_4]$ and $[C_4C_1im][PF_6]$, see Scheme 3.14.

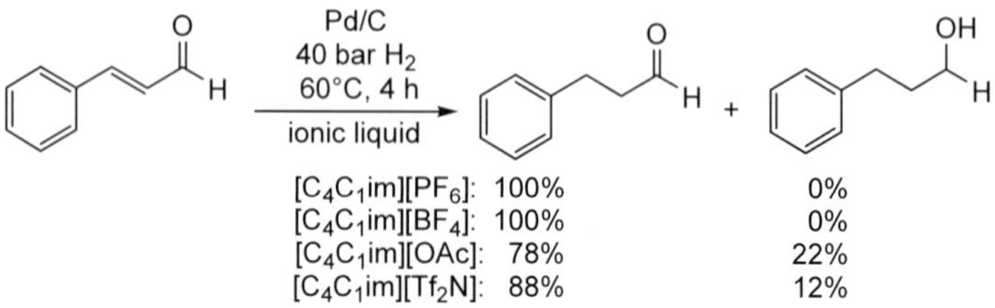

Scheme 3.14: Reduction of cinnamaldehyde with palladium on activated carbon

With citral as substrate, complete selectivity towards the C=C bond was obtained in several ionic liquids, but at the cost of rather low conversion. However, 99% conversion and selectivity were achieved in $[C_4C_1im][BF_4]$, see Table 3.7. Extraction of the product at the end of reaction with diethyl ether enabled facile catalyst recycling without the need of a filtration step.

Table 3.7: Conversion and selectivity in the reduction of citral as a function of the ionic liquid after 4 hours at 40 bar, 60°C with 10 mol% Pd/C

Ionic Liquid	Conversion [%]	Selectivity [%]
$[C_4C_1im][PF_6]$	69	96
$[C_4C_1im][BF_4]$	99	99
$[C_8py][BF_4]$	84	99
$[C_2C_1im][Tf_2N]$	81	81
$[C_6C_1im][Tf_2N]$	75	81

However, there are also examples where the catalyst degrades quickly in the ionic liquid whereas the same particles are very stable when used in neat substrate.[75]

3.6 Supported Ionic Liquid Catalysis

The concept of solid-liquid phase supported catalysts probably originated in 1989 when a water-soluble rhodium catalyst immobilised in a thin film on porous silica was employed in hydroformylation[77] and this methodology has since been adapted to encompass ionic liquids.[78] Using this approach a homogeneous catalyst is immobilised in a thin film of ionic liquid which itself is confined to the surface of a solid support. Whereas catalysts, which are covalently bound to solid supports often afford lower selectivity, these disadvantages are not observed if the catalyst is immobilised in a liquid film. Another significant advantage is the drastically reduced amount of ionic liquid required. Obviously, only co-solvents that do not dissolve the ionic liquid can be used, although in many reactions these are not required. A summary of SILP-hydrogenation reactions is provided in Table 3.8.

Table 3.8: SILP-hydrogenation reactions in ionic liquids

Ionic Liquid	Catalyst	Comments	Ref
$[C_4C_1im][PF_6]$	$[RuCl(\eta^6\text{-}p\text{-cymene})(BINAP)]Cl$	40 bar, 60°C, 97 % *ee*; 2-propanol as solvent; poly(diallyldimethylammonium) as support; reuse possible; markedly lower TOF than in homogeneous system.	[79]
$[C_4C_1im][PF_6]$	$RhCl_2(PPh_3)_3$	5 bar, 30°C; various solvents; poly (diallyldimethylammonium) as support; higher activity than in non-supported biphasic system; reuse possible.	[79]
$[C_4C_1im][PF_6]$	$[Rh(nbd)(PPh_3)_2][PF_6]$	2-40 bar, 22-50°C; silica gel as support; much higher TOF than in normal biphasic system; effect of H_2-pressure on activity studied; no significant loss of activity after 18 cycles; leaching below detection limits.	[80]
$[C_4C_1im][PF_6]$	Pd/C	Catalytic membrane; poly(vinylidene fluoride)-hexafluoropropylene copolymer as support.	[81]
guanidinium lactate (TMGL)	Pd	20-40°C; nanoparticles immobilised on molecular sieves by guanidinium-based ionic liquid; very high activity; particle size unchanged after five cycles.	[82]

A catalytically active membrane was obtained by mixing palladium on carbon, $[C_4C_1im][PF_6]$ and a poly(vinylidene fluoride) co-polymer.[81] The presence of the ionic liquid increases the flexibility of the system and thus the permeability of the otherwise rigid polymer. The membrane was successfully used for the hydrogenation of propene with 70% conversion after two hours.

A rhodium catalyst immobilised on silica gel/[C_4C_1im][PF_6] has been successfully applied to the hydrogenation of simple olefins like 1-hexene and cyclohexadiene.[80] Preparation of the supported ionic liquid catalyst involved dissolving [Rh(nbd)(PPh_3)$_2$][PF_6] in a mixture of [C_4C_1im][PF_6] and acetone, followed by addition of silica gel. After removal of the acetone under vacuum, an ionic liquid layer of *ca.* 0.6 nm thickness remains. The resulting solid was active for at least 18 batch runs without any significant decrease in activity. Turnover frequencies of the supported catalyst in neat olefin were significantly higher than in either an ionic liquid biphasic or homogeneous system employing acetone as solvent. This observation was attributed to two factors; first the absence of any coordinating solvent, and second higher catalyst concentration as the reaction could be performed in neat substrate.

Nanoparticles have the tendency to agglomerate which is detrimental to their catalytic activity and different methods have been developed to help prevent aggregation (*vide supra*). Excellent results were obtained in the solvent-free hydrogenation of simple alkenes with palladium nanoparticles immobilised on molecular sieves with the ionic liquid tetramethylguanidinium lactate.[82] After five catalytic cycles, no change in the particle size of 1-2 nm was detected. The thickness of the ionic liquid layer was estimated to be *ca.* 0.4 nm, which implies that the palladium particles are not fully immersed in the ionic liquid, as shown schematically in Figure 3.5. The combination of small particle size, high surface area and exposure of the particles led to high turnover frequencies of 1,200-4,000 $mol \cdot mol^{-1} \cdot h^{-1}$ at room temperature at very low catalyst loadings.

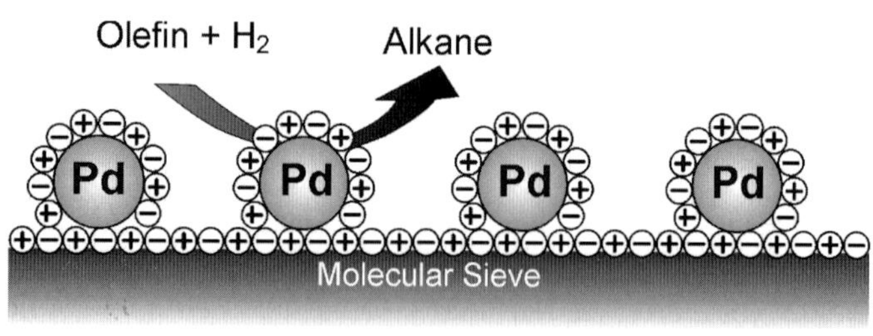

Figure 3.5: Immobilisation of palladium(0)-nanoparticles with a guanidinium-lactate ionic liquid at the surface of a molecular sieve

A polymeric support, poly(diallyldimethylammonium chloride), has been used in combination with either $RhCl(PPh_3)_3$ or [$RuCl_2(\eta^6$-

p-cymene)(BINAP)]Cl as catalysts for the hydrogenation of 2-cyclohexene-1-one, 1,3-cyclooctadiene and methyl acetoacetate.[79] In this case however, a large amount of ionic liquid relative to the polymer was used (4:1 by weight) so that the polymer serves as a sponge for the ionic liquid rather than as a support for a thin film. In comparison to the normal biphasic process, catalyst activity is markedly higher. In fast reactions like the enantioselective hydrogenation of methyl acetoacetate with $[RuCl_2(\eta^6\text{-}p\text{-cymene})(BINAP)]Cl$, mass transfer becomes critical for the polymeric system and reaction rates remain below those observed under homogeneous conditions. Reuse of the catalyst was possible without deterioration of activity in the second run.

Apart from these examples, nanoparticle-containing ionic liquids can also be used to impregnate a solid support from which the liquid is finally removed.[83] A solution of palladium colloids in $[C_4C_1im][Tf_2N]$ was added to an orthosilicate and allowed to gelate. Afterwards, the ionic liquid was extracted with acetonitrile and the resulting aerogel used to reduce cinnamaldehyde.

3.7 Outlook

Although hydrogenation reactions in ionic liquids are still in many ways in their infancy, as are those in other alternative solvents such as perfluorinated solvents and supercritical fluids, some comparisons can be made. Water has now been extensively evaluated as an immobilisation solvent for biphasic hydrogenation reactions and while there are many excellent systems, water is limited by substrate solubility and the fact that many catalysts are unstable in water. Clearly, ionic liquids overcome this limitation and in general the differences in polarity between a substrate and the related hydrogenated product result in a decrease in polarity and hence a decrease in solubility of the product in the ionic liquid and improved separation. There may be some cases, for example, the hydrogenation of certain functional groups, which do not lend themselves to ionic liquids and fluorous solvents might be better suited. Cost is sometimes cited as an inhibitory factor for employing ionic liquids, but apart from continual reductions in cost, the fact that many catalysts, such as cationic rhodium complexes bearing chiral phosphine ligands, do not require any modification to be used in ionic liquids is also advantageous.

3.8 References

[1] Buchanan, A. C.; Dworkin, A. S.; Smith, G. P. *J. Org. Chem.* **1982**, *47*, 603-607.

[2] Adams, C. J.; Earle, M. J.; Seddon, K. R. *Chem. Commun.* **1999**, 1043-1044.

[3] Chauvin, Y.; Mussmann, L.; Olivier, H. *Angew. Chem. Int. Ed. Engl.* **1996**, *34*, 2698-2700.

[4] Chauvin, Y. *Actual Chimique* **1996**, 44-46.

[5] Navarro, J.; Sagi, M.; Sola, E.; Lahoz, F. J.; Dobrinovitch, I. T.; Katho, A.; Joo, F.; Oro, L. A. *Adv. Synth. Catal.* **2003**, *345*, 280-288.

[6] Müller, L. A.; Dupont, J.; de Souza, R. F. *Macromol. Rap. Commun.* **1998**, *19*, 409-411.

[7] Berger, A.; de Souza, R. F.; Delgado, M. R.; Dupont, J. *Tetrahedron: Asymmetry* **2001**, *12*, 1825-1828.

[8] Anthony, J. L.; Maginn, E. J.; Brennecke, J. F. *J. Phys. Chem. B* **2002**, *106*, 7315-7320.

[9] Anthony, J. L.; Maginn, E. J.; Brennecke, J. F. In *Ionic Liquids, Industrial Applications for Green Chemistry*, Rogers, R. D., Seddon, K. R. Eds.; ACS Symposium Series 818, **2002**, 260-269.

[10] Dyson, P. J.; Laurenczy, G.; Ohlin, C. A.; Vallance, J.; Welton, T. *Chem. Commun.* **2003**, 2418-2419.

[11] Solinas, M.; Pfaltz, A.; Cozzi, P. G.; Leitner, W. *J. Am. Chem. Soc.* **2004**, *126*, 16142-16147.

[12] Brown, R. A.; Pollet, P.; McKoon, E.; Eckert, C. A.; Liotta, C. L.; Jessop, P. G. *J. Am. Chem. Soc.* **2001**, *123*, 1254-1255.

[13] Jessop, P. G.; Stanley, R. R.; Brown, R. A.; Eckert, C. A.; Liotta, C. L.; Ngo, T. T.; Pollet, P. *Green Chem.* **2003**, *5*, 123-128.

[14] Noyori, R. *Asymmetric Catalysis in Organic Synthesis*; John Wiley and Sons: New York, **1994**.

[15] Steines, S.; Wasserscheid, P.; Driessen-Hölscher, B. *J. Prak. Chem.-Chem. Ztg.* **2000**, *342*, 348-354.

[16] Mann, B. E.; Guzman, M. H. *Inorg. Chim. Acta* **2002**, *330*, 143-148.

[17] Suarez, P. A. Z.; Dullius, J. E. L.; Einloft, S.; DeSouza, R. F.; Dupont, J. *Polyhedron* **1996**, *15*, 1217-1219.

[18] Dupont, J.; Suarez, P. A. Z.; Umpierre, A. P.; de Souza, R. F. *Catal. Lett.* **2001**, *73*, 211-213.

[19] Dupont, J.; Suarez, P. A. Z.; Umpierre, A. P.; de Souza, R. F. *J. Braz. Chem. Soc.* **2000**, *11*, 293-297.

[20] Consorti, C. S.; Umpierre, A. P.; de Souza, R. F.; Dupont, J.; Suarez, P. A. Z. *J. Braz. Chem. Soc.* **2003**, *14*, 401-405.

[21] Suarez, P. A. Z.; Dullius, J. E. L.; Einloft, S.; deSouza, R. F.; Dupont, J. *Inorg. Chim. Acta.* **1997**, *255*, 207-209.

[22] Zhao, D. B.; Dyson, P. J.; Laurenczy, G.; McIndoe, S. *J. Mol. Catal. A* **2004**, *214*, 19-25.

[23] Daguenet, C.; Dyson, P. J. *Organometallics* **2004**, *23*, 6080-6083.

[24] Currao, A.; Feiken, N.; Macchioni, A.; Nesper, R.; Pregosin, P. S.; Trabesinger, G. *Helv. Chim. Acta* **1996**, *79*, 1587-1591.

[25] Dyson, P. J.; Ellis, D. J.; Welton, T. *Can. J. Chem.* **2001**, *79*, 705-708.

[26] Zhao, F.; Ikushima, Y.; Arai, M. *Green Chem.* **2003**, *5*, 656-658.

[27] Suarez, T.; Fontal, B.; Reyes, M.; Bellandi, F.; Contreras, R. R.; Ortega, J. M.; Leon, G.; Cancines, P.; Castillo, B. *React. Kinet. Catal. Lett.* **2004**, *82*, 325-331.

[28] Liu, F. C.; Abrams, M. B.; Baker, R. T.; Tumas, W. *Chem. Commun.* **2001**, 433-434.

[29] de Souza, R. F.; Rech, V.; Dupont, J. *Adv. Synth. Catal.* **2002**, *344*, 153-155.
[30] Monteiro, A. L.; Zinn, F. K.; DeSouza, R. F.; Dupont, J. *Tetrahedron: Asymmetry* **1997**, *8*, 177-179.
[31] Ngo, H. L.; Hu, A. G.; Lin, W. B. *Chem. Commun.* **2003**, 1912-1913.
[32] Hu, A. G.; Ngo, H. L.; Lin, W. B. *Angew. Chem. Int. Ed.* **2004**, *43*, 2501-2504.
[33] Ngo, H. L.; Hu, A. G.; Lin, W. B. *Tetrahedron Lett.* **2005**, *46*, 595-597.
[34] Berthod, M.; Joerger, J. M.; Mignani, G.; Vaultier, M.; Lemaire, M. *Tetrahedron: Asymmetry* **2004**, *15*, 2219-2221.
[35] Dyson, P. J.; Ellis, D. J.; Henderson, W.; Laurenczy, G. *Adv. Synth. Catal.* **2003**, *345*, 216-221.
[36] Boxwell, C. J.; Dyson, P. J.; Ellis, D. J.; Welton, T. *J. Am. Chem. Soc.* **2002**, *124*, 9334-9335.
[37] MacLeod, S.; Rosso, R. J. *Adv. Synth. Catal.* **2003**, *345*, 568-571.
[38] Yinghuai, Z.; Carpenter, K.; Bun, C. C.; Bahnmueller, S.; Ke, C. P.; Srid, V. S.; Kee, L. W.; Hawthorne, M. F. *Angew. Chem. Int. Ed.* **2003**, *42*, 3792-3795.
[39] Guernik, S.; Wolfson, A.; Herskowitz, M.; Greenspoon, N.; Geresh, S. *Chem. Commun.* **2001**, 2314-2315.
[40] Pugin, B.; Studer, M.; Kuesters, E.; Sedelmeier, G.; Feng, X. *Adv. Synth. Catal.* **2004**, *346*, 1481-1486.
[41] Lee, S. G.; Zhang, Y. J.; Piao, J. Y.; Yoon, H.; Song, C. E.; Choi, J. H.; Hong, J. *Chem. Commun.* **2003**, 2624-2625.
[42] Giernoth, R.; Krumm, M. S. *Adv. Synth. Catal.* **2004**, *346*, 989-992.
[43] Widegren, J. A.; Finke, R. G. *J. Mol. Catal. A* **2003**, *198*, 317-341.
[44] Widegren, J. A.; Bennett, M. A.; Finke, R. G. *J. Am. Chem. Soc.* **2003**, *125*, 10301-10310.
[45] Dyson, P. J.; Ellis, D. J.; Parker, D. G.; Welton, T. *J. Chem. Soc., Chem. Commun.* **1999**, 25-26.
[46] Plasseraud, L.; Süss-Fink, G. *J. Organomet. Chem.* **1997**, *539*, 163-170.
[47] Ogo, S.; Abura, T.; Watanabe, Y. *Organometallics* **2002**, *21*, 2964-2969.
[48] Dwars, T.; Oehme, G. *Adv. Synth. Catal.* **2002**, *344*, 239-260.
[49] Benyei, A.; Joo, F. *J. Mol. Catal.* **1990**, *58*, 151-163.
[50] Geldbach, T. J.; Dyson, P. J. *J. Am. Chem. Soc.* **2004**, *126*, 8114-8115.
[51] Kawasaki, I.; Tsunoda, K.; Tsuji, T.; Yamaguchi, T.; Shibuta, H.; Uchida, N.; Yamashita, M.; Ohta, S. *Chem. Commun.* **2005**, 2134-2136.
[52] Comyns, C.; Karodia, N.; Zeler, S.; Andersen, J. A. *Catal. Lett.* **2000**, *67*, 113-115.
[53] Berthold, H.; Schotten, T.; Honig, H. *Synthesis* **2002**, 1607-1610.
[54] Haack, K.-J.; Hashiguchi, S.; Fujii, A.; Ikariya, T.; Noyori, R. *Angew. Chem. Int. Ed.* **1997**, *36*, 285-288.
[55] Aiken, J. D.; Finke, R. G. *J. Mol. Catal. A* **1999**, *145*, 1-44.
[56] Roucoux, A.; Schulz, J.; Patin, H. *Chem. Rev.* **2002**, *102*, 3757-3778.
[57] Reetz, M. T.; Winter, M.; Breinbauer, R.; Thurn-Albrecht, T.; Vogel, W. *Chem. Eur. J.* **2001**, *7*, 1084-1094.
[58] Rao, C. N. R.; Kulkarni, G. U.; Thomas, P. J.; Edwards, P. P. *Chem. Soc. Rev.* **2000**, *29*, 27-35.
[59] Mevellec, V.; Roucoux, A.; Ramirez, E.; Philippot, K.; Chaudret, B. *Adv. Synth. Catal.* **2004**, *346*, 72-76.
[60] Hornstein, B. J.; Finke, R. G. *Chem. Mater.* **2003**, *15*, 899-909.
[61] Ozkar, S.; Finke, R. G. *J. Am. Chem. Soc.* **2002**, *124*, 5796-5810.
[62] Aiken, J. D.; Finke, R. G. *Chem. Mater.* **1999**, *11*, 1035-1047.

[63] Wei, G. T.; Yang, Z. S.; Lee, C. Y.; Yang, H. Y.; Wang, C. R. C. *J. Am. Chem. Soc.* **2004**, *126*, 5036-5037.
[64] Kim, K. S.; Demberelnyamba, D.; Lee, H. *Langmuir* **2004**, *20*, 556-560.
[65] Itoh, H.; Naka, K.; Chujo, Y. *J. Am. Chem. Soc.* **2004**, *126*, 3026-3027.
[66] Zhao, D. B.; Fei, Z. F.; Scopelliti, R.; Dyson, P. J. *Inorg. Chem.* **2004**, *43*, 2197-2205.
[67] Rossi, L. M.; Machado, G.; Fichtner, P. F. P.; Teixeira, S. R.; Dupont, J. *Catal. Lett.* **2004**, *92*, 149-155.
[68] Silveira, E. T.; Umpierre, A. P.; Rossi, L. M.; Machado, G.; Morais, J.; Soares, G. V.; Baumvol, I. L. R.; Teixeira, S. R.; Fichtner, P. F. P.; Dupont, J. *Chem. Eur. J.* **2004**, *10*, 3734-3740.
[69] Huang, J.; Jiang, T.; Han, B. X.; Gao, H. X.; Chang, Y. H.; Zhao, G. Y.; Wu, W. Z. *Chem. Commun.* **2003**, 1654-1655.
[70] Fonseca, G. S.; Umpierre, A. P.; Fichtner, P. F. P.; Teixeira, S. R.; Dupont, J. *Chem. Eur. J.* **2003**, *9*, 3263-3269.
[71] Dupont, J.; Fonseca, G. S.; Umpierre, A. P.; Fichtner, P. F. P.; Teixeira, S. R. *J. Am. Chem. Soc.* **2002**, *124*, 4228-4229.
[72] Fonseca, G. S.; Scholten, J. D.; Dupont, J. *Synlett* **2004**, 1525-1528.
[73] Anderson, K.; Goodrich, P.; Hardacre, C.; Rooney, D. W. *Green Chem.* **2003**, *5*, 448-453.
[74] Mu, X. D.; Evans, D. G.; Kou, Y. A. *Catal. Lett.* **2004**, *97*, 151-154.
[75] Scheeren, C. W.; Machado, G.; Dupont, J.; Fichtner, P. F. P.; Texeira, S. R. *Inorg. Chem.* **2003**, *42*, 4738-4742.
[76] Gallezot, P.; Richard, D. *Catal. Rev.* **1998**, *40*, 81-126.
[77] Arhancet, J. P.; Davies, M. E.; Merola, J. S.; Hanson, B. E. *Nature* **1989**, *339*, 454.
[78] Mehnert, C. P. *Chem. Eur. J.* **2005**, *11*, 50-56.
[79] Wolfson, A.; Vankelecom, I. F. J.; Jacobs, P. A. *Tetrahedron Lett.* **2003**, *44*, 1195-1198.
[80] Mehnert, C. P.; Mozeleski, E. J.; Cook, R. A. *Chem. Commun.* **2002**, 3010-3011.
[81] Carlin, R. T.; Fuller, J. *J. Chem. Soc., Chem. Commun.* **1997**, 1345-1346.
[82] Huang, J.; Jiang, T.; Gao, H. X.; Han, B. X.; Liu, Z. M.; Wu, W. Z.; Chang, Y. H.; Zhao, G. Y. *Angew. Chem. Int. Ed.* **2004**, *43*, 1397-1399.
[83] Anderson, K.; Fernandez, S. C.; Hardacre, C.; Marr, P. C. *Inorg. Chem. Commun.* **2004**, *7*, 73-76.

Chapter 4

HYDROFORMYLATION

4.1 Introduction

The transformation of an alkene into an aldehyde by addition of CO and H_2 (syngas) across an olefinic double bond represents one of the world's most important, homogeneously catalysed processes.[1-5] Unwanted side-reactions include isomerisation and hydrogenation and, in the case of internal olefins, numerous products may arise as shown in Scheme 4.1 for methyl-3-pentenoate.

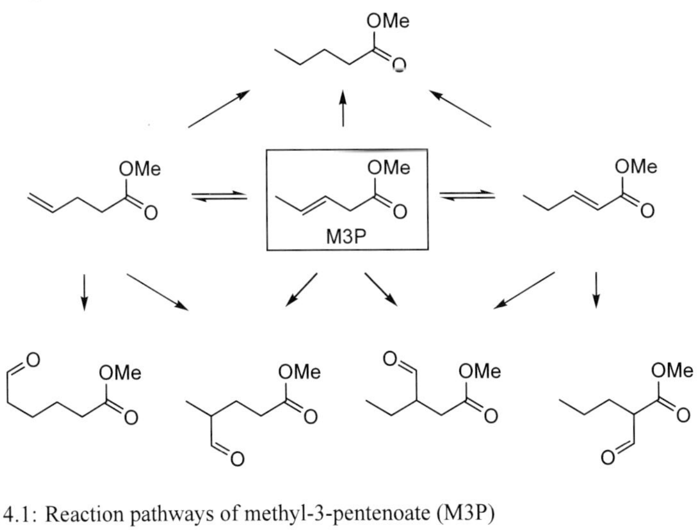

Scheme 4.1: Reaction pathways of methyl-3-pentenoate (M3P)

Various approaches have been used for the immobilisation of hydroformylation catalysts. These include (i) anchoring the catalyst to a dendrimer,[6-10] polymer[11-13] or inorganic solid;[14,15] (ii) fluorous biphasic systems;[16-18] (iii) aqueous biphasic systems[19,20] or use of supercritical CO_2 for the extraction of the reaction mixture.[17,18] Aqueous biphasic catalysis is employed on an industrial scale in the Ruhrchemie-RhônePoulenc process for the conversion of propene and syngas to *n*-butyraldehyde, with less than 4% *i*-butyraldehyde produced. The active catalyst is formed *in situ* from a rhodium salt and the sulfonated phosphine ligand TPPTS, thus achieving immobilisation of the complex in the aqueous layer. This allows the product to be isolated simply by decantation of the organic layer. Turnover frequencies up to 10,000 mol·mol^{-1}·h^{-1} have been achieved for the hydroformylation of propene using the water soluble BINAS ligand, a sulfonated binaphthyl diphosphine.[21] However, apart from trying to achieve high turnover frequencies, the main interest is the design of catalysts with high selectivity for the, usually more commercially valuable, linear products over branched products.

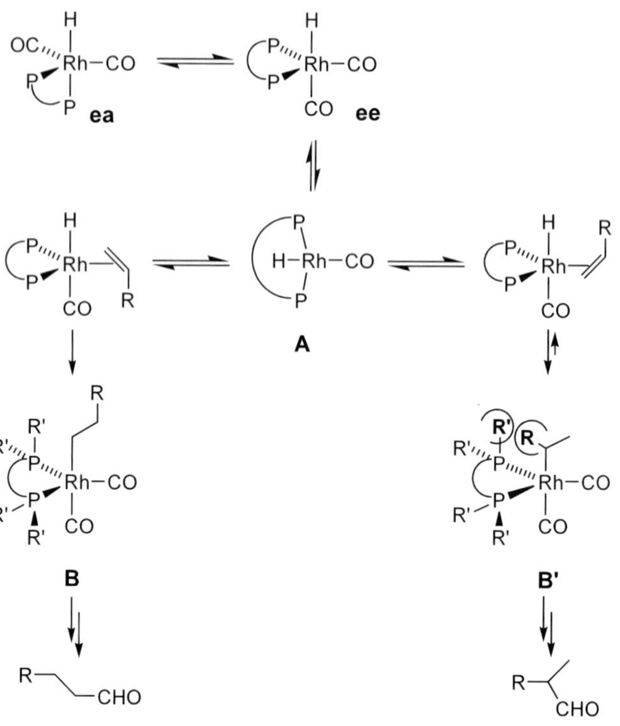

Scheme 4.2: Rationalisation for the formation of linear over branched hydroformylation products

Rhodium complexes with hydrophilic phosphines generally appear to afford higher linear-to-branched ratios (*l*:*b*) than hydrophobic ones.[1,22] This is probably a consequence of greater steric bulk due to the presence of e.g. sulfonate groups. Furthermore, intramolecular association of sulfonated substituents via hydrogen bonding may be of importance for monodentate phosphines. These interactions may lead to an increased dissociation energy of the phosphine ligands thereby reducing the formation of the di-carbonyl complex $RhH(CO)_2(PR_3)$ from the catalytically active species $RhH(CO)(PR_3)_2$.[23] NMR measurements indicate that at low syngas pressure (1 bar) $[C_4C_1im][BF_4]$ solutions of a mixture of $Rh(acac)(CO)_2$ and an excess of the sulfonated phosphine ligand TPPMS contain mostly the triphosphine species $RhH(CO)(TPPMS)_3$, while at higher pressure (135 bar) $RhH(CO)_2(TPPMS)_2$ is predominantly formed.[24]

With bidentate phosphine ligands, evidence for a dynamic equilibrium between the **ee** and the **ea** isomers under hydroformylation conditions has been established. In the presence of phosphines with a large bite angle, the *trans* intermediate **A** is stabilised (Scheme 4.2). During catalysis, following coordination of the alkene and hydride migration, conformer **B** is favoured over **B'** due to steric bulk arising from the backbone substituents, affording preferentially the linear product.[25]

Due to the decreasing solubility of olefins in water with increasing molecular mass, aqueous biphasic catalysis is not suitable for higher olefins ($>C_5$). For example, the water-solubility of 1-pentene at room temperature is 148 ppm, whereas only 50 ppm 1-hexene and 2.7 ppm 1-octene are dissolved.[26] Accordingly, higher olefins are transformed in industry with cobalt catalysts under rather forced conditions. Ionic liquids, on the other hand, offer a large flexibility in terms of tuning the solubility of the olefin by changing the length of the alkyl substituents on the cation and/or the nature of the anion.

4.2 Gas and Substrate Solubility in Ionic Liquids

Some experimental data are available with respect to alkene solubility in ionic liquids. For imidazolium-type ionic liquids, solubility increases in the order $[BF_4]^- < [PF_6]^- < [CF_3SO_3]^- < [CF_3CO_2]^- < [Tf_2N]^-$. As an example, the concentration of 1-hexene was found to be roughly four times higher in $[C_4C_1im][Tf_2N]$ as compared to $[C_4C_1im][BF_4]$.[27] The solubility of 1-octene in $[C_4C_1im][PF_6]$ was reported to be 0.025 mol·mol^{-1} at 25°C[28] and this value increases to 4.7 mol·mol^{-1} when the perfluorinated anion $[B\{C_6H_4(SiMe_2CH_2CH_2C_6F_{13})\text{-}p\}_4]^-$ is employed instead.[29] Halide-free anions like $[n\text{-}C_8H_{17}SO_4]^-$ may even lead to monophasic reaction conditions. This is reflected in a marked increase in activity and turnover frequencies for the hydroformylation of 1-octene being almost three-fold faster in

$[C_4C_1im][n\text{-}C_8H_{17}SO_4]$ relative to $[C_4C_1im][BF_4]$.[30] In addition, longer alkyl substituents on the cation also markedly improve the solubility of alkenes.[31]

The H_2-solubility in common ionic liquids appears to be almost independent of the salt composition. It varies between 0.62 $mmol{\cdot}l^{-1}{\cdot}bar^{-1}$ in $[C_8C_1im][BF_4]$ and 0.98 $mmol{\cdot}l^{-1}{\cdot}bar^{-1}$ in $[C_4C_1im][CF_3CO_2]$, comparable to water, i.e. 0.81 $mmol{\cdot}l^{-1}{\cdot}bar^{-1}$, and markedly lower than in simple alcohols or hydrocarbon solvents.[32] In contrast to these findings, the concentration of CO may change by a factor of more than six and was found to range between 0.66 $mmol{\cdot}l^{-1}{\cdot}bar^{-1}$ in $[C_4C_1py][CF_3CO_2]$ and 4.20 $mmol{\cdot}l^{-1}{\cdot}bar^{-1}$ in $[C_8C_1im][Tf_2N]$. This increased scale in solubility may be due to the permanent dipolar moment of CO, which makes this gas more susceptible to changes in the composition of the solvent. Figure 4.1 shows the solubility of 1-hexene, 1-octene, H_2 and CO in $[C_4C_1im]^+$ based ionic liquids with different anions, as well as in water and toluene.[32,33] Since the solubility of CO and H_2 is markedly higher in the organic substrate than in the ionic liquid the overall gas concentration in the reaction mixture may be highly dependent on the amount of substrate dissolved in the ionic liquid.

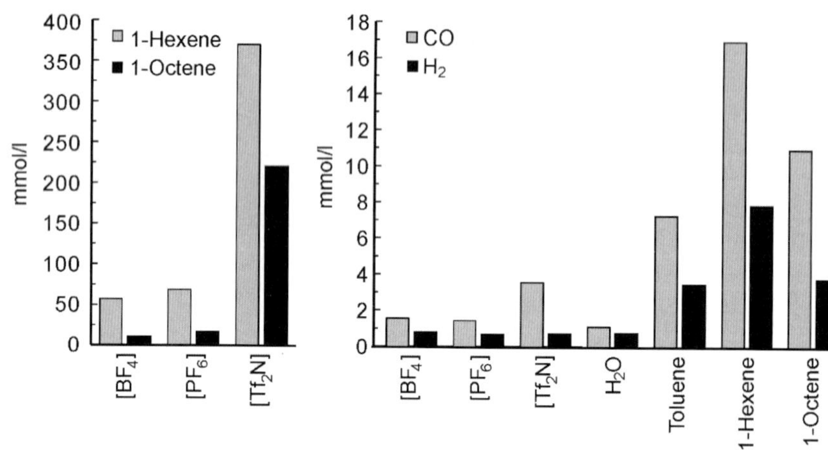

Figure 4.1: Substrate solubility in $[C_4C_1im]^+$ based ionic liquids with different anions, and in selected organic solvents for comparison reasons

4.3 Hydroformylation in Ionic Liquids

Hydroformylation in an ionic liquid was first described by Parshall in 1972, who reported on the reaction of ethylene with $PtCl_2$ as catalyst in liquid $[(C_2)_4N]Cl\text{-}SnCl_2$ (mp. 78°C).[34] In 1987, Knifton employed molten $[(C_4)_4P]Br$ (mp. 103-105°C) as solvent for the reaction of internal octenes with ruthenium and cobalt complexes.[35] The phosphonium salt matrix was found to stabilise the active ruthenium carbonyl species. Accordingly, the products could be obtained by distillation and the catalyst recycled without loss of activity. The first paper to report on rhodium-catalysed hydroformylation in a room temperature ionic liquid was published by Chauvin *et al.*[36] and nearly all subsequent research activity has focused on this metal. With 1-pentene as substrate and $[C_4C_1im][PF_6]$ as ionic liquid, catalyst leaching could be successfully suppressed when charged sulfonated phosphine ligands were used. Although promising results from biphasic aqueous catalysis with cobalt have been reported,[37] this metal, which is still extensively used in industrial hydroformylation, is practically absent in biphasic ionic liquid catalysis. An overview on non-rhodium catalysis is provided in Table 4.1, and reports of rhodium-catalysed hydroformylation are summarised in Table 4.2.

Table 4.1: Hydroformylation reactions with metals other than rhodium

Ionic Liquid	Catalyst	Substrate	TOF	*l:b*	Ref
$[(C_4)_4P]Br$	$Co_2(CO)_8$	int. Olefins			[35]
$[(C_4)_4P]Br$	RuO_2 + various phosphines	int. Olefins			[35]
$[C_2C_1im][BF_4]$	$[RuH(CO)(NCCH_3)_2(PPh_3)_3]^+$	1-Pentene	113		[36]
$[C_xC_1im]Cl$	$Ru_3(CO)_{12}$	1-Hexene		~1	[38]
$[(C_2)_4N]Cl\text{-}SnCl_2$	$PtCl_2$	Ethylene			[34]
$[C_2C_1im][BF_4]$	$Pt(Cl)_2(cod)$	1-Pentene	113		[36]
$[C_4C_1im]Cl\text{-}SnCl_2$ $[C_4C_1py]Cl\text{-}SnCl_2$	$PtCl_2(PPh_3)_2$ + 4 PPh_3	1-Octene	103-126	24	[39]
$[C_4C_1im]Cl\text{-}SnCl_2$ $[C_4C_1py]Cl\text{-}SnCl_2$	$PtCl_2(PPh_3)_2$ + 4 PPh_3	Methyl-3-pentenoate	31-37	1.3	[39]

In general, in rhodium-catalysed hydroformylation reactions conducted in ionic liquids the catalyst is generated *in situ*, typically from catalyst precursors such as $Rh(OAc)_4$, or more commonly $Rh(acac)(CO)_2$, together with an excess of an appropriate phosphine ligand. An induction period is often observed with the full activity only being reached after several catalytic runs. This has been attributed to sluggish conversion of the catalyst precursor to the active hydride species,[40-42] and to impurities present in the added ligand that are only removed after several product-extraction steps.[40]

Table 4.2: Hydroformylation with rhodium-catalysts

Ionic liquid	Ligand	Substrate	TOF (*l*:*b*)	Comments	Ref
$[C_1C_1C_1im][OTf]$ $[C_2C_1C_1im][OTf]$	- -	1-Hexene 1-Dodecene	23-33 (1-2) 27-30 (0.5-0.7)	Chiral Rh-COD complex as catalyst precursor, no phosphine ligands present; more side-reactions than with toluene as solvent.	[43]
			monodentate phosphines		
$[C_4C_1im][PF_6]$	PPh_3	1-Pentene	227,333 (3)	Some of the catalyst is extracted into the organic phase.	[36] [44]
$[C_1C_1C_1im][OTf]$ $[C_2C_1C_1im][OTf]$	PPh_3	1-Hexene 1-Dodecene	30-32 (1.7-2.0) 33 (2.5)	$RhCl(PPh_3)_3$ as catalyst precursor; no additional phosphine ligands added; more side-reactions than with toluene as solvent.	[43]
$[C_4C_1im][PF_6]$	PPh_3	1-Octene	100-680 (2.8)	Recycled 3 times; TOF drops considerably due to catalyst loss.	[40]
$[C_4C_1im][PF_6]$	PPh_3	Methyl-3-pentonate	107 (0.1)	Recycled 4 times; TOF drops from 169 in the first to 5 in the fifth run; product isolated *via* distillation.	[45]
$[C_4C_1im][PF_6]$	$P(OPh)_3$	1-Hexene	400 (6.1)	High ligand excess (150 per Rh); continuous flow catalysis with *sc*-CO_2.	[46] [47]
$[C_4C_1im][PF_6]$	TTPMS	1-Pentene	59 (3.9)	THF as co-solvent; no significant leaching observed.	[44]
$[C_4C_1im][PF_6]$	TPPMS	1-Nonene	160-320 (3.7)	High ligand excess (15000 per Rh); active for 12 runs; leaching $<$ 0.06 %; product extracted with *sc*-CO_2; ligand is slowly oxidised.	[46] [47]
$[C_8C_1im][Tf_2N]$	TPPDS	1-Octene	517 (3.8)	Continuous-flow process; catalyst stable for at least three days.	[47]
$[C_2C_1im][BF_4]$	TPPTS	1-Pentene	103 (2.8)	Heptane as co-solvent; no significant leaching observed.	[44]
$[C_4C_1im][BF_4]$ $[C_4C_1im][PF_6]$ $[C_4C_1C_1im][PF_6]$	TPPTS	1-Hexene	180-4,500 (1.8-2.6)	Activity *ca.* one order of magnitude lower than in toluene with PPh_3 as ligand; high-pressure NMR studies of the catalytic system; TOF decreases continuously upon recycling.	[24]

Ionic liquid	Ligand	Substrate	TOF (*l:b*)	Comments	Ref
$[C_4C_1im][PF_6]$	TPPTS	1-Octene	78-80 (4.9)	Recycled 3 times; TOF remains stable.	[28] [40]
Polyether melt	TPPTS	1-Tetradec.	16-95 (0.4-3.9)	$RhCl_3$ as catalyst precursor; leaching 0.5 %, recycled 6 times; TOF decreases only slowly after induction period.	[48]
$[C_4C_1im][PF_6]$	TPPOTS	1-Hexene	256 (12.6)	Recycled twice, considerable lower conversion after the second cycle.	[27]
Triazine-ionic liquids	DPP-$[C_1C_1py]$	1-Octene	20-207 (1.2-19)	Recycled twice, TOF remains relatively stable; significant increase in selectivity as a function of the ionic liquid.	[49]
$[C_4C_1im][PF_6]$	TPPM-$[C_1C_1im]$	1-Octene	51 (2.8)	No significant catalyst leaching; product isolated by phase-separation.	[50]
$[C_4C_1im][BF_4]$	TPPM-Guanidine	1-Hexene	190-205 (3.7-4.0)	With [TPPM-Guanidine]$[BF_4]$, catalyst leaching below detection limit.	[27]
$[C_4C_1im][PF_6]$	TPPD-Guanidine	1-Octene	276-330 (1.7-2)	Recycled 3 times; TOF stable after induction period.	[40]
$[C_4C_1im][C_8\text{-}SO_4]$	TPPD-Guanidine	1-Octene	862-892 (2.5-2.9)	Halide-free ionic liquid; monophasic system, with addition of cyclohexene biphasic; high activity reflection of high octane solubility.	[30]
$[C_4C_1im][PF_6]$	DPP-$[C_4C_1im]$	1-Octene	552 (1.1)	No significant catalyst leaching; product isolated by phase-separation.	[50]
$[C_4C_1im][BF_4]$	DPP-$[C_2C_1py]$	1-Hexene	232 (2.6)	Catalyst leaching of ca. 2% observed.	[27]
$[C_4C_1im][PF_6]$	MPP-$[C_2C_2im]_2$	1-Octene	32 (2.8)	No significant leaching observed.	[51]

Ionic liquid	Ligand	Substrate	TOF (*l*:*b*)	Comments	Ref
			Bidentate phosphines		
$[C_4C_1im][PF_6]$	DPP^{i}Pr-Cobaltocene	1-Octene	66 (2.6)	Catalyst leaching <0.2%; same activity in the second cycle.	[28]
$[C_4C_1im][PF_6]$	DPP-Cobaltocene	1-Octene	810 (16.2)	Catalyst leaching <0.2%; same activity in the second cycle.	[28]
$[C_4C_1im][PF_6]$	Modified neutral binaphthylphosphit	Methyl-3-pentonate	80-180 (1)	Recycled 10 times, TOF relatively stable; product isolated via distillation.	[45]
$[C_4C_1im][PF_6]$	Xantphos	1-Octene	245 (2.1)	Most of the Rh-complex lost to the organic phase after first cycle.	[52]
$[C_4C_1im][PF_6]$	Sulfoxantphos	1-Olefins	7-47 (1.7-61)	TOF stable for at least four cycles; low selectivity for 1-octene (4.8), high selectivity for 1-decene (61) and 1-octadecene (49).	[52]
$[C_4C_1im][PF_6]$	Sulfoxantphos	1-Octene	382-411 (6-7)	High pressure IR and NMR studies.	[53]
$[C_4C_1im][PF_6]$	Guanidine-Xantphos	1-Octene	15-58 (3.9)	Recycled 10 times; TOF stable after induction period.	[40]
$[C_4C_1im][PF_6]$ $[C_6C_1im][PF_6]$	POP-Xantphos	1-Hexene 1-Octene 2-Octene	4,000-8,900 (54-58) 65-10,100 (38-65) 1,200 (0.2-0.7)	TOF stable after induction period and highly dependent on catalyst concentration, H_2 partial pressure and stirring rate; leaching <0.07% with 1-octane but high with 1-hexane; 2-octane leads to decomposition; catalyst stable in ionic liquid for more than 14 days under air.	[41] [42]

4.4 Ligand Design

The success of biphasic hydroformylation has been largely due to ligand design. As the catalytically active rhodium species, $RhH(CO)(PR_3)_2$ and $RhH(CO)_2PR_3$ are neutral, phosphines bearing ionic substituents are crucial to immobilise the catalyst in the solvent phase. A large variety of different ligands have been synthesised for this purpose and some of these are shown in Figure 4.2 and Figure 4.3, respectively.

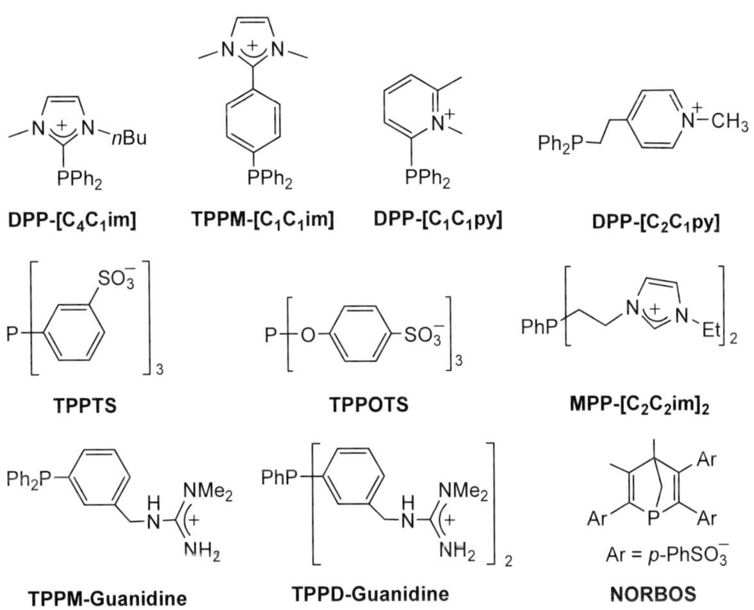

Figure 4.2: Charged monodentate phosphine ligands used in biphasic hydroformylation

With neutral phosphines, the reaction takes place both in the organic and the ionic liquid phase and much of the catalyst is lost during product isolation. Sulfonated phosphines such as the commercially available NaTPPTS, initially developed for aqueous biphasic catalysis, effectively prevent the catalyst from leaching into the product. Other approaches involve introduction of NR_3^+-fragments like guanidinium or use of charged metallocenes. Although immobilisation of the catalyst is achieved fairly easily in this manner with leaching below ICP-OES detection limits, high turnover frequencies and good linear to branched ratios are harder to obtain. In aqueous media simple monodentate NaTPPTS affords decent linear to branched ratios, whereas results in ionic liquids are rather disappointing (>15 *vs.* <5) and it appears that more elaborate ligands are required to increase the quantity of linear products. One exception involves the monodentate phosphite TPPOTS (*l:b* = 12.6) – a ligand that cannot be used in aqueous biphasic catalysis as it undergoes hydrolysis.[27] Otherwise, good *l:b* ratios have so far only been achieved with bidentate ligands, especially those from the Xantphos family. Here, the large bite angle of ca. 110° shifts the dynamic equilibrium to the **ee**-isomer, affording predominantly the linear product with up to 98% selectivity as demonstrated utilising the phenoxaphosphino-modified POP-Xantphos ligand introduced by van Leeuwen and co-workers.[41]

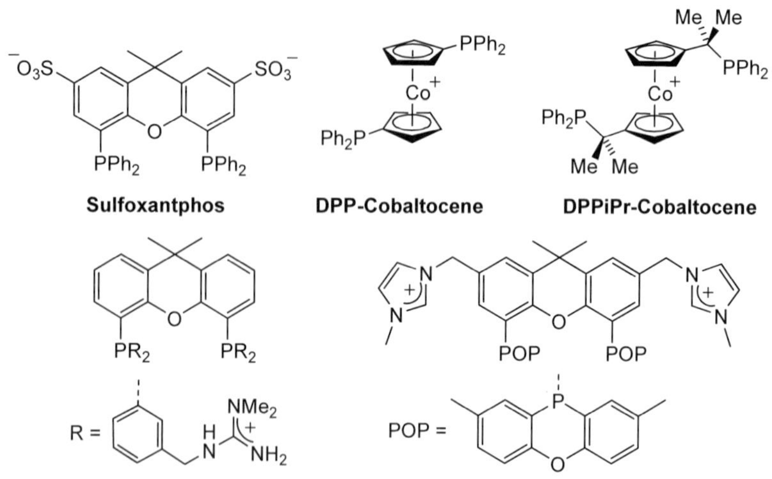

Figure 4.3: Charged bidentate phosphine ligands used in biphasic hydroformylation

Whereas the activity with the neutral version of the POP-Xantphos ligand in common solvents is as high as 1,700 mol·mol^{-1}·h^{-1}, only moderate TOFs around 100 mol·mol^{-1}·h^{-1} are observed in ionic liquids under comparable conditions. This decrease in activity is most likely due to formation of a dimeric rhodium species, see Scheme 4.3.[54] A higher partial hydrogen pressure and lower catalyst concentration may help to shift the equilibrium to the monomer. A tremendous increase in activity was observed as a result of lowering the catalyst concentration from 6.4 mM to 1.7 mM and depending on the stirring rate, turnover frequencies greater than 6,000 mol·mol^{-1}·h^{-1} are achieved. At such low catalyst concentration, catalyst degradation occurs upon exposure to air and the ionic liquid solution must be stored under a pressure of CO/H_2 in order to retain its activity.

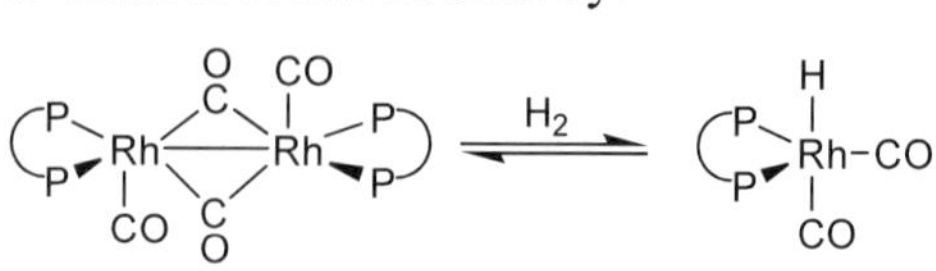

Scheme 4.3: Equilibrium between dinuclear and mononuclear rhodium species

As always, the correct interplay of steric and electronic effects determines the success of a given ligand. The latter is nicely demonstrated comparing the two bidentate ligands DPP-cobaltocene and DPP^{i}Pr-cobaltocene: a small spacer between the phosphine and the cationic metallocene, thereby separating the charge from the phosphorous atom, is all that is needed to decrease both activity (TOF 810 *vs.* 66) and *l:b* ratio (16.2 *vs.* 2.6) drastically.[28] The performance of a number of ligands in the hydroformylation of 1-octene are compared in Table 4.3.

Table 4.3: Performance of different ligands in the hydroformylation of 1-octene at 100°C in $[C_4C_1im][PF_6]$

Ligand;	L:Rh	p [bar]	TOF	*l:b*	leaching [%]	Ref
POP-Xantphos	4	60	6,200	44	< 0.07	[42]
Guanidine-Xantphos	2	30	52	21	< 0.07	[40]
DPP-Cobaltocene	2	10	810	16	< 0.2	[28]
Sulfoxantphos	4	15	34	14.8	not reported	[52]
Sulfoxantphos	4	10	382	7	not reported	[53]
dppf	2	10	828	3.8	significant	[28]
dppe	2	10	35	3.0	significant	[28]
MPP-$[C_2C_2im]_2$	2	30	32	2.8	not significant	[51]
DPP-$[C_4C_1im]$	2	30	51	2.8	not reported	[50]
DPP^{i}Pr-Cobaltocene	2	10	66	2.6	< 0.2	[28]
TPPTS	2	30	80	2.6	not significant	[40]

4.5 Influence of the Solvent

Whereas the rate of the reaction might markedly depend on the chosen reaction medium, the linear-to-branched ratio does not appear to be affected by the nature of the solvent. There is only one example to date, where a marked increase of the *l:b* ratio was reported as a function of the ionic liquid.[49] At high concentration (Rh:P = 1:20) of a charged pyridinium-phosphine ligand, DPP-$[C_1C_1py]^+$, *l:b* ratios up to 19 were achieved in the rather exotic triazine based ionic liquid tris[2-(oxoethyl-*N*-2'-fluoroethylimidazolium)]triazine (see Figure 4.4), whereas in 2-oxobutyl-3',3',3'-trifluoroproply imidazolium *l:b*-ratios of 4 were obtained under otherwise comparable reaction conditions. No explanation for this observation was given and further studies are necessary to evaluate whether the ionic liquid remains innocent or, which seems more likely, is involved in the formation of a new catalytically active species.

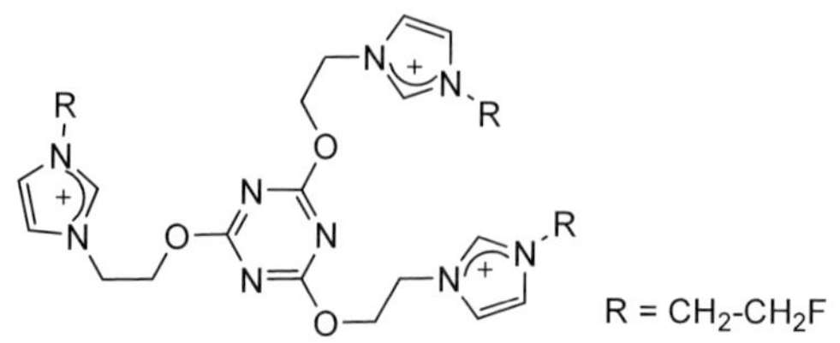

Figure 4.4: Structure of the tris[2-(oxoethyl-*N*-2'-fluoroethylimidazolium)]triazine cation

Ionic liquids based on the imidazolium cation, often utilising $[PF_6]^-$ as the anion, have been extensively used in hydroformylation reactions. This choice is probably a reflection of commercial availability rather than these being the most suitable reaction media. Under the reaction conditions of hydroformylation, $[PF_6]^-$ is susceptible towards hydrolysis to form HF and phosphoric acid, a process that is also observed to a lesser extent with $[BF_4]^-$, to afford HF and boric acid.[55] Accordingly, water-free conditions are necessary to avoid catalyst deactivation via formation of stable metal-fluorine bonds and to prevent corrosion of the reaction vessel. Hydrolysis is not an issue with perfluorinated anions like $[Tf_2N]^-$ which are, however, currently considerably more expensive than either $[PF_6]^-$ or $[BF_4]^-$. In the longer term, fluorine free anions are desirable and the suitability of ammonium tosylates,[48] phosphonium tosylates[56] and octylsulfate[30] has recently been demonstrated. The latter is especially interesting as its precursor, Na[n-$C_8H_{17}SO_4$] is a cheap bulk chemical. However, due to the presence of the long alkyl chain, a co-solvent might be necessary to induce phase separation and allow for facile isolation of the reaction products. Imidazolium chlorides have been employed successfully in the ruthenium catalysed hydroformylation of 1-hexene using CO_2 as the carbon source.[38] Normally, the presence of chloride is detrimental to catalysis, but in this case chloride facilitates deprotonation of a ruthenium hydride to form an active species which is capable of converting CO_2 to CO, as shown in Scheme 4.4.[57]

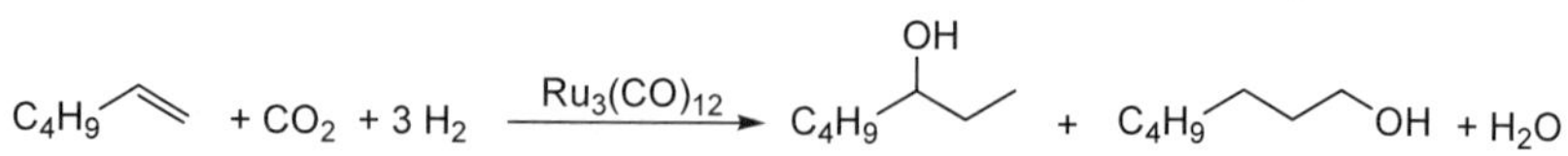

Scheme 4.4: Ruthenium-catalysed hydroformylation with CO_2 as carbon source

Chlorostannate liquids such as $[C_4C_1im][Sn_2Cl_5]$ have also been shown to be a potential alternative in *i.e.* platinum-catalysed hydroformylation, where they activate the catalyst precursor in a Lewis-acid/base reaction, as depicted in Scheme 4.5.[39]

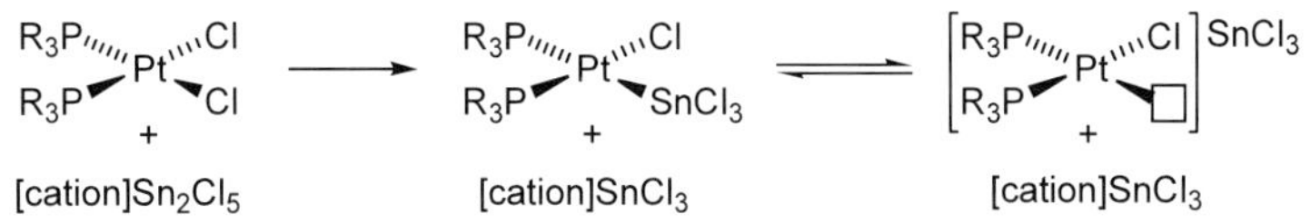

Scheme 4.5: Proposed mechanism for the activation of $PtCl_2(PR_3)_2$ in an acidic chlorostannate ionic liquid

As hydroformylation is performed at elevated temperatures ranging typically from 80 to 140°C, ionic liquids that melt above ambient temperature may also be considered as potential solvents, especially as these might further facilitate product separation once they have solidified at ambient conditions. Ionic liquids derived from pyridinium salts may be problematic as alkyl transfer from the nitrogen to the phosphine can take place under the reaction conditions.[58]

No detailed studies on catalyst degradation pathways in ionic liquids have yet been reported. Due to the reaction of $[PF_6]^-$ with water, transformation of $P(OR)_3$ to $P(OR)_nF_{3-n}$ has been observed.[47] In the case of aqueous biphasic catalysis with TPPTS as ligand, it is known that migration of a phenyl ring and ultimately elimination of sulfonated phenyl derivatives can take place.[59,60] However, when suitable (ionic) ligands are employed, the catalyst can remain active for more than 10 cycles and may be stable for more than 14 days under air.[41]

4.6 Special Processes

The recycling potential of hydroformylation catalysts in ionic liquids is highly dependent on the way the product is isolated. In terms of catalyst lifetime, simple decantation is certainly the method of choice. However at an industrial level, distillation is the most common separation technique and evaluation of catalyst recyclability under somewhat more stressful conditions has been determined.[45]

Whereas extraction with an organic solvent is not desirable from an environmental point of view, supercritical CO_2 represents an environmentally benign and clean alternative as demonstrated.[46] If the process is run continuously, as depicted in Figure 4.5, it can even compete with the industrial cobalt-based process.[47,61,62] Careful tuning is, however, required and the effects of different ionic liquids, temperature, gas composition and substrate flow rate are all important.

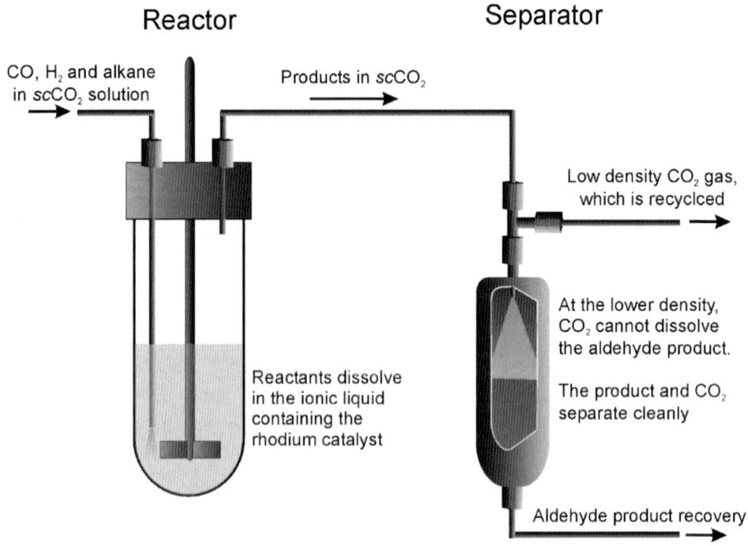

Figure 4.5: Diagram of a continuous-flow hydroformylation setup employing ionic liquids and supercritical CO_2

Silica-supported ionic liquid-phase (SILP) catalysis has been developed as an alternative approach to address the problem of product isolation, a methodology well known from aqueous catalysis, and an overview of SILP-hydroformylation reactions in ionic liquids is given in Table 4.4.[63]

Table 4.4: Supported ionic liquid phase hydroformylation in ionic liquids

Ionic Liquid	Description	Ref
$[C_4C_1im][PF_6]$ $[C_4C_1im][BF_4]$	Surface of silica gel modified with covalently anchored ionic liquid fragment; 1-hexene as substrate; TPPTS as ligand; leaching 0.07% ($[C_4C_1im]PF_6]$) - 2.1% ($[C_4C_1im]BF_4]$). TOF 3,360-3,900, *l:b* 2.4.	[64]
$[C_4C_1im][PF_6]$ $[C_4C_1im][BF_4]$ TMGL	MCM-41 mesoporous silica as support; 1-hexene as substrate; TPPTS as ligand; catalyst isolation by centrifugation; activity remained stable for at least 11 runs. TOF: 87-500; *l:b*: 0.7-3.5.	[65]
$[C_4C_1im][PF_6]$ $[C_4C_1im][C_8H_{17}SO_4]$	Continuous fixed-bed gas-phase hydroformylation with non-modified and partially dehydroxynated silica gel; propene as substrate; sulfoxantphos as ligand; with dehydroxynated support, activity and selectivity remained stable for > 60 hours. TOF: 1.5-40.8; *l:b*: 1.7-23.7	[66] [67]
$[C_4C_1im][PF_6]$	Continuous fixed-bed liquid-phase hydroformylation with non-modified silica gel; 1-octene as substrate; NORBOS as ligand. TOF: 44; *l:b*: 2.6	[68]

Surface-modified silica gel containing covalently anchored ionic liquid fragments together with a catalyst/ionic liquid solution has been used for batch hydroformylation of 1-hexene, and the system is illustrated in Figure 4.6.[64] Furthermore, unmodified silica can be used as demonstrated in the continuous fixed-bed hydroformylation of propene[66,67] and 1-octene.[68] Applying a partially dihydroxylated silica support (heating at 500°C for 15 hours) was found to significantly improve the catalyst stability.[67] In a system comprised of a rhodium-sulfatoxantphos ligand and with $[C_4C_1im][C_8H_{17}SO_4]$ as support phase, the catalyst was stable and active in the continuous flow propene stream for at least 60 hours.

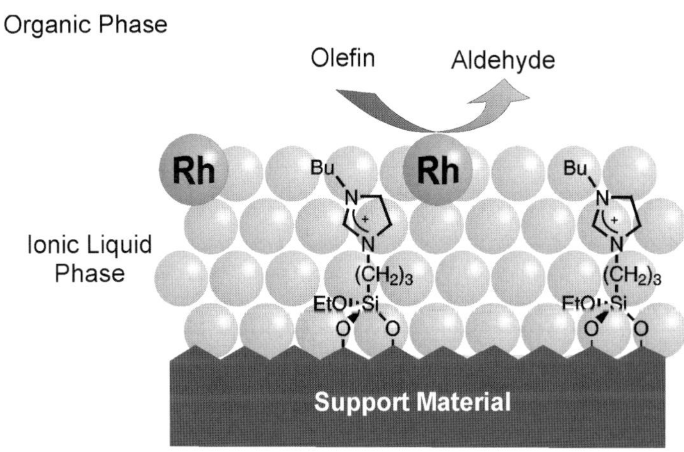

Figure 4.6: Silica-supported ionic liquid phase hydroformylation

4.7 Outlook

Hydroformylation in ionic liquids has made considerable progress in the past decade and catalysts are now available that achieve excellent linear-to-branched ratios together with good turnover frequencies and negligible leaching. Surprisingly, enantioselective hydroformylation[69-72] in ionic liquids has hardly been studied[43] (attempts to determine *ee*s have not been made), and this would certainly be an interesting expansion to the field.

The beauty of the industrial aqueous biphasic process, where a sufficiently soluble gas is converted into an insoluble liquid will, however, be hard to match. Ionic liquids that are designed to dissolve higher olefins will also dissolve some of the formed, more polar product and separation problems are likely to occur, although elegant approaches to overcome such problems have been reported.

The POP-Xantphos ligand, which at the present affords the best turnover frequencies and *l:b* ratios is considerably more expensive than the rather simple TPPTS ligand employed in aqueous biphasic hydroformylation. Compared to the classical rhodium-phosphine process, manufacturing costs in the aqueous biphasic process are about 10% lower. Accordingly, requirements for a fairly expensive catalyst will be exceptional if it should be considered in an industrial process, especially if one bears the additional costs for the required ionic liquid in mind. Nevertheless, the prospect of biphasic ionic liquid hydroformylation looks very promising.

4.8 References

[1] Trzeciak, A. M.; Ziolkowski, J. J. *Coord. Chem. Rev.* **1999**, *190-192*, 883-890.

[2] Frohning, C. D.; Kohlpaintner, C. W. In *Applied Homogeneous Catalysis with Organometallic Compounds*; Cornils, B., Herrmann, W. A., Eds.; VCH: Weinheim, **1996**; Vol. 1, pp 27-104.

[3] van Leeuwen, P. W. N. M.; Claver, C., Eds. *Rhodium Catalyzed Hydroformylation*; Kluwer Academic Publishers: Dordrecht, **2000**.

[4] Cornils, B. *Top. Curr. Chem.* **1999**, *206*, 133.

[5] Joó, F. *Aqueous Organometallic Chemistry*; Kluwer Academic Publishers: Dordrecht, **2001**.

[6] Reetz, M. T.; Lohmer, G.; Schwickardi, R. *Angew. Chem. Int. Ed.* **1997**, *36*, 1526-1529.

[7] Lu, S. M.; Alper, H. *J. Am. Chem. Soc.* **2003**, *125*, 13126-13131.

[8] Findeis, R. A.; Gade, L. H. *Eur. J. Inorg. Chem.* **2003**, 99-110.

[9] Ropartz, L.; Haxton, K. J.; Foster, D. F.; Morris, R. E.; Slawin, A. M. Z.; Cole-Hamilton, D. J. *Dalton Trans.* **2002**, 4323-4334.

[10] Ropartz, L.; Morris, R. E.; Foster, D. F.; Cole-Hamilton, D. J. *J. Mol. Catal. A* **2002**, *182*, 99-105.

[11] Shibahara, F.; Nozaki, K.; Matsuo, T.; Hiyama, T. *Bioorg. Med. Chem. Lett.* **2002**, *12*, 1825-1827.

[12] Shibahara, F.; Nozaki, K.; Hiyama, T. *J. Am. Chem. Soc.* **2003**, *125*, 8555-8560.

[13] Yoneda, N.; Minami, T.; Shiroto, Y.; Hamato, K.; Hosono, Y. *J. Jpn. Pet. Inst.* **2003**, *46*, 229-239.

[14] van Leeuwen, P.; Sandee, A. J.; Reek, J. N. H.; Kamer, P. C. J. *J. Mol. Catal. A* **2002**, *182*, 107-123.

[15] Yuan, Y. Z.; Zhang, H. B.; Yang, Y. Q.; Zhang, Y.; Tsai, K. R. *Catal. Today* **2002**, *74*, 5-13.

[16] Chen, W. P.; Xu, L. J.; Xiao, J. L. *Chem. Commun.* **2000**, 839-840.

[17] Mathivet, T.; Monflier, E.; Castanet, Y.; Mortreux, A.; Couturier, J. L. *C. R. Chim.* **2002**, *5*, 417-424.

[18] Osuna, A. M. B.; Chen, W. P.; Hope, E. G.; Kemmitt, R. D. W.; Paige, D. R.; Stuart, A. M.; Xiao, J. L.; Xu, L. J. *Dalton Trans.* **2000**, 4052-4055.

[19] Cornils, B.; Herrmann, W. A., Eds. *Aqueous-Phase Organometallic Catalysis*; Wiley-VCH: Weinheim, **1998**.

[20] Bischoff, S.; Kockritz, A.; Kant, M. *Top. Catal.* **2000**, *13*, 327-334.

[21] Bahrmann, H.; Bach, H.; Frohning, C. D.; Kleiner, H. J.; Lappe, P.; Peters, D.; Regnat, D.; Herrmann, W. A. *J. Mol. Catal. A* **1997**, *116*, 49-53.

[22] Breitt, B. *Acc. Chem. Res.* **2003**, *36*, 264-275.

[23] Horvath, I. T.; Kastrup, R. V.; Oswald, A. A.; Mozeleski, E. J. *Catal. Lett.* **1989**, 85-90.

[24] Mehnert, C. P.; Cook, R. A.; Dispenziere, N. C.; Mozeleski, E. J. *Polyhedron* **2004**, *23*, 2679-2688.

[25] Kamer, P. C. J.; Van Leeuwen, P. W. N. M. *Acc. Chem. Res.* **2001**, *34*, 895-904.

[26] McAuliffe, C. *J. Phys. Chem.* **1966**, *70*, 1267.

[27] Favre, F.; Olivier-Bourbigou, H.; Commereuc, D.; Saussine, L. *Chem. Commun.* **2001**, 1360-1361.

[28] Brasse, C. C.; Englert, U.; Salzer, A.; Waffenschmidt, H.; Wasserscheid, P. *Organometallics* **2000**, *19*, 3818-3823.

[29] van den Broeke, J.; Winter, F.; Deelman, B. J.; van Koten, G. *Org. Lett.* **2002**, *4*, 3851-3854.

[30] Wasserscheid, P.; van Hal, R.; Bösmann, A. *Green Chem.* **2002**, *4*, 400-404.

[31] Waffenschmidt, H. *Zur Hydroformylierung in Ionischen Flüssigkeiten*; Dissertation RWTH Aachen, **2000**.

[32] Dyson, P. J.; Laurenczy, G.; Ohlin, C. A.; Vallance, J.; Welton, T. *Chem. Commun.* **2003**, 2418-2419.

[33] Ohlin, C. A.; Dyson, P. J.; Laurenczy, G. *Chem. Commun.* **2004**, 1070-1071.

[34] Parshall, G. W. *J. Am. Chem. Soc.* **1972**, *94*, 8716 - 8719.

[35] Knifton, J. F. *J. Mol. Catal.* **1987**, *43*, 65-78.

[36] Chauvin, Y.; Olivier, H.; Mussmann, L. European Patent *EP0776880A1*, **1996**.

[37] Beller, M.; Krauter, J. G. E. *J. Mol. Catal. A* **1999**, *143*, 31-39.

[38] Tominaga, K.; Sasaki, Y. *Chem Lett* **2004**, *33*, 14-15.

[39] Wasserscheid, P.; Waffenschmidt, H. *J. Mol. Catal. A* **2000**, *164*, 61-67.

[40] Wasserscheid, P.; Waffenschmidt, H.; Machnitzki, P.; Kottsieper, K. W.; Stelzer, O. *Chem. Commun.* **2001**, 451-452.

[41] Bronger, R. P. J.; Silva, S. M.; Kamer, P. C. J.; van Leeuwen, P. *Chem. Commun.* **2002**, 3044-3045.

[42] Bronger, R. P. J.; Silva, S. M.; Kamer, P. C. J.; van Leeuwen, P. W. N. M. *Dalton Trans.* **2004**, 1590-1596.

[43] Stenzel, O.; Raubenheimer, H. G.; Esterhuysen, C. *Dalton Trans.* **2002**, 1132-1138.

[44] Chauvin, Y.; Mussmann, L.; Olivier, H. *Angew. Chem. Int. Ed.* **1996**, *34*, 2698-2700.

[45] Keim, W.; Vogt, D.; Waffenschmidt, H.; Wasserscheid, P. *J. Catal.* **1999**, *186*, 481-484.

[46] Sellin, M. F.; Webb, P. B.; Cole-Hamilton, D. J. *Chem. Commun.* **2001**, 781-782.

[47] Webb, P. B.; Sellin, M. F.; Kunene, T. E.; Williamson, S.; Slawin, A. M. Z.; Cole-Hamilton, D. J. *J. Am. Chem. Soc.* **2003**, *125*, 15577-15588.

[48] Kong, F. Z.; Jiang, J. Y.; Jin, Z. L. *Catal. Lett.* **2004**, *96*, 63-65.

[49] Omotowa, B. A.; Shreeve, J. M. *Organometallics* **2004**, *23*, 783-791.

[50] Brauer, D. J.; Kottsieper, K. W.; Liek, C.; Stelzer, O.; Waffenschmidt, H.; Wasserscheid, P. *J. Organomet. Chem.* **2001**, *630*, 177-184.
[51] Kottsieper, K. W.; Stelzer, O.; Wasserscheid, P. *J. Mol. Catal. A* **2001**, *175*, 285-288.
[52] Dupont, J.; Silva, S. M.; de Souza, R. F. *Catal. Lett.* **2001**, *77*, 131-133.
[53] Silva, S. M.; Bronger, R. P. J.; Freixa, Z.; Dupont, J.; van Leeuwen, P. *New J. Chem.* **2003**, *27*, 1294-1296.
[54] Castellanos-Páez, A.; Castillón, S.; Claver, C.; van Leeuwen, P. W. N. M.; de Lange, W. G. J. *Organometallics* **1998**, *17*, 2543-2552.
[55] Swatloski, R. P.; Holbrey, J. D.; Rogers, R. D. **2003**, *5*, 361-363.
[56] Karodia, N.; Guise, S.; Newlands, C.; Andersen, J. A. *Chem. Commun.* **1998**, 2341-2342.
[57] Tominaga, K.; Sasaki, Y.; Hagihara, K.; Watanabe, T.; Saito, M. *Chem. Lett.* **1994**, 1391.
[58] March, J.; *Advanced Organic Chemistry 4^{th}ed.*; Wiley: New York, **1992**, p 413.
[59] Herrmann, W. A.; Kohlpaintner, C. W. *Angew. Chem. Int. Ed.* **1993**, *32*, 1524-1544.
[60] Beller, M.; Cornils, B.; Frohning, C. D.; Kohlpaintner, C. W. *J. Mol. Catal. A* **1995**, *104*, 17-85.
[61] Cole-Hamilton, D. J.; Webster, J. M. *Chem. Commun.* **2004**, 612-613.
[62] Cole-Hamilton, D. J. *Science* **2003**, *299*, 1702-1706.
[63] Arhancet, J. P.; Davis, M. E.; Merola, J. S.; Hanson, B. E. *Nature* **1989**, *339*, 454-455.
[64] Mehnert, C. P.; Cook, R. A.; Dispenziere, N. C.; Afeworki, M. *J. Am. Chem. Soc.* **2002**, *124*, 12932-12933.
[65] Yang, Y.; Lin, H.; Deng, C.; She, J.; Yuan, Y. *Chem. Lett.* **2005**, *34*, 220-221.
[66] Riisager, A.; Wasserscheid, P.; van Hal, R.; Fehrmann, R. *J. Catal.* **2003**, *219*, 452-455.
[67] Riisager, A.; Fehrmann, R.; Flicker, S.; van Hal, R.; Haumann, M.; Wasserscheid, P. *Angew. Chem. Int. Ed.* **2005**, *44*, 815-819.
[68] Riisager, A.; Eriksen, K. M.; Wasserscheid, P.; Fehrmann, R. *Catal. Lett.* **2003**, *90*, 149-153.
[69] Dieguez, M.; Pamies, O.; Claver, C. *Tetrahedron: Asymmetry* **2004**, *15*, 2113-2122.
[70] Clarkson, G. J.; Ansell, J. R.; Cole-Hamilton, D. J.; Pogorzelec, P. J.; Whittell, J.; Wills, M. *Tetrahedron: Asymmetry* **2004**, *15*, 1787-1792.
[71] Ren, Y. L.; Zhao, W. C.; Lu, S. J. *Prog. Chem.* **2004**, *16*, 266-272.
[72] Yan, M.; Li, X. S.; Chan, A. S. C. *Chin. J. Chem.* **2004**, *22*, 203-206.

Chapter 5

OXIDATION REACTIONS

Ionic liquids were initially developed as solvents for electrochemical applications. The electrochemical window of clean ionic liquids can be huge,[1] allowing for a wide range of redox reactions.[2,3] It has further been demonstrated that they are also suitable solvents for enzymatic oxidations[4-7] but both topics are beyond the scope of this book. Only transformations that involve the metal-catalysed addition of oxygen to unsaturated carbon bonds as well as the oxidation of alcohols, aldehydes and ketones to their corresponding ketones, carboxylic acids and esters shall be discussed in this chapter.

The first metal catalysed oxidation in an ionic liquid originated in 2000, describing the oxidation of aromatic aldehydes to their corresponding carboxylic acids.[8] In the few years that have since passed, several dozen papers have appeared that deal with transition metal catalysed oxidations and of these epoxidation and dihydroxylation reactions have attracted most interest. In addition there are also numerous examples where ionic liquids act as catalyst or co-catalyst in oxidation reactions without serving as the reaction medium. These include the oxidative carbonylation of amines with imidazolium and phosphonium alkylselenites,[9] nickel-catalysed oxidation of ethylbenzene with $[R_4N][BF_4]$ and $[C_4C_1im][PF_6]$ as co-catalyst,[10,11] direct oxidative carboxylation of styrene with $[(C_4)_4N]Br$[12] and the $PdCl_2$-catalysed oxidation of styrene with $[C_4C_1im]^+$-based ionic liquids.[13] However, those properties that render ionic liquids a special class of solvents are not really exploited in these reactions and accordingly no detailed description will be given here.

5.1 Oxidants

There are no differences between ionic liquids and molecular solvents with respect to the reagents used to oxidise a given substrate, but they may display different properties, especially in terms of their solubility. Oxidants employed can roughly be divided into three categories, gaseous, aqueous and solid, and a brief summary of the most common reagents is given in Table 5.1.

A comparative study concerning the solubility of O_2 in ionic liquids is not yet available. It seems, however, that at least in imidazolium-type ionic liquids the oxygen solubility is much lower than initially anticipated. From photochemical quenching experiments, the O_2 concentration in $[C_4C_1im][PF_6]$ was estimated to be smaller than 0.2 $mmol \cdot l^{-1}$ at atmospheric pressure.[14] A very low oxygen solubility was also observed in $[C_4C_1im][Tf_2N]$, as measured using a gravimetric microbalance.[15] Molecular oxygen has been predominantly used in ionic liquids to oxidise alcohols, aldehydes or ketones. To overcome the intrinsic low solubility and to minimise mass transfer limitations, efficient mixing of the liquid and gaseous phase must be ensured. Addition of a suitable co-solvent may, however, be required nevertheless. Perfluorinated solvents exhibit very high oxygen solubility and combination of ionic and perfluorinated solvents may help to increase the reaction rate.[16] Alternatively, ionic liquids bearing perfluorinated alkyl chains could probably be used, but little is yet known about their physical properties.

Table 5.1: Oxidants currently used in ionic liquids

Oxidant	Comment
O_2	Cheap and environmentally friendly; poorly soluble in common ionic liquids.
H_2O_2	Cheap and environmentally friendly; water may be unwanted; exact concentration not obvious.
NaOCl	Cheap and environmentally friendly; water may be unwanted; exact concentration not obvious.
$(CH_3)_3C$-OOH (TBHP)	Relatively cheap; available as aqueous and anhydrous solution.
$PhI(OAc)_2$	Relatively expensive but easy to dose.
H_2N-CO-$NH_2 \cdot H_2O_2$ (UHP)	Water-free peroxide source; easy to handle; dissolves much more effectively in ionic liquids than in common organic solvents.
N-morpholine-*N'*-oxide (NMO)	Expensive; residual *N*-methyl morpholine may be problematic.

As aqueous oxidants, hydrogen peroxide, *tert*-butyl hydrogenperoxide (TBHP) and sodium hypochloride are frequently used. Their main disadvantage is that their exact concentration is not always reliable (unless assayed directly before use) and therefore over-oxidation of the substrate and/or catalyst may occur. Also, the presence of relatively large quantities of water may lead to further, unwanted reactions such as the ring-opening of epoxides. Aqueous oxidants have been used in combination with both hydrophilic and hydrophobic ionic liquids although no general statement can be made, with respect to which is the better choice.

The class of solid oxidants comprises phenyliodide diacetate, urea hydrogenperoxide (UHP), and *N*-morpholine-*N'*-oxide (NMO). These water-free reagents are easy to dose, but are also significantly more expensive than aqueous oxidants. While UHP is only sparingly soluble in solvents like dichloromethane or acetonitrile, it readily dissolves in ionic liquids like $[C_2C_1im][BF_4]$, affording higher reaction rates relative to conventional solvents. Obviously, one mole by-product per mole target molecule is produced with these organic oxidants, which may, for example, complicate the isolation of the desired product or interfere with the catalyst. Nevertheless, positive effects like modulation of the pH of the reaction mixture by formed urea have also been observed.

5.2 Ionic Liquids for Oxidation Reactions

It is of course essential that the ionic liquid is stable in the presence of the oxidant, which excludes ionic liquids with metallic anions such as chlorocuprates. In many oxidation reactions water is present as co-solvent, reagent or it is produced in the course of the reaction, which further eliminates the use of chloroaluminates. However, less care has to be taken with respect to drying the ionic liquid compared to other catalytic reactions when aqueous oxidants are used. Common imidazolium and ammonium based ionic liquids are neither water-nor oxygen sensitive and thus well suited and may even act as co-catalyst in the oxidation reaction. Some anions like $[BF_4]^-$ and $[PF_6]^-$ are, however, susceptible towards hydrolysis. Where the ionic liquid is highly viscous, a co-solvent such as dichloromethane might be necessary to afford acceptable reaction rates, especially when molecular oxygen is used as reagent.

Those catalysts that are employed in oxidation reactions are often relatively tolerant with respect to the presence of halide impurities. This is underlined from the fact that NaOCl or $[C_4C_1im]Br$ are used as oxidant or solvent, respectively. However, some catalysts are able to oxidise halides to their corresponding hypohalous acids, which in turn catalyse the disproportionation of hydrogen peroxide to water and molecular oxygen.[17]

In methyltrioxorhenium-catalysed reactions, residual pyridine, pyrazole or imidazole may deactivate the active catalyst and the ionic liquid should be purified accordingly.[18,19] The nature of the ionic liquid may decide the success or failure of a given reaction and, for example, Ru-catalysed alcohol oxidations appear to be very sensitive to the type and quality of ionic liquid used.

5.3 Oxidation of Alkenes

5.3.1 Racemic Epoxidation

The epoxidation of alkenes, both racemic and enantioselective, has been extensively studied in ionic liquids and Table 5.3 provides a summary of these reactions. Although no transition metal is involved, the base-catalysed epoxidation of electrophilic alkenes in ionic liquids is worth briefly mentioning. Depending on the substrate and the reaction conditions, *α,β*-unsaturated carbonyl compounds were oxidised with aqueous H_2O_2 in $[C_4C_1im][BF_4]$ or $[C_4C_1im][PF_6]$ within minutes under mild conditions and no hydrolysis products were observed.[20-22] The extraction of the products with *sc*CO_2 instead of conventional solvents was demonstrated to be feasible, albeit on a small scale.[23]

Methyltrioxorhenium (MTO) is an extensively used catalyst for olefin epoxidation[24] and a number of reviews are available on its application.[25-28] The complex reacts with hydrogen peroxide to form the mono η^2-peroxo species **mpMTO** and the diperoxide **dpMTO**. Both complexes are capable of transferring an oxygen atom to olefins or other substrates, as shown in Scheme 5.1, but the diperoxide species is about five times more reactive than the monoperoxo complex.[17] Care has to be taken that halide impurities are absent from the solvent as these are oxidised much faster than olefins.[17]

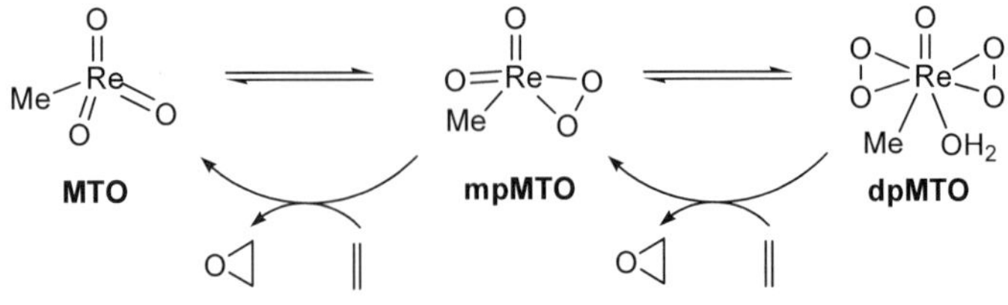

Scheme 5.1: Catalytically active species in **MTO**-catalysed epoxidation reactions

In conventional solvents addition of acid, which in return may lead to ring-opening of the epoxide, is required to stabilise the active complexes, whereas no such additives are usually necessary in an ionic liquid. Good conversion and selectivity was obtained in the MTO-catalysed epoxidation of cyclic alkenes with UHP as oxidant in $[C_2C_1im][BF_4]$.[29,30] In contrast only rather poor results were observed with 1-decene as substrate, which may be a reflection of the very low solubility in the ionic liquid.[29] A beneficial effect of using UHP as oxidant is that due to the absence of water solvolysis of the epoxide to the corresponding diol is generally not observed whereas with H_2O_2 as oxidant the corresponding diol is often the main product. In addition, urea liberated during the reaction may have a beneficial effect in the epoxidation reaction as it modulates the pH and thus prevents acid-catalysed ring opening. The catalyst appeared to be highly stable in the ionic liquid and remained active for at least 72 hours, but no attempts were made to recycle the catalyst.[29] Good results under similar reaction conditions were also observed in the domino epoxidation-nucleophilic ring opening of glycols with UHP as oxidant and dibutylphosphate (DBP) as nucleophile (Scheme 5.2).[31]

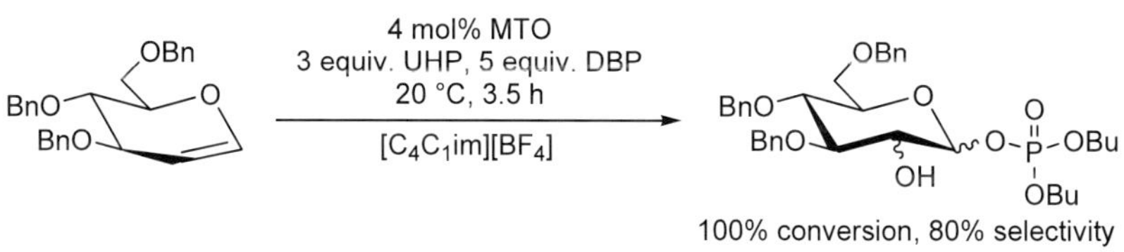

Scheme 5.2: Domino epoxidation-nucleophilic ring opening

A large range of different ionic liquids have been screened in the epoxidation of cyclooctene with dioxomolybdenum(VI) complexes and *tert*-butyl hydrogenperoxide as oxidant, as shown in Table 5.2.[32] With the diazabutadiene complex, **48**, as catalyst, inferior turnover frequencies were observed relative to the reaction in neat substrate or in dichloromethane and the recycling potential of the catalyst turned out to be only limited. Catalyst immobilisation was better with the cationic tris(methylaminomethyl)ethane complex, **49**, however at the expense of selectivity. Of the ionic liquids tested, $[C_4C_1im][Tf_2N]$ gave the best results for both molybdenum complexes.

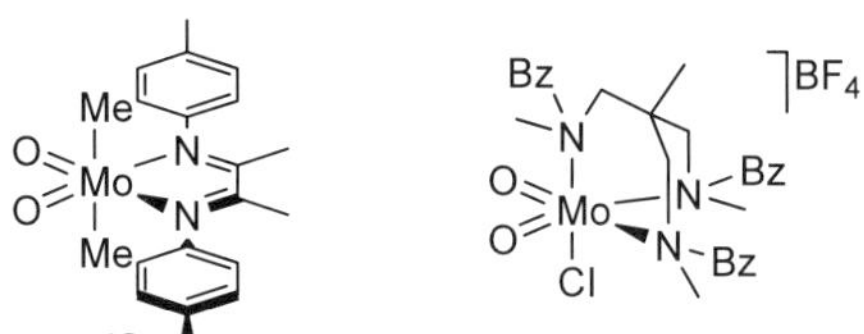

48 **49**

Figure 5.1: Dioxomolybdenum(VI) complexes used in epoxidation reactions

Comparable results were further observed when neutral cyclopentadienyl-molybdenum(II) compounds like $MoCpCl(CO)_3$ were used as catalysts in that immobilisation of the catalyst in an ionic liquid afforded lower turnover frequencies relative to the reaction in neat substrate.[33]

Table 5.2: Epoxidation of cyclooctene with TBHP as oxidant and Mo-complexes **48** and **49** as catalysts

	Complex **48**		Complex **49**	
Solvent	TOF$^{a)}$	Conversion$^{b)}$	TOF$^{a)}$	Conversion$^{b)}$
neat cyclooctene	73	100	48	64
1,2-dichloroethane	53	100	19	71
$[C_4C_1im][PF_6]$	24	90	40	71
$[C_4C_1im][Tf_2N]$	34	96	47	93
$[C_8C_1im][PF_6]$	25	94		
$[C_4C_1C_1im][Tf_2N]$	7	53		
$[(C_5O_2)C_1im][PF_6]$	2	14		

$^{a)}$ Calculated after 1 hour reaction; $^{b)}$ after 30 hours at 55°C

Macrocyclic ligands are well established in olefin epoxidation reactions. The Mn(porphyrin) catalyst, **50**, was employed in the epoxidation of various alkenes in a homogeneous $[C_4C_1im][PF_6]$-CH_2Cl_2 mixture with good to excellent conversion and selectivity, as shown in Scheme 5.3.[34] From UV-VIS spectroscopy it appears that the catalytically active Mn-oxo intermediate is stabilised in this solvent system, thus affording superior results relative to neat dichloromethane solutions. Significantly better results were obtained with $PhI(OAc)_2$ as oxidant compared to PhIO. This was attributed to the greater Lewis basicity of the former reagent, which may lead to a faster coordination to the electron deficient metalloporphyrin. Recycling of the catalyst was thus possible ten times with only a minor deterioration of its activity. If, however, an excess of oxidant was present, degradation of the catalyst was observed. Aryl substituents with electron-withdrawing groups, as in the case of the sulfonated and fluorinated complexes **50b** and **50d**, respectively, led to a slight decrease in both activity and selectivity, relative to the phenyl-substituted complex **50a**.[34,35]

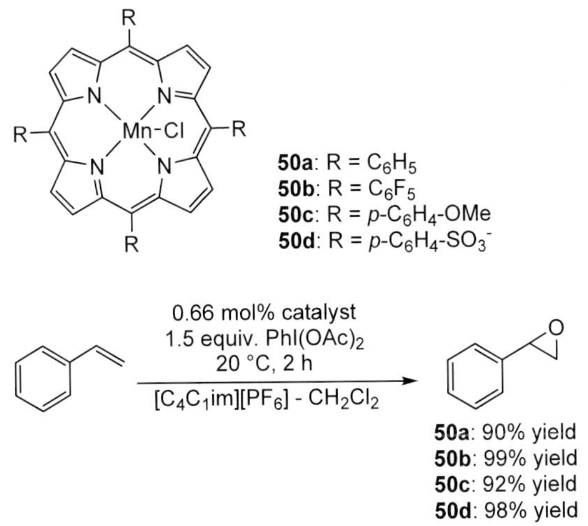

Scheme 5.3: Epoxidation of styrene with manganese-porphyrin catalysts

Similar results were also achieved with the highly charged iron(III) porphyrin catalyst **51**, using H_2O_2 as the oxidant.[36] In a biphasic $[C_4C_1im]Br$-CH_2Cl_2 system yields decreased from 74% in the first run to 62% in the fifth run, see Scheme 5.4. As in the previous example, catalyst degradation was observed in the presence of excess oxidant. Depending on the substrate, aldehydes and alcohols were observed as by-products in addition to the desired epoxide, which may be due to the rearrangement of an initially formed iron-oxo-alkene complex.[37]

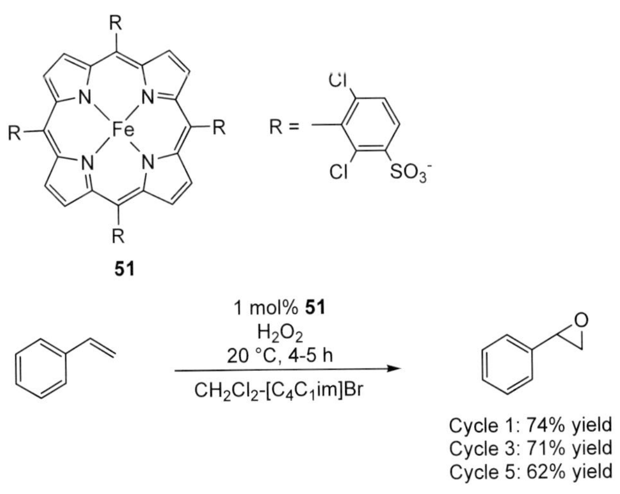

Scheme 5.4: Epoxidation of styrene with an iron-porphyrin catalyst

The racemic epoxidation of olefins has further been investigated using a peroxotungstenate catalyst, $[\{W(=O)(O_2)_2(H_2O)\}_2(\mu\text{-}O)]^{2-}$, supported on ionic liquid modified silica.[38] The solid support was reacted with 1-octyl-3-(3-triethoxysilylpropyl)-4,5-dihydroimidazolium, affording a SiO_2 surface on which the ionic liquid is covalently bound, as depicted in Figure 5.2. This heterogeneous catalyst was successfully used to epoxidise olefins using H_2O_2 as oxidant and reaction rates were comparable rates to those under homogeneous conditions.

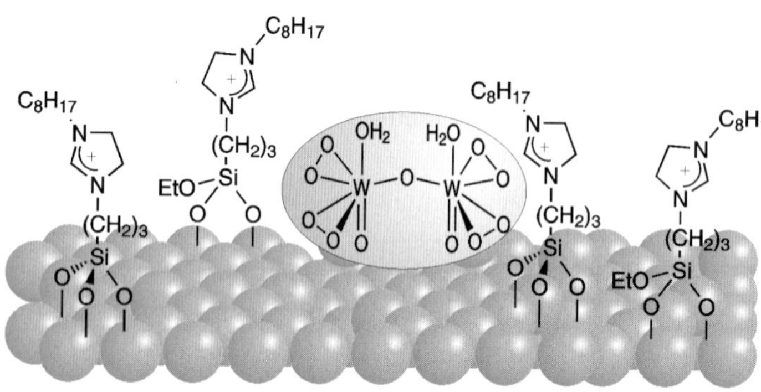

Figure 5.2: Tungstenate complex immobilised on ionic liquid modified silica

5.3.2 Enantioselective Epoxidation, Ring-Opening and Kinetic Resolution of Epoxides

The catalytic asymmetric epoxidation of alkenes offers a powerful strategy for the synthesis of enantiomerically enriched epoxides and enantioselective oxidation reactions in ionic liquids have been summarised previously.[39] Complexes based on chiral salen ligands – usually with manganese(III) as the coordinated metal – often afford excellent yields and enantioselectivities and the catalytic cycle for the reaction is depicted in Scheme 5.5.[40]

Various strategies have been pursued in order to immobilise chiral epoxidation catalysts and these encompass covalent attachment to solid supports,[41] steric occlusion in nanosized cages of zeolites,[42-44] entrapment in a polydimethylsiloxane membrane[45,46] and fluorous biphasic systems.[47] However, these approaches frequently require tedious ligand modifications and often lead to a marked decrease in both selectivity and activity of the transition metal catalyst.

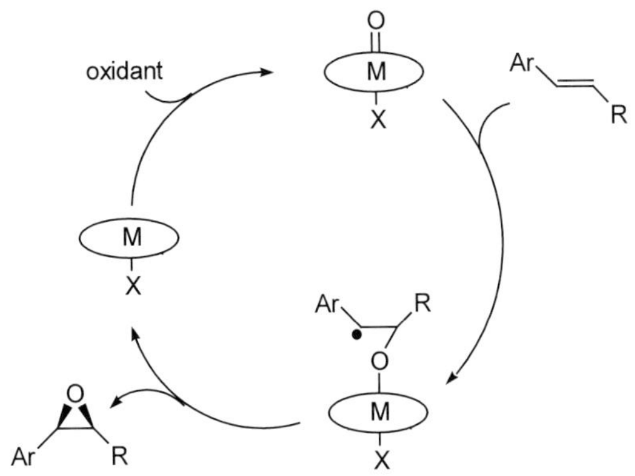

Scheme 5.5: Catalytic cycle for enantioselective epoxidation with salen complexes

Good to excellent enantioselectivity was achieved in the epoxidation of mainly cyclic olefins with the chiral salen-catalyst **52** immobilised in $[C_4C_1im][PF_6]$, but selectivity deteriorated upon catalyst recycling, see Scheme 5.6.[48] Relative to molecular solvents, higher reaction rates were observed even under biphasic conditions when the epoxidation reaction was carried out in the presence of an ionic liquid. UV-VIS spectroscopic[34] and cyclovoltammetric[49] studies suggest that the commonly observed superior reaction rates are a reflection of the solvent's ability to stabilise the active metalla-oxo intermediate.

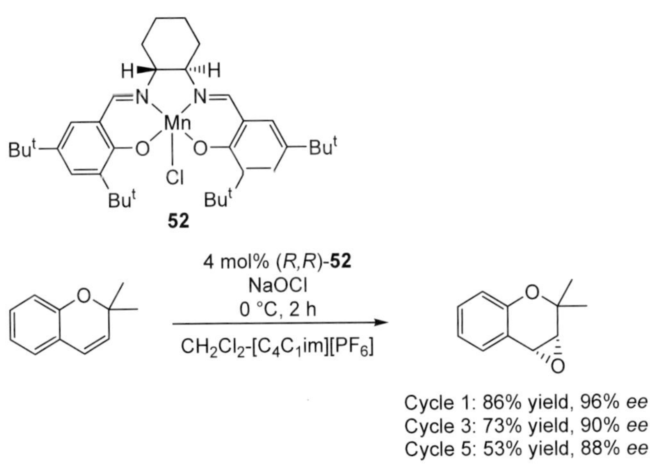

Scheme 5.6: Epoxidation of 2,2-dimethylchromene with a chiral manganese-salen catalyst

Comparable results were achieved in the asymmetric epoxidation of dihydronaphthalene with the Katsuki-type catalyst **53** in $[C_4C_1im][PF_6]$.[50] Again, higher reaction rates were observed in the presence of the ionic liquid and the overall stability of the catalyst appeared to be superior to that with catalyst **52**. The presence of a co-ligand, 4-phenylpyridine *N*-oxide (4-PPNO), was found to be crucial, in order to provide high activity and selectivity (see Scheme 5.7).

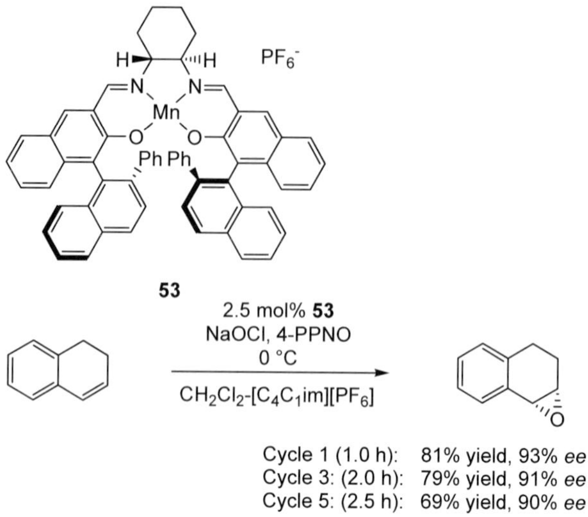

Scheme 5.7: Epoxidation of 1,2-dihydronaphtalene with a chiral Katsuki-type catalyst

Although not truly oxidation reactions, the kinetic resolution and ring opening of epoxides with cobalt and chromium salen complexes is worth mentioning here. Chiral Co(salen) complexes have been successfully used for the kinetic resolution of racemic epichlorohydrin.[51] Notably, the Co(II) complex **54a**, which is inactive in conventional solvents, effectively catalysed the resolution of epichlorohydrin in $[C_4C_1im]^+$-based ionic liquids. It was found that complex **54a** was oxidised *in situ* to the active Co(III) species without the need to add any acid. Acidic impurities in the ionic liquids may account for this observation, but no investigations have yet been undertaken. The catalytic system involving $[C_4C_1im][Tf_2N]$ as solvent was reusable up to ten times without any loss of activity and enantioselectivity (>99% *ee*), as shown in Scheme 5.8.

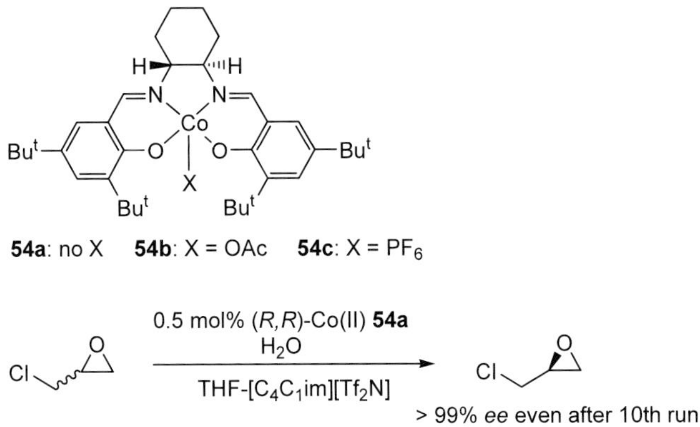

Scheme 5.8: Kinetic resolution of racemic epichlorohydrin

The asymmetric ring-opening of *meso*-epoxides with trimethylsilyl azide, $TMSN_3$, catalysed by the chromium complex **55** was investigated in $[C_4C_1im]^+$ based ionic liquids with different anions (see Scheme 5.9).[52] Catalyst immobilisation was most efficient when the anion is $[BF_4]^-$ or $[OTf]^-$, but in these hydrophilic liquids, yield and selectivity were very poor (<5 %). In contrast, good yields (up to 86 %) and *ee*'s (up to 97 %) were obtained when hydrophobic anions such as $[PF_6]^-$ or $[SbF_6]^-$ were employed. A 5:1 mixture of $[C_4C_1im][OTf]$ and $[C_4C_1im][BF_4]$ offered the best compromise between catalyst retention and performance, such that the catalyst could be reused at least four times.

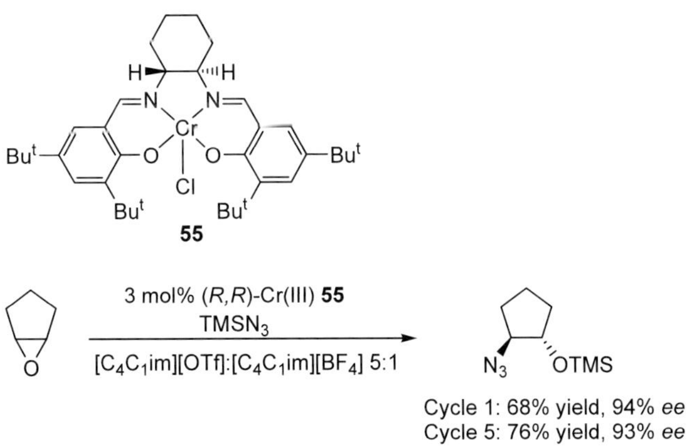

Scheme 5.9: Enantioselective ring-opening of cyclopentene oxide

Table 5.3: Summary of epoxidation reactions in ionic liquids

Ionic liquid	Catalyst	Oxidant	Comments	Ref
$[C_4C_1im]$[various]	OH^-	H_2O_2	Epoxidation of electrophilic olefins; products extracted with *sc*CO_2.	[23]
$[C_4C_1im][BF_4]$ $[C_4C_1im][PF_6]$	OH^-	H_2O_2	Epoxidation of electrophilic olefins; products extracted with ethyl acetate; only epoxide products observed.	[20]
$[C_nC_1im][PF_6]$	OH^-	H_2O_2	Epoxidation of α,β-unsaturated carbonyl compounds; products extracted with toluene. Ionic liquid/water biphasic system; superior activity in the presence of an ionic liquid; ionic liquid can be reused at least 7 times.	[21]
$[C_4C_1im][BF_4]$	OH^- ReO_3Me	H_2O_2	Epoxidation of chromones and flavonoids; product extracted with Et_2O; very low conversion with ReO_3Me as catalyst.	[22]
$[C_4C_1im][BF_4]$	ReO_3Me	UHP	Domino epoxidation-nucleophilic ring opening of glycols with dibutylphosphate; product extracted with Et_2O.	[17] [31]
$[C_2C_1im][BF_4]$	ReO_3Me	UHP H_2O_2	Catalyst appears to be stable for at least 72 hours in the ionic liquid; product extracted with Et_2O; with H_2O_2 as oxidant, diols rather than epoxides are obtained; detailed kinetic studies.	[17] [29] [53]
silylated dihydro-imidazolium	W(VIII) peroxide	H_2O_2	Supported ionic liquid phase (SILP) catalyst with SiO_2 as solid support; similar rates as for the homogeneous reaction.	[38]
large variety	Dioxomolybdenum complexes	TBHP	Epoxidation of cyclooctene; activity under biphasic conditions markedly lower compared to homogeneous solutions; significant decrease in activity upon recycling, probably due to leaching of the catalyst; Tf_2N-ionic liquids are better suited for recycling.	[32]
$[C_4C_1im][BF_4]$ $[C_4C_1im][PF_6]$ $[C_4C_1im][Tf_2N]$	MoCp-complexes	TBHP	Epoxidation of cyclooctene; product extracted with hexane; activity under biphasic conditions markedly lower compared to homogeneous solutions; best results in $[Tf_2N]^-$-based ionic liquids.	[33]

Ionic liquid	Catalyst	Oxidant	Comments	Ref
$[C_4C_1im]Br$	Iron-porphyrin complex	H_2O_2	CH_2Cl_2 as co-solvent; product isolated by decantation; catalyst recycled 4 times, slow decrease in activity; excess oxidant leads to degradation of the catalyst; epoxide-selectivity is substrate-dependent.	[36]
$[C_4C_1im][PF_6]$	Mn(porphyrin) complex	$PhI(OAc)_2$	Epoxidation of styrene; CH_2Cl_2 as co-solvent; better results in presence of an ionic liquid; effect of ionic liquid concentration and oxidant concentration studied; products extracted with hexane; catalyst active for at least 10 runs without significant decrease in activity; $MnTPPS_4$ is more active than MnTPPCl.	[34] [35]
$[C_4C_1im][PF_6]$	Mn(salen) complex	NaOCl	Asymmetric epoxidation; CH_2Cl_2 as co-solvent; reaction proceeds faster in the ionic liquid relative to conventional solvents; product extracted with hexane; catalyst was recycled 4 times, activity and selectivity slowly decrease.	[48]
$[C_4C_1im][BF_4]$ $[C_4C_1im][PF_6]$	Mn(salen) complex	NaOCl	Asymmetric epoxidation; CH_2Cl_2 or ethyl acetate as co-solvent; reaction proceeds faster in the ionic liquid relative to conventional solvents; product extracted with hexane; catalyst was recycled 8 times, activity and selectivity decrease steadily.	[50]
$[C_4C_1im][PF_6]$ $[C_4C_1im][Tf_2N]$	Co(salen) complex		Kinetic resolution of racemic epichlorohydrin; product isolated by distillation and extraction with water; THF as co-solvent; very high enantioselectivity; Co(II)-can be used directly as catalyst precursor; catalyst active for at least 10 runs without decrease in activity or selectivity; activity increases with time.	[51]

5.3.3 Dihydroxylation

The osmium-catalysed oxidation of olefins to their corresponding diols represents a reaction that is widely used in organic synthesis. However, due to toxicity, high cost performance and volatility of the reagent, few large-scale industrial applications have evolved. Furthermore, product contamination with toxic osmium poses another obstacle. These obstacles have been overcome to some extent by, for example, microencapsulation of OsO_4,[54-57] polymer-bound catalysts[58,59] or by use of ion-exchangers,[60] allowing facile separation and recycling.

Dihydroxylation reactions in ionic liquids have been undertaken by several research groups and a summary of that work is provided in Table 5.4. Imidazolium based liquids have been used almost exclusively and it appears that the volatility and thereby the acute toxicity of OsO_4 is greatly suppressed in the presence of ionic liquids. For example, after standing for three months, no OsO_4 could be detected on the septum of a reaction vessel containing OsO_4 in an ionic liquid.[61] Olefins were readily oxidised by OsO_4 with NMO as oxidant in neat ionic liquid. The ionic liquid-catalyst solution was active for at least five cycles, but rather high catalyst loadings (5 mol%) were required (Scheme 5.10).[61]

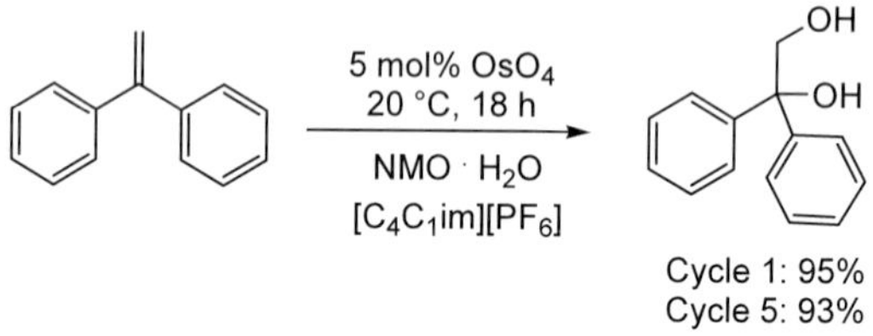

Scheme 5.10: Dihydroxylation of 1,1-diphenylethene

Under the frequently used Upjohn-conditions, which employ a solvent mixture of H_2O and *tert*-butanol, loss of OsO_4 from the ionic liquid phase can be considerable. Addition of simple amines such as 4-dimethylamino pyridine (DMAP) significantly reduces catalyst leaching due to the formation of a zwitterionic amine-osmium adduct (see Figure 5.3), which is preferably retained in the ionic liquid phase.[62]

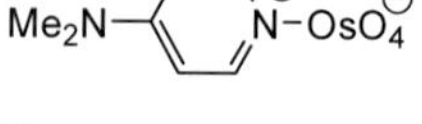

Figure 5.3: The DMAP-osmate adduct

Table 5.4: Summary of dihydroxylation reactions in ionic liquids

Ionic liquid	Catalyst	Oxidant	Comments	Ref
$[C_2C_1im][BF_4]$	OsO_4	NMO	Dihydroxylation of olefins; other oxidants than NMO give only poor results; catalyst reused 4 times, activity basically unchanged; high catalyst concentration (5 mol%).	[61]
$[C_4C_1im][PF_6]$	$K_2[OsO_4]$	NMO/H_2O_2	Dihydroxylation of olefins with H_2O_2 and catalytic amount of NMO; NMO regenerated with the help of an electron transfer mediator; biomimetic system; catalyst active for at least 5 runs.	[63]
$[C_4C_1im][PF_6]$ $[C_8C_1im][PF_6]$ and others	$K_2[OsO_2(OH)_4]$	NMO $K_3[Fe(CN)_6]$ K_2CO_3	Enantioselective dihydroxylation of olefins with H_2O and *t*-BuOH as co-solvents; detailed study on reaction parameters; product extracted with Et_2O; activity and selectivity stable for 9 runs, then considerable decrease; osmium contamination in the product < 7 ppb.	[64] [65]
$[C_4C_1im][PF_6]$	OsO_4	NMO	Enantioselective dihydroxylation of olefins with H_2O and *t*-butanol as co-solvents; products extracted with Et_2O; addition of an amine ligand significantly improves catalyst retention; catalyst recycled 5 times, activity decreases only slowly.	[62]
$[C_4C_1im][PF_6]$	OsO_4	NMO	Enantioselective dihydroxylation of olefins with H_2O and acetone as co-solvents; products extracted with *tert*-butyl methyl ether. Use of a charged chiral PHAL-ligand.	[66]
$[C_4C_1im][PF_6]$ $[C_4C_1im][SbF_6]$	OsO_4	NMO	Enantioselective dihydroxylation of aryl olefins with acetone-H_2O as co-solvent; 2.5-5 equivalents of chiral ligand added; products extracted with Et_2O; catalyst reused four times, marked decrease in activity after the third run.	[67]
$[C_4C_1im][PF_6]$	$K_2[OsO_2(OH)_4]$	NMO	Enantioselective dihydroxylation of *trans*-stilbene with H_2O and acetone as co-solvents and $(DQ)_2PHAL$ or $(DHQ)_2PHAL$ as ligand; products extracted with diethyl ether.	[68]
$[C_4C_1im][Tf_2N]$ $[C_4C_1C_1im][Tf_2N]$	$K_2[OsO_2(OH)_4]$	NMO	Enantioselective dihydroxylation of olefins with 2 equivalents $(DHQD)_2PHAL$ as ligand; product extracted with either Et_2O or *sc*CO_2.	[69]

Biomimetic dihydroxylation represents an interesting methodology in that hydrogen peroxide, which is cheap and environmentally friendly, is used as oxidant.[63] The key step in the system is an electron transfer mediator (ETM) which reoxidises *N*-methyl morpholine (NMM) to *N*-morpholine-*N*'-oxide, as shown in Scheme 5.11. Compounds that can act as an ETM include, for example, vanadium or rhenium complexes in high oxidation states or flavin. Catalytic amounts of K_2OsO_4, NMM and an electron transfer mediator were immobilised in $[C_4C_1im][PF_6]$ together with the substrate and acetone as co-solvent and 0.5 equivalents of both DMPA and tetraethylamine acetate (TEAA) were added. The latter reagent serves to accelerate the hydrolysis of the osmate ester that forms during the catalytic cycles.

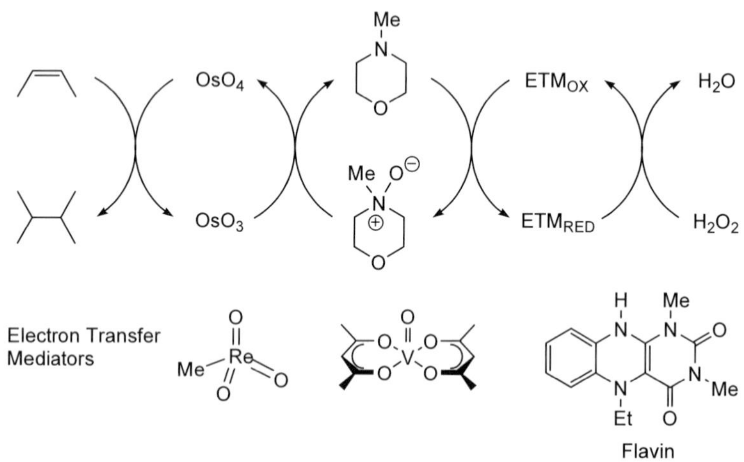

Scheme 5.11: Biomimetic dihydroxylation of olefins with electron transfer mediators

A variety of olefins could be oxidised using this approach and best results were obtained with flavin as ETM. Activity of the catalyst was maintained for at least five cycles and both the mediator as well as the *N*-morpholine-*N*'-oxide remain in the ionic liquid, thus only H_2O_2 and co-solvent need to be recharged.

Of particular interest are asymmetric dihydroxylation (AD) reactions. Depending on the oxidant employed, which is commonly either NMO or $K_3[Fe(CN)_6]/K_2CO_3$, two different mechanisms may apply (see Schemes 5.12 and 5.13).[70] One essential difference is that *N*-morpholine-*N*'-oxide is typically used in a homogeneous acetone-water solvent mixture while a biphasic reaction mixture is usually observed with aqueous $K_3[Fe(CN)_6]/K_2CO_3$ as oxidant. In the latter case the olefin is oxidised (osmylation) in the organic phase and the hydrolysis takes place at the organic-aqueous interface.

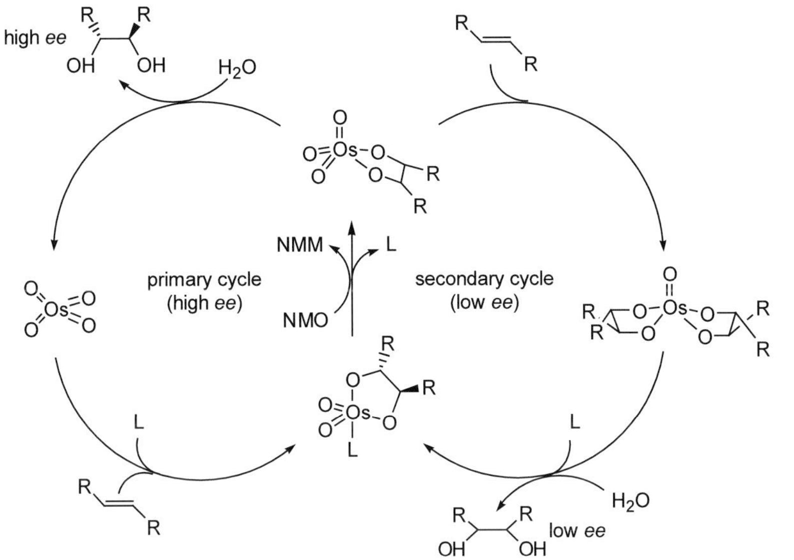

Scheme 5.12: Catalytic cycle for AD-reaction with *N*-methylmorpholine-*N*-oxide as co-oxidant

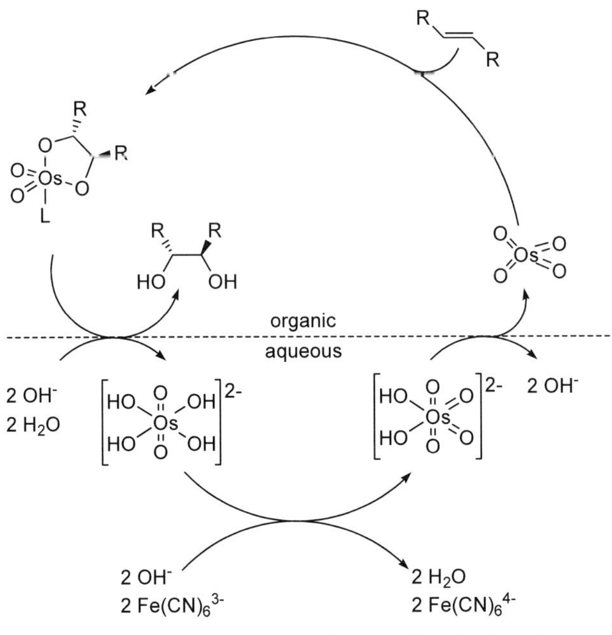

Scheme 5.13: Catalytic cycle for AD-reaction with potassium ferricyanide as co-oxidant

That situation may change in the presence of an ionic liquid. Depending on the co-solvent used, homogeneous or heterogeneous conditions may result. Although $[C_4C_1im][PF_6]$ is essentially immiscible with either water or *tert*-butanol, homogeneous mixtures were observed, once substrate and NMO were added.[61,62] Monophasic and biphasic reaction conditions were compared for reactions employing NMO as well as $K_3[Fe(CN)_6]/K_2CO_3$ as oxidant and good yield and selectivity was observed for both reaction systems.[65]

The combination of $K_2[OsO_2(OH)_4]$ or OsO_4 with chiral alkaloid ligands affords highly recyclable systems for the dihydroxylation of olefins. Bis-*Chinona* alkaloid ligands containing phthalazine or pyrimidine backbones, as shown in Figure 5.8, provide superb enantioselectivity and different models have been devised to rationalise the chirality transfer.[71,72]

With two equivalents of **$(DHQD)_2$PHAL** as ligand very high recyclability was reported when the catalyst was immobilised in $[C_4C_1im][Tf_2N]$.[69] In the dihydroxylation of hexane a yield and *ee* of 92% were observed even after the catalyst had been reused already 13 times. When the product was extracted with diethyl ether, 1-2% of the initial amount of the osmium catalyst was lost from the ionic liquid. With *sc*CO_2 as extractant, osmium-leaching was as low as 0.34% if the CO_2-pressure was not too high (100 bar).

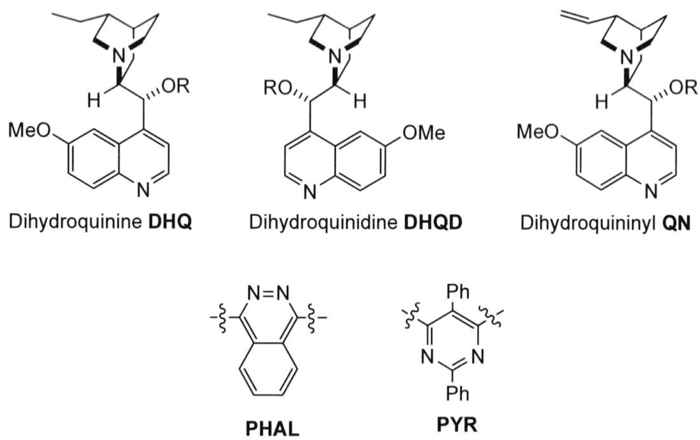

Figure 5.8: Stuctures of *Chinchona* alkaloid derived ligands

Complex formation between OsO_4 and the alkaloid ligand occurs reversibly *via* the quinuclidine nitrogen. Use of an alkaloid ligand bearing polar or even charged groups should help to immobilise the metal complex and minimise leaching of toxic osmium during the product extraction step. An elegant solution lies in the use of 1,4-bis(9-*O*-quininyl)phtalazid, **(QN)$_2$PHAL**, as ligand: Under AD-reaction conditions this ligand is converted to its more polar derivative **(QN)$_2$PHAL(4-OH)** which leads to a significantly better catalyst retention in the ionic liquid, see Scheme 5.14.[67] While up to 80% of **(DHQ)$_2$PHAL** were detected in the organic phase of a diethyl ether - [C$_4$C$_1$im][PF$_6$] mixture, less then 1% ligand leaching was detected with **(QN)$_2$PHAL(4-OH)** as ligand.[68] With 0.1 mol% of catalyst a total turnover number of 2,370 was achieved after three cycles.

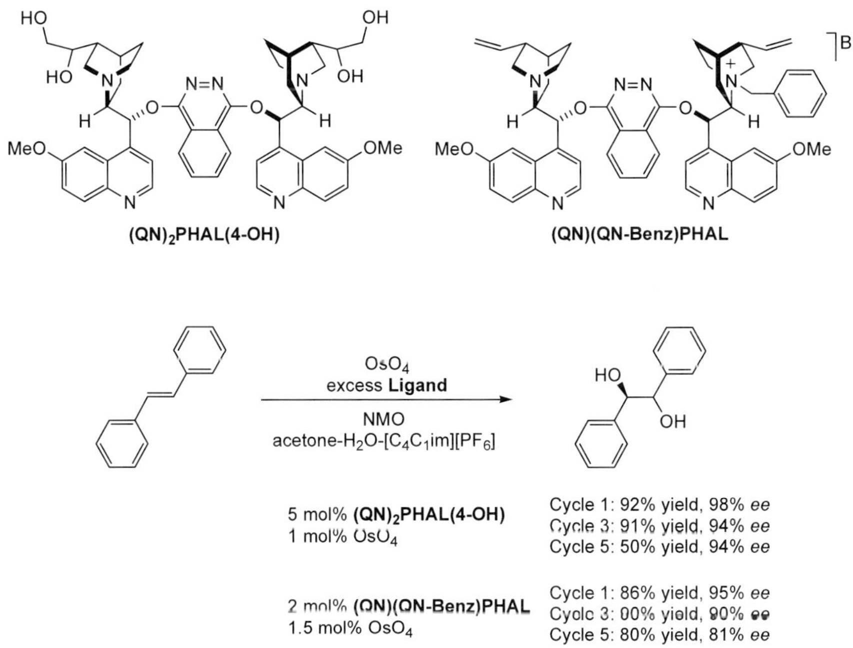

Scheme 5.14: Enantioselective dihydroxylation of *trans*-stilbene

Catalyst retention can be further optimised if a charged derivative of the **(QN)$_2$PHAL**-ligand is used. Quaternisation of one of the nitrogen atoms with benzyl bromide affords the cationic ligand **(QN)(QN-Benz)PHAL**, and indeed recyclability was improved in the presence of this ligand, however at the cost of lower *ee*, as shown in Scheme 5.14.[66] The addition of one equivalent of tetraethylammonium acetate was found to improve the selectivity .[60]

With $K_2OsO_2(OH)_4$ as catalyst and $K_3Fe(CN)_6/K_2CO_3$ as oxidant, 1-hexene was converted to the corresponding diol with good conversion and selectivity.[64] Only after the ninth run was a marked decrease in catalyst performance observed, (see Scheme 5.15).

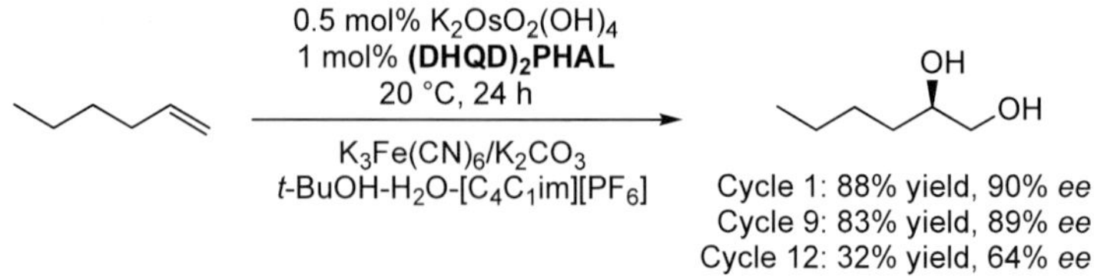

Scheme 5.15: Enantioselective dihydroxylation of 1-hexene

For this system, a detailed study of the various reaction parameters was undertaken, using styrene as a model substrate.[65] Variations in the temperature were found to have a strong effect on the catalytic activity, with best results being obtained between 25°C and 45°C whereas the selectivity remained essentially unaffected over this temperature range.

5.3.4 Formation of Ketones

Very few examples of the oxidation of olefins to ketones in ionic liquids have been reported. In one case, $[C_4C_1im][BF_4]$ or $[C_4C_1im][PF_6]$ were used in the palladium-catalysed oxidation of styrene to acetophenone with H_2O_2 as oxidant, however the concept of biphasic catalysis was not exploited and no attempts were made to recycle the catalyst.[13] The ionic liquid serves the purpose of a co-catalyst rather than that of a reaction medium.

A solvent combination of *sc*CO_2-$[C_4C_1im][PF_6]$ was found to afford superior selectivity in the Wacker-type oxidation of 1-hexene with $PdCl_2$-CuCl as catalyst to afford 2-hexanone as the main product as shown in Scheme 5.16.[73] In the absence of either *sc*CO_2 or ionic liquid, considerably lower selectivity for 2-hexanone was observed. Catalyst solutions were recycled five times with a low, but steady, decrease in activity.

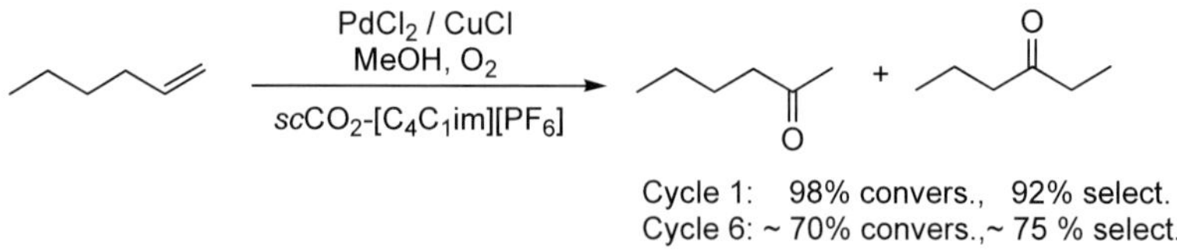

Scheme 5.16: Wacker-type oxidation of 1-hexene

5.4 Oxidation of Alcohols and Thiols

The palladium catalysed dehydrogenation of benzylic alcohols was performed in molten $[(C_4)_4N]Br$ without addition of base or need for oxygen (see Scheme 5.17).[74] Hydrogen gas liberated during the reaction has to be removed from the reaction mixture in order to achieve good conversion. The product was extracted with diethyl ether and recycling of the catalyst phase was possible, however yields decreased with each run.

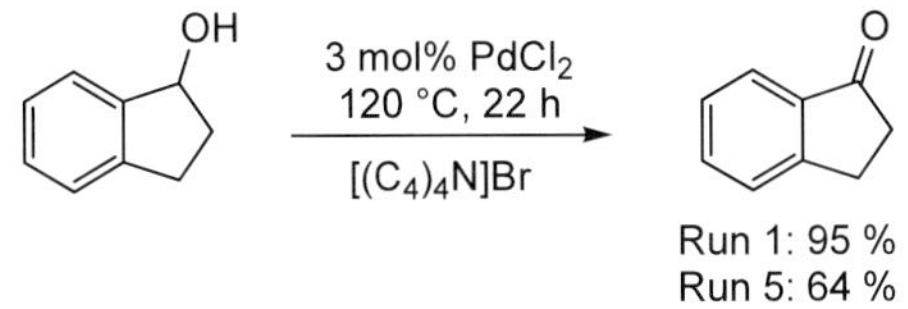

Scheme 5.17: Dehydrogenation of indan-1-ol

The aerobic oxidation of benzylic and allylic alcohols to their corresponding carbonyl compounds in $[C_4C_1im][PF_6]$ with TEMPO-CuCl (TEMPO = 2,2,6,6-tetramethylpiperidinyl-1-oxy) as catalyst was found to proceed at higher rates than that of aliphatic alcohols, which is in agreement with results in classical solvents.[75] After product extraction with diethyl ether, the ionic liquid was washed with water and dried at 70°C, prior to the next run. In that manner catalyst activity remained relatively stable for 8 cycles.

A number of different ionic liquids have been screened in the ruthenium-catalysed oxidation of secondary alcohols (see Scheme 5.18). Three different ruthenium compounds, $RuCl_3$, $RuCl_2(PPh_3)_3$ and $[RuCl_2(p\text{-cymene})]_2$ were compared and best results were obtained with $RuCl_2(PPh_3)_3$.[76] While imidazolium-based ionic liquids gave only poor results (anion = Cl) or suppressed the reaction completely (anion = $[BF_4]^-$ or $[PF_6]^-$), tetraalkylammonium-based solvents such as Aliquat 336 (tricaprylmethylammonium chloride) or tetramethylammonium hydroxide afforded much better yields.

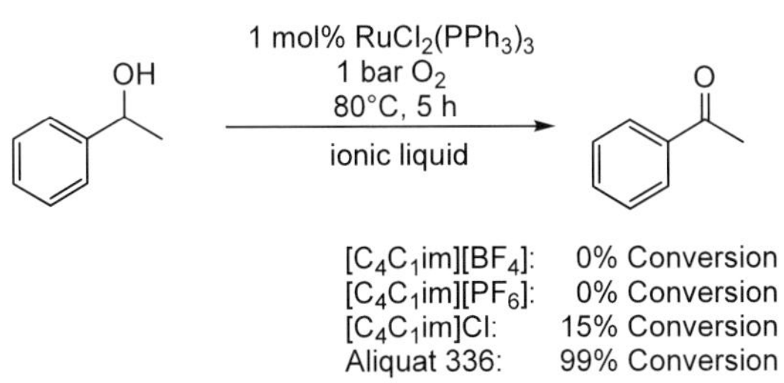

Scheme 5.18: Aerobic oxidation of 1-phenylethanol to acetophenone

A range of primary and secondary alcohols were oxidised to their corresponding aldehydes and ketones in a variety of different ionic liquids with high catalytic loadings (5-10 mol%) of tetra-*N*-propylammonium perruthenate, $[(C_3)_4N][RuO_4]$ and either NMO or air as oxidant.[77,78] The catalyst was found to be active for at least five cycles without loss of activity when simple substrates were employed. With more complex substrates, as for example in the oxidation of cinnamyl alcohol, activity decreased rapidly. When molecular oxygen was employed as oxidant best results were obtained when the oxygen was bubbled through the solution and CuCl was added as co-catalyst. In most cases addition of aminopyridine did not have a beneficial effect on the outcome of the reaction.[78]

A range of aliphatic and aromatic thiols may be oxidised with O_2 to their corresponding disulfides in neat $[C_4C_1im][BF_4]$ in good to excellent yield with cobalt(II) phtalocyanine catalysts **56a** and **56b**, shown in Scheme 5.19.[79] In comparison to common organic solvents, higher yields and turnover frequencies were observed and product separation was facile.

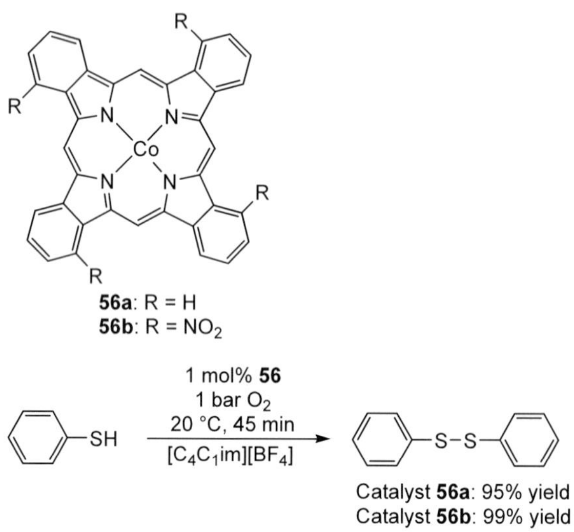

Scheme 5.19: Oxidation of thiophenol with cobalt(II) phtalocyanine complexes

5.5 Other Oxidation Reactions

5.5.1 Oxidation of Aldehydes and Ketones

The first catalytic oxidation to be carried out in an ionic liquid was probably the oxidation of aromatic aldehydes with $Ni(acac)_2$ and molecular oxygen to the corresponding carboxylic acids, reported by Howarth in 2000 (Scheme 5.20).[8] Overall yields were only moderate, which was attributed to the low oxygen solubility in the ionic liquid, but the catalyst containing ionic liquid was successfully reused without deterioration in activity.

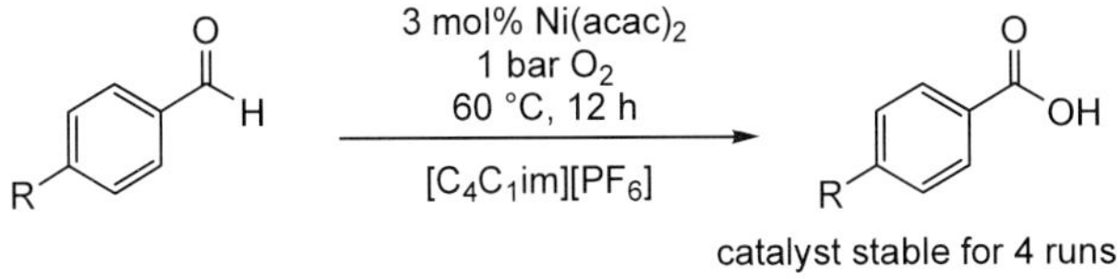

Scheme 5.20: Aerobic oxidation of aromatic aldehydes

The Baeyer–Villiger oxidation of a range of cyclic ketones was undertaken with ReO_3Me in $[C_4C_1im][BF_4]$ with H_2O_2 as oxidant.[80] The ring-strained cyclobutanone was oxidised quantitatively at room temperature to γ-butyrolactone within one hour and the activity of the ionic liquid phase only deteriorated after the fourth cycle. Even at elevated temperatures only moderate yields were observed with unactivated ketones as substrate. The oxidation of the methylated flavonone naringenin (Scheme 5.21) proceeded more efficiently and under milder conditions in an ionic liquid as compared to a system in which the catalyst was immobilised onto a polymer support.[81]

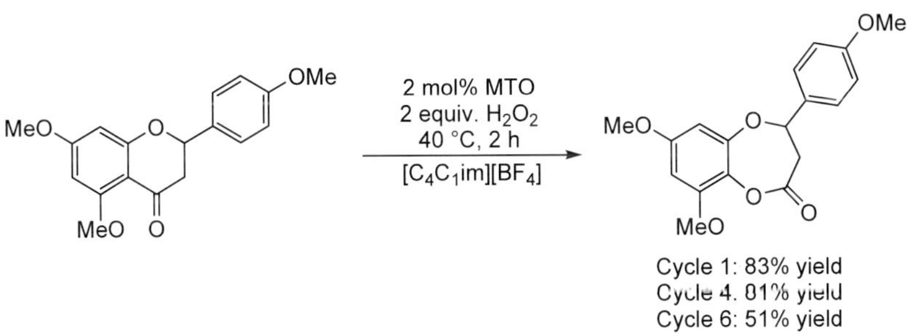

Scheme 5.21: Baeyer-Villiger oxidation of naringenin

Baeyer-Villiger oxidations in ionic liquids have also been reported using either *m*-chloroperbenzoic acid[82] or H_2O_2 as the oxidant.[83] In the latter case Sn-β-molecular sieves served as the catalyst.

5.5.2 Oxidation of Alkanes

The oxidation of cyclic alkanes has been investigated in a mixture of $[C_4C_1im][PF_6]$ and dichloromethane using the manganese-porphyrin catalysts **3a-d** and $PhI(OAc)_2$ as oxidant. Results from stopped-flow UV-VIS spectroscopy suggest that the oxygen atom transfer occurs *via* the *in situ* formation of a Mn(V) = O oxo-manganese complex. In the absence of the ionic liquid, only a less active Mn(IV) = O intermediate was detected.[35,84] Depending on the substrate, the presence of a nitrogen base that acts as an axial ligand to the manganese porphyrin complex can have a beneficial effect on the catalytic activity. The addition of dichloromethane as a co-solvent was found to be necessary to reduce the viscosity of the reaction mixture and such a solvent mixture afforded higher yields than either of the neat solvents alone. The substituents on the porphyrin ligand have a major influence on both the activity and selectivity of the complex, as shown in Scheme 5.22 for the oxidation of tetralin.[35]

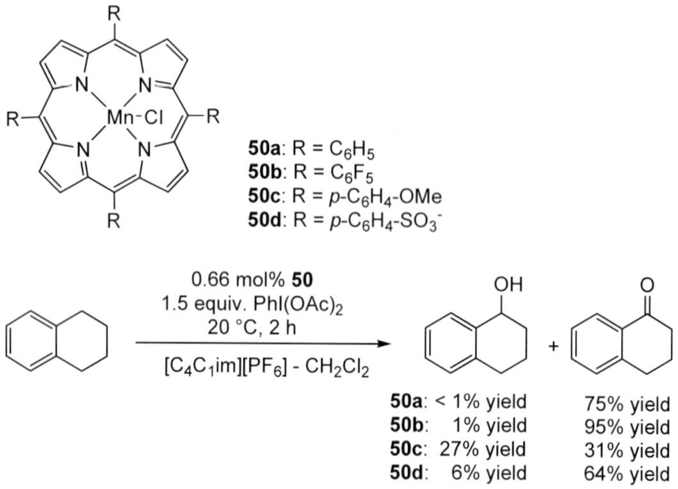

Scheme 5.22: Oxidation of tetralin with the manganese-porphyrine catalyst **50**

With molecular oxygen as the oxidant, addition of perfluorinated solvents can help to increase the reaction rate as demonstrated in the $Co(acac)_2$-catalysed oxidation of ethylbenzene to acetophenone.[16] The function of the perfluorohexane solvent is to increase the otherwise low oxygen concentration in the ionic liquid phase.

5.5.3 Oxidation of Arenes

Benzene may be oxidised to phenol in an ionic liquid-aqueous biphasic system employing dodecanesulfonate salts of Fe^{3+}, Fe^{2+}, Cu^{2+}, Co^{2+} and Ni^{2+} as catalyst and hydrogen peroxide as oxidant.[85] Ionic liquids with long alkyl chains like $[C_8C_1im][PF_6]$ form one phase with benzene and at the same time dissolve the catalyst. As the phenol product is preferably soluble in the aqueous phase product isolation was facile. Yet, reaction rates were only moderate and after six hours at 50°C conversion of benzene was in the range of 35-55%.

5.6 Conclusion

For many oxidation reactions, the presence of ionic liquids appears to have a beneficial effect in that increased reaction rates are observed and stabilisation of catalytically active intermediates in these solvents has been demonstrated by means of *in situ* spectroscopic methods in several cases. And while the oxygen solubility is very poor in common ionic liquids solid oxidants are often dissolved more efficiently, resulting in increased reaction rates, ever under biphasic conditions.

It is probably in asymmetric dihydroxylation, where the use of ionic liquids appears to be most promising. The decreased acute toxicity of osmium tetroxide due to its suppressed volatility certainly represents a great benefit for those who work with this reagent and its derivatives. Furthermore, high cost of both the osmium catalyst, as well as the chiral ligand, make recycling of the catalyst-ionic liquid particularly attractive. On the other hand, disposal of osmium-contaminated ionic liquids in an environmentally benign manner has yet to be addressed.

5.7 References

[1] Bonhôte, P., Dias, A.-P.; Papageorgiou, N.; Kalyanasundaram, K.; Grätzel, M. *Inorg. Chem.* **1996**, *35*, 1168-1178.

[2] Evans, R. G.; Klymenko, O. V.; Hardacre, C.; Seddon, K. R.; Compton, R. G. *J. Electroanal. Chem.* **2003**, *556*, 179-188.

[3] AlNashef, I. M.; Matthews, M. A.; Weidner, J. W. In: *Ionic Liquids as Green Solvents: Progress and Prospects*, ACS Symposium Series 856, **2003**, pp 509-525.

[4] Hinckley, G.; Mozhaev, V. V.; Budde, C.; Khmelnitsky, Y. L. *Biotechnol. Lett.* **2002**, *24*, 2083-2087.

[5] Okrasa, K.; Guibe-Jampel, E.; Therisod, M. *Tetrahedron: Asymmetry* **2003**, *14*, 2487-2490.

[6] Laszlo, J. A.; Compton, D. L. *J. Mol. Catal. B* **2002**, *18*, 109-120.

[7] Walker, A. J.; Bruce, N. C. *Chem. Commun.* **2004**, 2570-2571.

[8] Howarth, J. *Tetrahedron Lett.* **2000**, *41*, 6627-6629.

[9] Kim, H. S.; Kim, Y. J.; Bae, J. Y.; Kim, S. J.; Lah, M. S.; Chin, C. S. *Organometallics* **2003**, *22*, 2498-2504.

[10] Alcántara, R.; Canoira, L.; Guilherme-Joao, P.; Santos, A. S.; Vázquez, I. *Appl. Catal. A* **2000**, *203*, 259-268.

[11] Alcantara, R.; Canoira, L.; Guilherme-Joao, P.; Perez-Mendo, J. P. *Appl. Catal. A* **2001**, *218*, 269-279.

[12] Sun, J.; Fujita, S.-I.; Bhanage, M.; Arai, M. *Catal. Commun.* **2004**, *5*, 83-87.

[13] Namboodiri, V. V.; Varma, R. S.; Sahle-Demessie, E.; Pillai, U. R. *Green Chem.* **2002**, *4*, 170-173.

[14] Álvaro, M.; Ferrer, B.; Garcia, H.; Narayana, M. *Chem. Phys. Lett.* **2002**, *362*, 435-440.

[15] Anthony, J. L.; Anderson, J. L.; Maginn, E. J.; Brennecke, J. F. *J. Phys. Chem. B* **2005**.

[16] Seddon, K. R.; Stark, A. *Green Chem.* **2002**, *4*, 119-123.

[17] Abu-Omar, M. M.; Owens, G. S.; Durazo, A. In: *Ionic Liquids as Green Solvents: Progress and Prospects*, ACS Symposium Series 856, **2003**, pp 277-288.

[18] Herrmann, W. A.; Fischer, R. W.; Rauch, M. U.; Scherer, W. *J. Mol. Catal.* **1994**, *86*, 243-266.

[19] Wang, W.; Espenson, J. H. *J. Am. Chem. Soc.* **1998**, *120*, 11335-11341.

[20] Bortolini, O.; Conte, V.; Chiappe, C.; Fantin, G.; Fogagnolo, M.; Maietti, S. *Green Chem.* **2002**, *4*, 94-96.

[21] Wang, B.; Kang, Y.-R.; Yang, L.-M.; Suo, J.-S. *J. Mol. Catal. A.* **2003**, *203*, 29-36.

[22] Bernini, R.; Mincione, E.; Coratti, A.; Fabrizi, G.; Battistuzzi, G. *Tetrahedron* **2004**, *60*, 967-971.

[23] Bortolini, O.; Campestrini, S.; Conte, V.; Fantin, G.; Fogagnolo, M.; Maietti, S. *Eur. J. Org. Chem.* **2003**, 4804-4809.

[24] Herrmann, W. A.; Fischer, R. W.; Marz, D. W. *Angew. Chem. Int. Ed.* **1991**, *30*, 1638-1641.

[25] Herrmann, W. A.; Kuhn, F. E. *Acc. Chem. Res.* **1997**, *30*, 169-180.

[26] Romao, C. C.; Kuhn, F. E.; Herrmann, W. A. *Chem. Rev.* **1997**, *97*, 3197-3246.

[27] Espenson, J. H. *Chem. Commun.* **1999**, 479-488.

[28] Owens, G. S.; Arias, J.; Abu-Omar, M. M. *Catal. Today* **2000**, *55*, 317-363.

[29] Owens, G. S.; Abu-Omar, M. M. *Chem. Commun.* **2000**, 1165-1166.

[30] Abu-Omar, M. M.; Owens, G. S.; Durazo, A. In: *Ionic Liquids as Green Solvents: Progress and Prospects*, ACS Symp. Ser. **2003**, *856*, 277-288.

[31] Soldaini, G.; Cardona, F.; Goti, A. *Tetrahedron Lett.* **2003**, *44*, 5589-5592.

[32] Valente, A. A.; Petrovski, Z.; Branco, L. C.; Afonso, C. A. M.; Pillinger, M.; Lopes, A. D.; Romao, C. C.; Nunes, C. D.; Goncalves, I. S. *J. Mol. Catal. A* **2004**, *218*, 5-11.

[33] Kühn, F. E.; Zhao, J.; Abrantes, M.; Sun, W.; Afonso, C. A. M.; Branco, L. C.; Gonçalves, I. S.; Pillinger, M.; Romao, C. C. *Tetrahedron Lett.* **2005**, *46*, 47-52.

[34] Li, Z.; Xia, C. G.; Ji, M. *Appl. Catal. A* **2003**, *252*, 17-21.

[35] Li, Z.; Xia, C. G. *J. Mol. Catal. A* **2004**, *214*, 95-101.

[36] Srinivas, K. A.; Kumar, A.; Chauhan, S. M. S. *Chem. Commun.* **2002**, 2456-2457.

[37] Groves, J. T.; Myers, R. S. *J. Am. Chem. Soc.* **1983**, *105*, 5791-5796.

[38] Yamaguchi, K.; Yoshida, C.; Uchida, S.; Mizuno, N. *J. Am .Chem. Soc.* **2005**, *127*, 530-531.

[39] Song, C. E. *Chem. Commun.* **2004**, 1033-1043.

[40] Jacobsen, E. N.; Wu, M. H. In: *Comprehensive Asymmetric Catalysis*; Jacobsen, E. N., Pfaltz, A., Yamamoto, H., Eds.; Springer Verlag: Berlin, **1999**; Vol. 2, p 649.

[41] Sabater, M. J.; Corma, A.; Domenech, A.; Fornes, V.; Garcia, H. *Chem. Commun.* **1997**, 1285-1286.

[42] Zhang, Y.; Yin, D. H.; Fu, Z. H.; Li, C. Z.; Yin, D. L. *Chin. J. Catal.* **2003**, *24*, 942-946.

[43] Chatterjee, D.; Mitra, A. *J. Mol. Catal. A* **1999**, *144*, 363-367.

[44] Piaggio, P.; McMorn, P.; Murphy, D.; Bethell, D.; Page, P. C. B.; Hancock, F. E.; Sly, C.; Kerton, O. J.; Hutchings, G. J. *Perkin Trans. 2,* **2000**, 2008-2015.

[45] Vankelecom, I. F. J.; Vercruysse, K. A. L.; Neys, P. E.; Tas, D. W. A.; Janssen, K. B. M.; Knops-Gerrits, P. P.; Jacobs, P. A. *Top. Catal.* **1998**, *5*, 125-132.

[46] Vankelecom, I. F. J.; Moens, N. M. F.; Vercruysse, K. A. L.; Parton, R. F.; Jacobs, P. A. *Stud. Surf. Sci. Catal.* **1997**, *108*, 437-444.

[47] Pozzi, G.; Cinato, F.; Montanari, F.; Quici, S. *J. Chem. Soc., Chem. Commun.* **1998**, 877-878.

[48] Song, C. E.; Roh, E. J. *Chem. Commun.* **2000**, 837-838.

[49] Gaillon, L.; Bedioui, F. *Chem. Commun.* **2001**, 1458-1459.

[50] Smith, K.; Liu, S.; El-Hiti, G. A. *Catal. Lett.* **2004**, *98*, 95-101.

[51] Oh, C. R.; Choo, D. J.; Shim, W. H.; Lee, D. H.; Roh, E. J.; Lee, S.; Song, C. E. *Chem. Commun.* **2003**, 1100-1101.

[52] Song, C. E.; Oh, C. R.; Roh, E. J.; Choo, D. J. *Chem. Commun.* **2000**, 1743-1744.

[53] Owens, G. S.; Durazo, A.; Abu-Omar, M. M. *Chem. Eur. J.* **2002**, *8*, 3053-3059.

[54] Song, C. E.; Yang, G. Y.; Ha, H.-J. *Tetrahedron: Asymmetry* **1997**, *8*, 841-844.

[55] Kobayashi, S.; Ishida, T.; Akiyama, R. *Org. Lett.* **2001**, *3*, 2649-2652.

[56] Ishida, T.; Akiyama, R.; Kobayashi, S. *Adv. Synth. Catal.* **2003**, *345*, 576-579.

[57] Chandrasekhar, S.; Narsihmulu, C.; Sultana, S. S.; Reddy, N. R. *Chem. Commun.* **2003**, 1716-1717.

[58] Bolm, C.; Gerlach, A. *Angew. Chem., Int. Ed.* **1997**, *36*, 741-743.

[59] Song, C. E.; Yang, G. Y.; Ha, H.-J.; Lee, S.-G. *Tetrahedron: Asymmetry* **1996**, *7*, 645-648.

[60] Choudary, M. B.; Chowar, S. N.; Jyothi, K.; Kantam, L. M. *J. Am. Chem. Soc.* **2002**, *124*, 5341-5349.

[61] Yanada, R.; Takemoto, Y. *Tetrahedron Lett.* **2002**, *43*, 6849-6851.

[62] Yao, Q. W. *Org. Lett.* **2002**, *4*, 2197-2199.

[63] Closson, A.; Johansson, M.; Bäckvall, J. E. *Chem. Commun.* **2004**, 1494-1495.

[64] Branco, L. C.; Afonso, C. A. M. *Chem. Commun.* **2002**, 3036-3037.

[65] Branco, L. C.; Afonso, C. A. M. *J. Org. Chem.* **2004**, *69*, 4381-4389.

[66] Jiang, R.; Kuang, Y. Q.; Sun, X. L.; Zhang, S. Y. *Tetrahedron: Asymmetry* **2004**, *15*, 743-746.

[67] Song, C. E.; Jung, D.-U.; Roh, E. J.; Lee, S.-G.; Chi, D. Y. *Chem. Commun.* **2002**, 3038-3039.

[68] Liu, Q.; Zhang, Z.; van Rantwijk, F.; Sheldon, R. A. *J. Mol. Catal. A* **2004**, *224*, 213-216.

[69] Branco, L. C.; Serbanovic, A.; Nunes da Ponte, M.; Afonso, C. A. M. *Chem. Commun.* **2005**, 107-109.

[70] Johnson, R. A.; Sharpless, K. B. In *Catalytic Asymmetric Synthesis*; Ojima, I., Ed.; Wiley-VCH: New York, **2000**, pp 357-398.

[71] Kolb, H. C.; Andersson, P. G.; Bennani, Y. L.; Crispino, G. A.; Jeong, K. S.; Kwong, H. L.; Sharpless, K. B. *J. Am. Chem. Soc.* **1993**, *115*, 12226-12227.

[72] Corey, E. J.; Noe, M. C. *J. Am. Chem. Soc.* **1996**, *118*, 11038-11053.

[73] Hou, Z. S.; Han, B. X.; Gao, L.; Jiang, T.; Liu, Z. M.; Chang, Y. H.; Zhang, X. G.; He, J. *New J. Chem.* **2002**, *26*, 1246-1248.

[74] Ganchegui, B.; Bouquillon, S.; Henin, F.; Muzart, J. *Tetrahedron Lett.* **2002**, *43*, 6641-6644.

[75] Ansari, I. A.; Gree, R. *Org. Lett.* **2002**, *4*, 1507-1509.

[76] Wolfson, A.; Wuyts, S.; De Vos, D. E.; Vankelecom, I. F. J.; Jacobs, P. A. *Tetrahedron Lett.* **2002**, *43*, 8107-8110.

[77] Ley, S. V.; Ramarao, C.; Smith, M. D. *Chem. Commun.* **2001**, 2278-2279.

[78] Farmer, V.; Welton, T. *Green Chem.* **2002**, *4*, 97-102.

[79] Chauhan, S. M. S.; Kumar, A.; Srinivas, K. A. *Chem. Commun.* **2003**, 2348-2349.

[80] Bernini, R.; Coratti, A.; Fabrizi, G.; Goggiamani, A. *Tetrahedron Lett.* **2003**, *44*, 8991-8994.

[81] Bernini, R.; Mincione, E.; Cortese, M.; Saladino, R.; Gualandi, G.; Belfiore, M. C. *Tetrahedron Lett.* **2003**, *44*, 4823-4825.

[82] Yadav, J. S.; Reddy, B. V. S.; Basak, A. K.; Narsaiah, A. V. *Chem. Lett.* **2004**, *33*, 248-249.

[83] Panchgalle, S. P.; Kalkote, U. R.; Niphadkar, P. S.; Joshi, P. N.; Chavan, S. P.; Chaphekar, G. M. *Green Chem.* **2004**, *6*, 308-309.

[84] Li, Z.; Xia, C. G.; Xu, C. Z. *Tetrahedron Lett.* **2003**, *44*, 9229-9232.

[85] Peng, J. J.; Shi, F.; Gu, Y. L.; Deng, Y. Q. *Green Chem.* **2003**, *5*, 224-226.

Chapter 6

CARBON-CARBON COUPLING REACTIONS

6.1 Introduction

The formation of new carbon-carbon bonds represents probably the most important synthetic step in organic chemistry and carbon-carbon cross coupling has had a huge impact over the last 30 years on the way complex molecules are prepared. This chapter almost exclusively deals with palladium-catalysed reactions with the only exceptions being a copper-mediated Heck-coupling as well as an example of a rhodium catalysed Suzuki-type coupling. Palladium-catalysed C-C coupling reactions in ionic liquids have been reviewed on a previous occasion, covering the literature until early 2003.[1]

Apart from the omnipresent imidazolium based ionic liquids, other classes of low melting salts have been successfully applied in carbon-carbon coupling reactions, notably tetraalkyl ammonium and phosphonium salts. The former include pyrrolidinium and piperidinium salts, but particularly tetrabutylammonium bromide, $[(C_4)_4N]Br$ (mp. 103-105°C), which has been evaluated quite extensively in Heck reactions and a remarkable increase in reactions rates is frequently observed with this solvent. Those examples where $[(C_4)_4N]Br$ acts merely as co-catalyst[2-6] rather than as reaction medium shall not be discussed here.

It is believed that the accelerated reaction rates that are observed in ammonium salts are due to their ability to stabilise nanoparticles, and especially when simple palladium salts are employed under ligand-free conditions, formation of colloidal palladium is highly likely.[7-9] As carbon-carbon coupling reactions are frequently conducted at elevated temperatures, typically between 120-150ºC, a wide range of low-melting salts may be considered as potential solvents for such transformations. Those salts which are solid at ambient temperature may actually facilitate the separation between product and catalyst, which is the case when the system is conducted under biphasic conditions in the presence of a co-solvent. In the absence of a co-solvent, extraction of a solid is required, which can be impractical.

Very promising results have also been observed using phosphonium-based ionic liquids in both Heck and Suzuki coupling reactions.[10-12] Notably, good activities were obtained with arylchlorides as substrate, which in molecular solvents are often found to react only sluggishly, if at all.

Finally, there are a few examples where task-specific ionic liquids have been employed. These examples encompass an imidazolium ionic liquid with a chiral alkyl chain used to induce chirality,[13] an imidazole-imidazolium liquid which acts as both ligand and solvent[14] and a nitrile-functionalised pyridinium ionic liquid which also serve as ligand and solvent.[15] The application of these specific solvents is discussed in greater detail below.

Though not a general rule, it is worthwhile pointing out that ionic liquids with halide anions often give good or superior activities than, for example, $[BF_4]^-$, $[PF_6]^-$ or $[Tf_2N]^-$ salts. In this respect C-C coupling reactions stand out from other catalysed transformations in ionic liquids where the presence of halide (often due to insufficient purification of the ionic liquid used) is frequently found to inhibit catalytic activity.

Despite the considerable interest on C-C coupling reactions in ionic liquids, very little attention has so far been paid to catalyst modification in order to improve their immobilisation in the ionic medium. This is probably in part due to the fact that the active species is often generated *in situ* from simple metal salts and that the ionic liquid is frequently non-innocent, acting as both solvent and ligand (see below). Nevertheless, further examples of well-defined molecular catalysts containing charged ligands would certainly be a valuable extension to the field. In this context it is also surprising that enantioselective coupling reactions are essentially absent from the literature. As chiral catalysts are often very expensive, probing their recyclability in an ionic liquid medium should be particularly attractive.

6.2 Catalyst Stability in Ionic Liquid Mediated C-C Coupling Reactions

The influence of the ionic liquid on the performance of a given catalyst can be relatively complex in C-C coupling reactions. Under the reaction conditions that usually prevail – presence of a base and elevated temperatures – generation of new catalytically active species from the reaction between the palladium salt and the ionic liquid cation can easily occur. Imidazolium-based ionic liquids are particularly reactive due to the presence of the acidic proton at the 2-position of the ring.[16,17] Indeed, the formation of *N,N*-heterocyclic carbene complexes from imidazolium based ionic liquids is relatively facile and depending on the specific circumstances, this can be either detrimental or advantageous. Various transition metal carbene complexes have been deliberately prepared in imidazolium-based ionic liquids and these compounds are often found to be active catalysts.[18-20]

It appears that palladium acetate is always transformed into a new species under the conditions present in C-C coupling reactions and the formation of carbene complexes from $Pd(OAc)_2$ and imidazolium based ionic liquids is particularly facile. However, the outcome depends strongly on the nature of the anion present.[21] As an example, $Pd(OAc)_2$ readily reacts with $[C_4C_1im]Br$ to afford imidazolylidene isomers of the composition $PdBr_2(C_4C_1imy)_2$, as shown in Scheme 6.1.[20] These complexes have been demonstrated to be highly efficient catalysts in Heck coupling reactions while their chloro-analogues were found to be less active, presumably due to the stronger palladium-chloride bond. In contrast, no carbene formation was observed between $Pd(OAc)_2$ and $[C_4C_1im][BF_4]$ in the absence of a halide source. Further, when $PdBr_2(C_4C_1imy)_2$ was immobilised in $[C_4C_1im][BF_4]$ instead of $[C_4C_1im]Br$, markedly lower reaction rates were detected and it has been suggested, that formation of a catalytically inactive tetrakis-carbene complex occurs.[22,23]

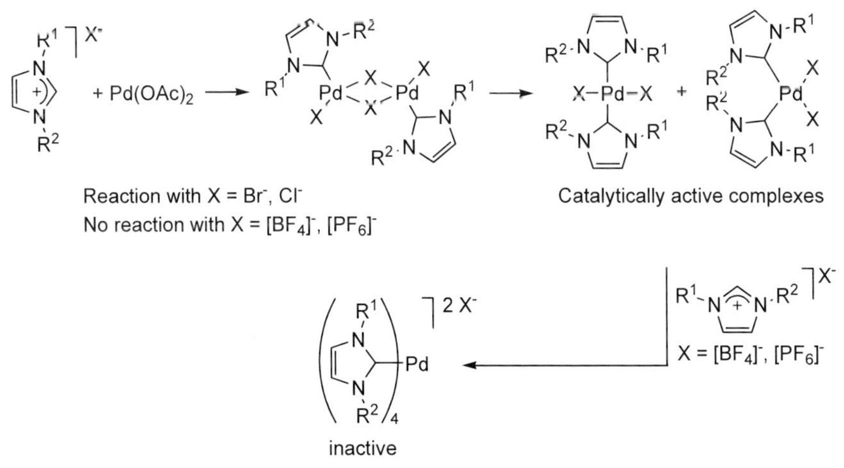

Scheme 6.1: Generation of active bis-imidazolylidene complexes and catalyst deactivation via formation of a tetrakiscarbene complex

The presence of a base is not necessarily mandatory for the formation of carbene complexes, as has been demonstrated in the reaction between $Pd(OAc)_2$ with [Mes_2im]Cl, shown in Scheme 6.2. However, instead of the expected symmetric carbene complex a zwitterionic species was isolated which turned out to be a significantly more active catalyst than the bis-carbene species.[24]

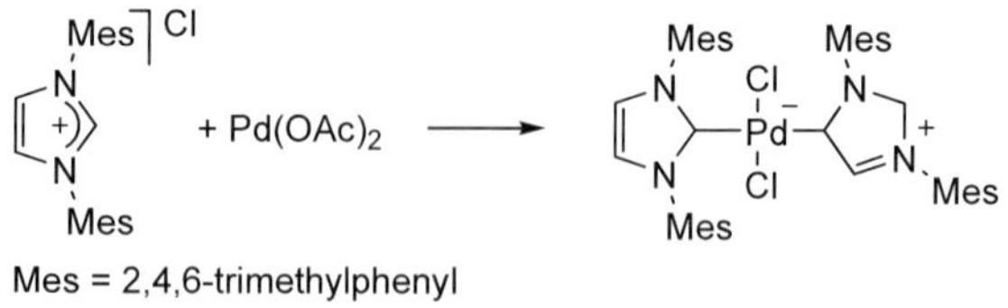

Scheme 6.2: Carbene-formation under neutral conditions

Imidazolium salts can further oxidatively add to, for example, nickel(0), palladium(0) and platinum(0) compounds, leading to stable metal-hydride complexes as shown in Scheme 6.3.[25-28]

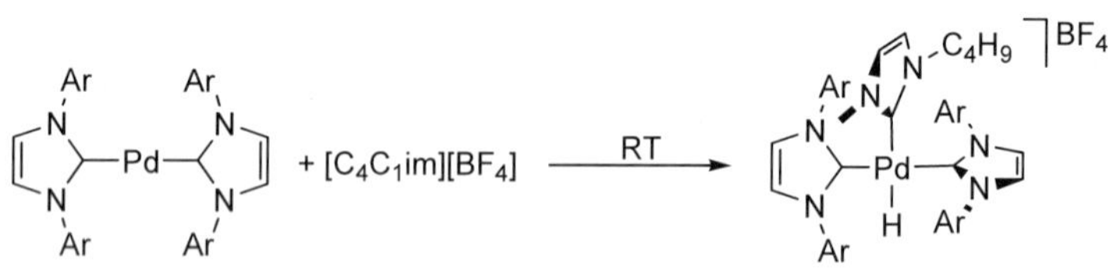

Scheme 6.3: Formation of a palladium(II) hydride complex via oxidative addition of [C_4C_1im][BF_4]

Finally, chemical or thermal-induced dealkylation (Hofmann elimination) from tetraalkylammonium salts can occur even in the absence of base as shown in Scheme 6.4. The formation of a palladium bis-imidazole complex from $[PdCl_4]^{2-}$ and [C_4C_1im][BF_4] has been reported to take place during the hydrodimerisation of butadiene[29] and an analogous product was obtained from a $[PdBr_4]^{2-}$ imidazolium salt in the presence of the radical initiator 2, 2'-azabisisobutyronitrile.[30]

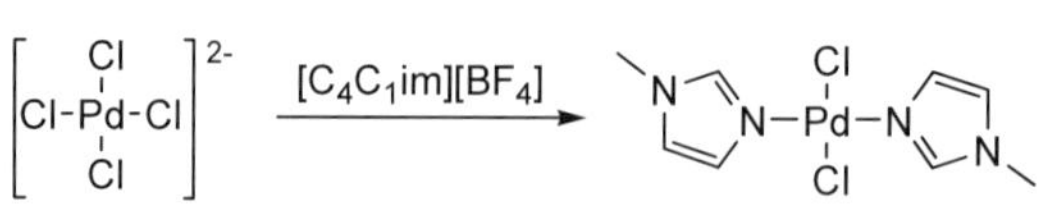

Scheme 6.4: Generation of a palladium bis-imidazole complex via alkyl elimination

Frequently, formation of palladium black is observed during C-C coupling reactions and it appears to be likely that palladium nanoparticles account at least for some of catalytic activity observed.[8,15,31] This is especially true when ligand-free conditions are employed and the stability and activity of nanoparticle catalysts/catalyst reservoirs can be significantly enhanced in the presence of an ionic liquid, which control their size and prevent particle agglomeration. Thus, the species formed during the reaction depends highly on the catalyst precursor, the ionic liquid used and the type of ligand, e.g. phosphine or arsine, employed. It would thus appear, that at least some of the active species may differ significantly in an ionic liquid from those present in molecular solvents.

6.3 Heck Coupling Reactions

6.3.1 General Considerations

The palladium catalysed arylation of alkenes with an organic halide was first reported Mizoroki[32] and Heck[33] in the early 1970's. As the reaction is very versatile and applicable to a wide range of substrates it has since become one of the cornerstones of palladium chemistry.[34-37]

The proposed mechanism for a standard Heck reaction is depicted in Scheme 6.5. Generally, a haloalkene or haloarene undergoes oxidative addition to an *in situ* generated, coordinatively unsaturated 14-electron palladium(0) complex, but other substrates such as tosylates, triflates or diazonium salts can also be applied. Subsequent *syn*-insertion into the C=C double bond of a complexed olefin yields a *σ-(β*-alkenyl) or *σ-(β*-aryl)alkylpalladium complex. If no hydrogen atom in a *pseudo* cis-position relative to the palladium is present, an internal rotation step is required prior to *syn*-elimination of the olefin to afford the *trans* olefin product and a palladium(II) hydride complex. The latter is restored to the initial Pd(0) species by base-induced reductive elimination.[37-40]

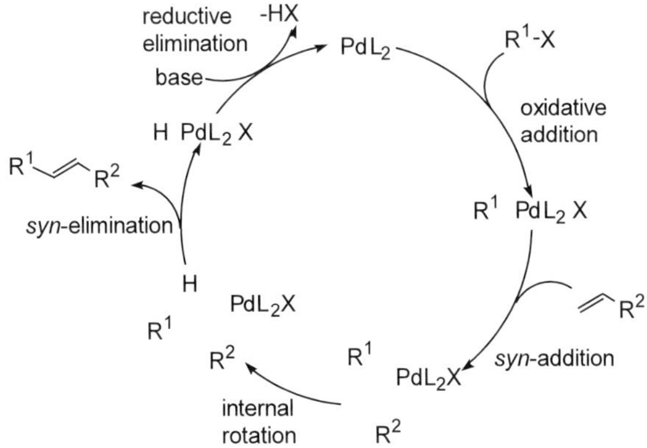

Scheme 6.5: General mechanism for the Heck reaction

Numerous homogeneous palladium catalysts (or palladium catalyst precursors) have been used as well as less well-defined nanoparticle pre-catalysts and an extensive literature is available on the topic.[34-37,41-45] A wide range of solvents can be used for Heck reactions, but polar solvents like DMF, dimethylacetamide (DMA), *N*-methyl-2-pyrrolidine or 1,4-dioxane are particularly common. It has further been known for some time that tetraalkylammonium salts exert a beneficial effect of on the activity and stability of palladium(II) catalysts.[46,47] This background may perhaps explain the relatively large number of publications dealing with Heck reactions in ionic liquids, but other approaches for the immobilisation and recovery of catalysts in the Heck reaction, such as using solid supports,[48-50] *sc*CO_2 as reaction medium[51-53] or biphasic fluorous[54-56] methods have also been evaluated.

A summary of Heck coupling reactions that have been conducted in ionic liquids is presented in Table 6.1. Due to the large number of examples available, only some selected cases will be discussed in detail. Yet, from the data available it emerges that the Heck reaction not only proceeds with higher activity in ionic liquids compared to molecular solvents, but in some cases the stability of the catalyst is also improved. Particularly good conversions are often obtained when simple palladium salts are used, often without the addition of co-ligands. As described above, *in situ* formation of carbene complexes that exhibit higher activities than the simple salts often employed in other solvents may account for such positive effects when imidazolium based ionic liquids are used. Another explanation may lie in the stabilisation of nanoparticles by the ionic liquid. The formation of palladium nanoparticles within the Heck reaction has been demonstrated,[57] and while often overlooked, the formation of nanoparticle catalyst reservoirs could frequently occur when simple palladium salts are used in the presence of a phase-transfer agent.[31] Since ionic liquids are structurally closely related to phase-transfer agents it is possible that many Heck reactions, and C-C coupling reactions in general, that take place in ionic liquids involve nanoparticle catalysts or catalyst precursors. It was, for example, observed that palladium black re-dissolves above 100°C in imidazolium based ionic liquids in the presence of Group 15 donor ligands.[58]

The fate of palladium acetate in different ionic liquids was investigated using *in situ* XAFS measurements.[59] With $[BF_4]^-$ and $[PF_6]^-$ salts of either imidazolium or pyridinium, formation of metallic palladium was observed. Conversely, no nanoparticles were formed when the corresponding chloride salts were used instead. Bis-carbene complexes were detected with $[C_4C_1im]Cl$ whereas with $[C_4C_1C_1im]Cl$, $[C_4py]Cl$ and $[C_6py]Cl$, that unlike $[C_4C_1C_1im]Cl$ do not contain an acidic proton, only $[PdCl_4]^{2-}$ was found. While addition of PPh_3 did not have a great effect with respect to the nature of the palladium species in these chloride salts, markedly less nanoparticle formation was observed in the $[BF_4]^-$ and $[PF_6]^-$ ionic liquids. It was also found that the formation of nanoparticle clusters is faster in the presence of substrate, *viz.* iodobenzene and butyl acrylate and nanoclusters with a diameter between 0.8 and 1.6 nm were shown to be the main species present during the reaction.[59]

Apart from palladium(0) and palladium(II) salts, such as $Pd_2(dba)_3$, $PdCl_2$ and $Pd(OAc)_2$, several well-defined Pd-carbene complexes have been employed in ionic liquids, as have some palladacycle complexes. However, ligands used are in most cases neutral and have not been included to facilitate catalyst retention in the ionic liquid.

Co-solvents are occasionally added to the ionic liquid, mainly to facilitate product isolation and/or to help dissolve inorganic bases. In some cases, triphasic (aqueous - ionic liquid - organic) systems result, either during the reaction or in the workup.[58,60] An organic phase (frequently diethyl ether or hexane) is used to absorb the product while any inorganic salt which has formed from the reaction of base with the palladium(II) complex may be taken up by the aqueous layer. Tetraalkylammonium or phosphonium ionic liquids usually form the middle phase.[10] As one example of such a system, coupling of ethyl acrylate with arylhalides proceeded with high yield and selectivity in an organic - Aliquat 336 - aqueous trisolvent system with Pd/C as catalyst.[60]

As an alternative to such triphasic procedures, washing the ionic liquid phase with water from time to time may be required to ensure recyclability. Otherwise, the physical properties of the ionic liquid – especially its viscosity – may change, and catalyst performance may also deteriorate, due to accumulated base and other residues in the solvent.

Some studies have been undertaken with respect to anion effects in imidazolium based ionic liquids. As illustrated above, $[C_4C_1im]Br$ may be a superior solvent to $[C_4C_1im][BF_4]$ for Heck reactions, as highly active carbene intermediates are more likely to arise in the former while with the latter these are either not formed or undergo transformation into a less active species as illustrated in Scheme 6.1.[20]

Table 6.1: Heck coupling reactions conducted in ionic liquids

Ionic liquid	Catalyst precursor	Base	Comments	Ref
$[(C_4)_4N][OAc]$ $[(C_4)_4N]Br$ and others	Large variety	NaOAc, $[(C_4)_4N]OAc$ and others	120-160 °C. Successful coupling or chlorobenzene with styrene using NaOAc as base; more active than the corresponding reactions in DMF; very high recyclability of the phospha-palladacyle catalyst in $[(C_4)_4N]Br$ with product isolated by distillation.	[61] [62]
$[(C_4)_4N]Br$	$PdCl_2$	$NaHCO_3$	80-120 °C. Ligand-free arylation of allylic alcohols; product extracted with Et_2O or hexane.	[63]
$[(C_4)_4N][BF_4]$ $[C_4C_4pyrr][BF_4]$	$PdCl_2$	K_2CO_3 Et_3N	90 °C. Ligand-free arylation of butyl acrylate with iodo- and bromo benzene; biphasic system with either water or toluene as co-solvent; product extracted with hexane.	[64]
$[C_4C_1im][BF_4]$ $[C_4C_1im]Br$	$Pd(OAc)_2$	NaOAc	90-125 °C. Phosphine-free arylation of acrylates; higher reaction rates in due to *in situ* formation of palladium carbene halides; product extracted with ethyl acetate.	[20]
$[C_4C_1im][BF_4]$	$Pd(OAc)_2$	Et_3N	80-120 °C. Arylation of butyl vinyl ether with arylhalides; dppp added as ligand; reaction proceeds with high regioselectivity (>99 %); no palladium black detected.	[65]
$[C_4C_1im][BF_4]$ $[C_4C_1im]Br$	$PdCl_2$ $Pd(OAc)_2$	NaOAc	30 °C. Ligand-free, ultrasound promoted arylation of alkenes and alkynes with aryliodides; palladium bis-carbenes and palladium nanoparticles (~1 nm) are identified after catalysis; product extracted with ethyl acetate/petrol ether.	[66]
$[(C_4)_4N]Br$	Palladium(0) carbene complexes	NaOAc	140-160 °C. Phosphine-free arylation of styrene with arylchlorides; excellent conversions with both activated and non-activated arylchlorides; product extracted with Et_2O.	[67]
$[(C_4)_4N]Br$ $[C_4C_1im]Br$	$Pd(OAc)_2$ Pd-benzthiazole carbene complex	$NaHCO_3$ Na_2CO_3 NaOAc	130 °C. Phosphine-free arylation of α-substituted acrylates, allylic alcohols and other substrates using arylhalides. Nanoparticles are identified and $[(C_4)_4N]OAc$ appears to play a crucial role in their formation; product extracted with cyclohexane.	[68-74]

Ionic liquid	Catalyst precursor	Base	Comments	Ref
$[C_4C_1im][PF_6]$ $[S\text{-}C_4C_1im][PF_6]$	$PdCl_2$ $Pd(OAc)_2$ $Pd(NCPh)_2Cl_2$	Ag_2CO_3	56-100 °C. Enantioselective Heck oxyarlyation; chiral induction with chiral ligands and/or chiral ionic liquid is very low; product extracted with toluene.	[13]
$[C_4C_1im][PF_6]$	$PdCl_2$ $Pd(OAc)_2$	Et_3N	180-220 °C. Microwave-accelerated arylation of butyl acrylate; best results with $PdCl_2$ and $P(o\text{-}Tol)_3$ as ligand; short reaction times (5-45 minutes); product isolated via distillation.	[75]
$[C_4C_1im][PF_6]$	$Pd(OAc)_2$		20-50 °C. Base-free arylation of olefins with arenediazonium salts; catalyst stable for at least 4 runs.	[76]
$[C_4C_1im][PF_6]$ $[C_4C_1im][Tf_2N]$	Pd-phosphine carbene complex	Pr_3N	130-150 °C. Arylation of butyl acrylate with iodobenzene in a microflow-system with continuous recycling of the ionic liquid-catalyst phase; product extracted with hexane.	[77]
$[C_4C_1im][PF_6]$ $[C_5C_1im]$[various] $[C_6py]Cl$	$PdCl_2$ $Pd(OAc)_2$	$NaHCO_3$ Et_3N	40-200 °C. Phosphines and arsines as ligands; arylation of acrylates; product extracted with hexane.	[58]
$[C_4C_1im][PF_6]$	Pd-bis-imidazole complexes	Et_3N NaOAc Na_2CO_3	120-160 °C. Phosphine-free arylation of butyl acrylate and other olefins with aryliodides and bromides; catalyst recycled five times without loss of activity; product extracted with Et_2O.	[78]
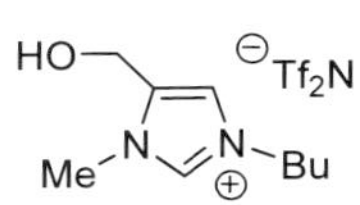	$Pd(OAc)_2$	Et_3N	100 °C. Phosphine-free arylation of methyl acrylate with aryliodides; presence of catalytic amounts of various halide ions were found to accelerate the reaction; excellent recycling of the catalyst demonstrated; product extracted with cyclohexane.	[21] [79]
$[(C_6)_3C_{14}P]Cl$ $[(C_4)_3C_1P]$[various]	$Pd_2(dba)_3$ $Pd(OAc)_2$	Et_3N NaOAc	50-100 °C. Coupling of activated and deactivated iodides and activated bromides; anion effects studied; triphasic aqueous workup; catalyst recycling possible; product extracted with hexane.	[10]

Ionic liquid	Catalyst precursor	Base	Comments	Ref
$[(C_4)_3C_{16}P]Br$ $[(C_4)_4N]Br$	$PdCl_2$ $Pd(OAc)_2$ $PdCl_2(PPh_3)_2$	Et_3N	100 °C. Arylation of butyl acrylate with arylhalides; addition of LiBr or NaOAc accelerate the reaction; product isolated by distillation.	[7]
imidazole-imidazolium IL	$PdCl_2$-IL complex	Na_2CO_3	100 °C. Coupling of aryliodides and chlorides; phosphine-free system; catalyst highly stable and reused 10 times; product extracted with Et_2O.	[14]
$[C_4C_1im][PF_6]$	$Pd(OAc)_2$	Et_3N	80-120 °C. Chelate phosphine ligand added; arylation of electron-rich olefins, vinyl ethers, enamides and allyltrimethylsilane with aryliodides and bromides; product extracted with CH_2Cl_2-H_2O	[80]
			Heterogeneous Catalysts	
Aliquiat 336	Pd/C	Et_3N, KOH, K_2CO_3, $NaHCO_3$	100 °C. Multiphasic reaction system; aryliodides react up to 10 times faster in the presence of Aliquat 336; product extracted with *iso*-octane.	[60]
$[(C_4)_4N]Br$	Supported Pd nanoparticles	Bu_3N	130 °C. Phosphine-free arylation of styrene with arylchlorides; palladium immobilised on layered double hydroxide; microwave irradiation leads to significant rate acceleration; product isolated by distillation.	[81]
$[(C_4)_4N]Br$	Pd nanoparticles	$[(C_4)_4N][OAc]$	130 °C. Nanoparticles immobilised on chitosan; products extracted with cyclohexane; high reaction rates for deactivated arylbromides and activated arylchlorides; highly recyclable catalyst system.	[82]
$[(C_4)_4N]Br$	Pd nanoparticles	$[(C_4)_4N][OAc]$	120 °C. Phosphine-free arylation of butyl methacrylate and α-methylstyrene with arylbromides; moderate selectivity for the terminal olefins; $Pd(OAc)_2$ as catalyst precursor.	[83]
$[(C_4)_4N]Br$	Copper bronze	$[(C_4)_4N][OAc]$	130 °C. Coupling of aryliodides and activated bromides; activity increases upon recycling due to formation of more active nanoparticles; catalysts remains stable for more than 15 runs; products extracted with cyclohexane.	[84]

Ionic liquid	Catalyst precursor	Base	Comments	Ref
$[C_4C_1im][PF_6]$	Pd nanoparticles	$Et(^iPr)_2N$	30-130 °C. Synthesis of ligand-free nanoparticles from palladacycles; size 1.7 nm; nanoparticles as reservoir for catalytically active species postulated.	[31]
$[C_4C_1im][PF_6]$	Silica-supported Pd nanoparticles	Et_3N	80 °C. Phosphine-free arylation of acrylates with aryliodide; $PdCl_2(NH_3)_2$ as catalyst precursor; palladium leaches from the support into the ionic liquid phase; product extracted with *tert*-butyl methyl ether.	[85]
$[C_4C_1im][PF_6]$	Pd/C	Et_3N	140 °C. Phosphine-free arylation of ethyl acrylate and other olefins with arylhalides; steady decrease in activity is observed with recycling, which is restored by washing the ionic liquid/catalyst solution with water; product extracted with hexane or Et_2O.	[86]
$[C_4C_1im][PF_6]$	$Pd(OAc)_2$ $Pd(PPh_3)_4$	Et_3N K_2CO_3	100-150 °C. Supported ionic liquid phase catalysis (SILP) with silica as solid phase; catalyst relatively stable for at least 5 runs; product decanted and solid phase washed with hexane; less than 0.24% catalyst loss.	[87]
$[C_4C_1im][PF_6]$	Pd/C	Et_3N	180-220 °C. Microwave-accelerated arylation of butyl acrylate; short reaction times (5-45 minutes); product isolated via distillation.	[75]
$[C_8C_1im][BF_4]$	Pd/C	Bu_3N	Microwave-accelerated arylation of olefins; very short reaction times (1.5-2 minutes) catalyst recycled five times with steady decrease in activity; product extracted with Et_2O.	[88]

6.3.2 Heck Reactions with Homogeneous Catalysts

The first example of a Heck reaction in a molten salt stems from as early as 1996 when tetraalkylammonium and phosphonium halides were used as reaction media for the coupling between arylhalides and *n*-butyl acrylate.[7] Particularly good results were achieved in trihexyl(tetradecyl)phosphonium chloride, both in terms of reactivity and ease of product isolation.

In molecular solvents catalytic activity tends to increase as the steric bulk of the phosphine increases whereas the effect of additional ligands with respect to their structure is not always obvious in an ionic liquid. A number of different ligands have been evaluated in the arylation of ethyl acrylate with 4-bromoanisole in $[C_4C_1im][PF_6]$ with $Pd(OAc)_2$ as the catalyst precursor.[58] In this system it was found that the relative activity of the ligands evaluated decreased in the order $P(o\text{-Tol})_3 > PPh_3 > AsPh_3 >$ dppf >> $P(O\text{-Ph})_3 > SbPh_3 >$ dppe. If, however, a pyridinium based ionic liquid is used instead, the presence of phosphine ligands appeared to hinder the reaction. For example, the coupling between iodobenzene and ethyl acrylate in $[C_6py]Cl$ afforded 98% of the envisaged product after 24 hours at 40°C, whereas only 82% was obtained when 4 mol% PPh_3 was added.

A range of different palladium compounds immobilised in $[(C_4)_4N]Br$ have been screened in the coupling between styrene and chlorobenzene, with or without additional co-ligands. Apart from various simple palladium(II) salts these compounds included the carbene complexes **16-18** and the phospha-palladacycles **19** and **20**, shown in Figure 6.1.[61,62]

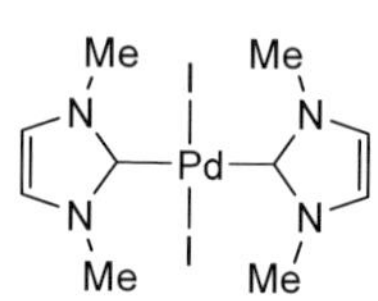

16

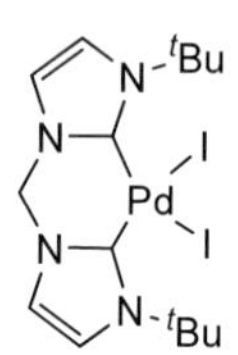

17

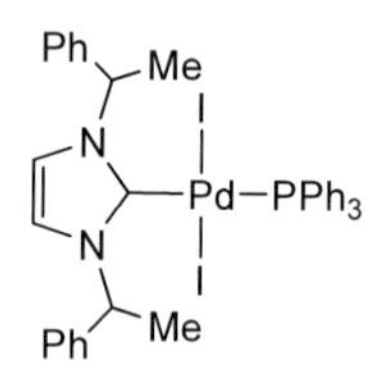

18

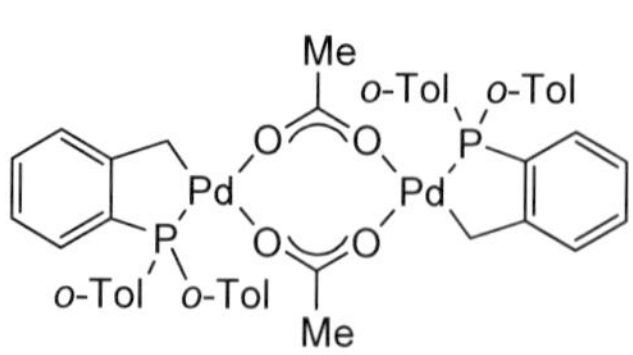

19

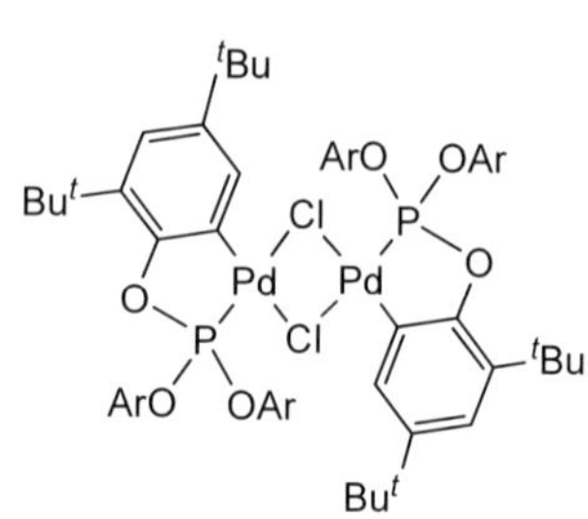

20

Figure 6.1: Carbene and phosphapalladacyle complexes evaluated for the coupling of styrene with chlorobenzene in $[(C_4)_4N]Br$

The yields obtained after two and 18 hours are displayed in Table 6.2 and are impressive given that chlorobenzene is often completely unreactive as a substrate in Heck coupling reactions, especially with catalysts like $PdCl_2$ and $Pd(OAc)_2$. All the tested catalysts gave increased yields relative to DMF. When $[(C_4)_4N][OAc]$ is used as base in the coupling of organobromides the ionic liquid solvent is generated as the by-product, improving the recyclability. Products were removed by distillation and depending on the catalyst, recycling was possible up to 12 times with only a minor reduction of activity.

Table 6.2: Heck olefination of chlorobenzene with styrene in $[(C_4)_4N]Br$ to afford 1,2-diphenylethene.[a)]

	Yield after 2 hours		Yield after 18 hours	
Catalyst	DMF	$[(C_4)_4N]Br$	DMF	$[(C_4)_4N]Br$
$PdCl_2$	0	7	0	50
$Pd_2(dba)_3$	0	1	2	8
$Pd_2(dba)_3$+ 2 eq. $P(t\text{Bu})_3$	56	86	72	92
$Pd(OAc)_2$ + 3 eq. $P(o\text{-Tol})_3$	3	11	29	46
$Pd(PPh_3)_4$	4	24	17	65
19 + 5 eq. $[AsPh_4]Cl$	19	37	41	84
16	0	4	3	51
20	1	11	5	49

a) Chlorobenzene (1.0 equiv.), styrene (1.5 equiv.), Na(OAc) (1.2 equiv.), Pd (2 mol %).

Comparable reaction conditions were applied in the coupling of activated and non-activated arylchlorides with styrene or 2-ethylhexyl acrylate, using the palladium carbene catalysts shown in Scheme 6.6. While compounds **22** and **23** were found to be highly active catalysts, complex **21** was thermally unstable and decomposed to palladium black during the catalysis.[67] The yield and selectivity were only moderate in DMA, but results improved markedly when the reaction was carried out in $[(C_4)_4N]Br$.

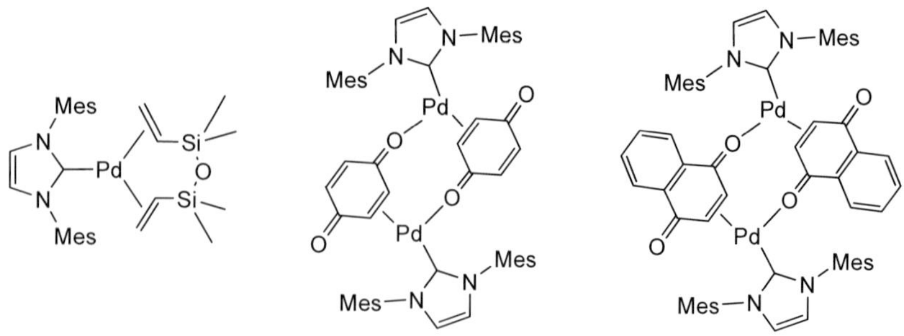

21 **22** **23**

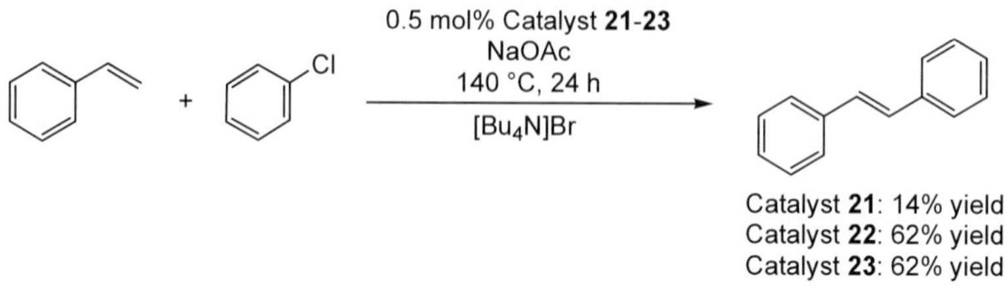

Scheme 6.6: Coupling of chlorobenzene and styrene catalysed by Pd carbene-olefin complexes

High reaction rates and very good selectivity were also observed in the reaction between aryliodides and acrylates or vinyl compounds with the bis-imidazole-complexes **24** and **25,** shown in Figure 6.2.[78]

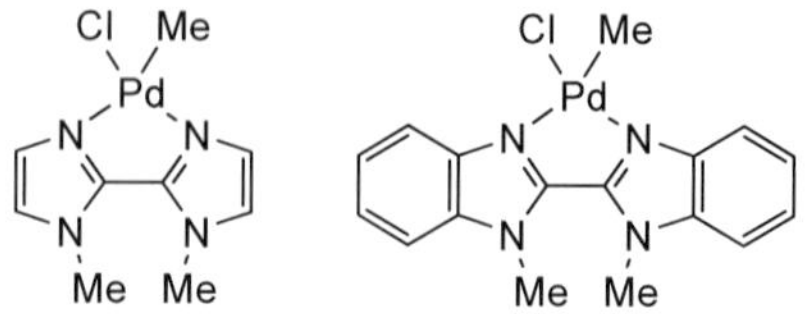

24 **25**

Figure 6.2: Palladium bis-imidazole complexes

A somewhat related, highly stable catalyst system was obtained from $PdCl_2$ immobilised in a imidazole-imidazolium ionic liquid which serves both as solvent and ligand. [14] In the reaction solution a bis-imidazole complex, **26**, is formed, which has also been synthesised and characterised separately (see Scheme 6.7). In this system catalyst immobilisation appears to be very efficient and after 11 runs with three different substrates high conversion was observed even with chlorobenzene as substrate.

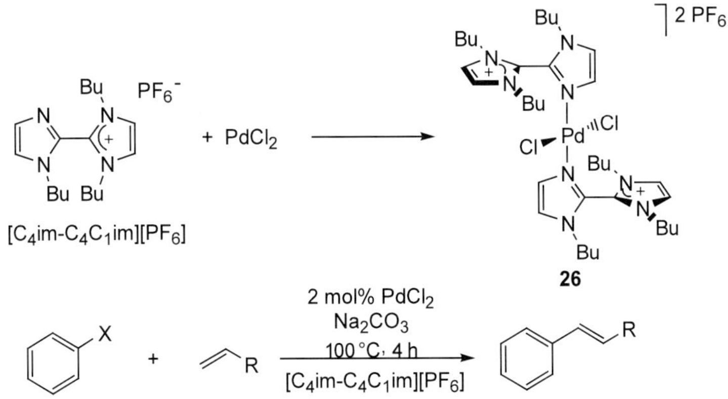

26

X	R	run	yield
I	CO_2Me	1	92
	CO_2Me	5	91
Cl	CO_2Me	6	70
	CO_2Me	8	75
Cl	Ph	9	86
	Ph	11	83

Scheme 6.7: *In situ* formation of an ionic liquid stabilised palladium complex in an imidazole-imidazolium ionic liquid and its performance in the Heck coupling of various substrates

Good catalyst immobilisation was also realised by attaching a charged imidazolium moiety onto a palladacycle complex, resulting in the cationic complex **27**, depicted in Figure 6.3. No palladium was detected in the organic phase after extensive extraction of a solution of $[C_4C_1im][PF_6]$ containing the catalyst with diethyl ether. Nevertheless, the catalytic activity was only disappointing and it was proposed that catalyst decomposition occurred in the presence of a base.

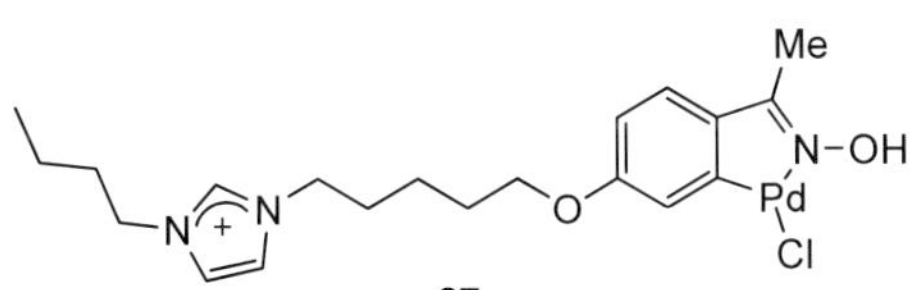

27

Figure 6.3: Carbapalladacycle tested in Heck and Suzuki-coupling reactions

Apart from improved reaction rates, facilitated product isolation or high recyclability, performing the reaction in an ionic liquid may result in improved selectivities. This is illustrated in the arylation of the electron rich olefin butyl vinyl ether in the presence of $[C_4C_1im][BF_4]$. In this solvent, a substantial improvement in the regioselectivity was observed relative to either DMF or DMSO, yet at somewhat lower reaction rates (Scheme 6.8).[65]

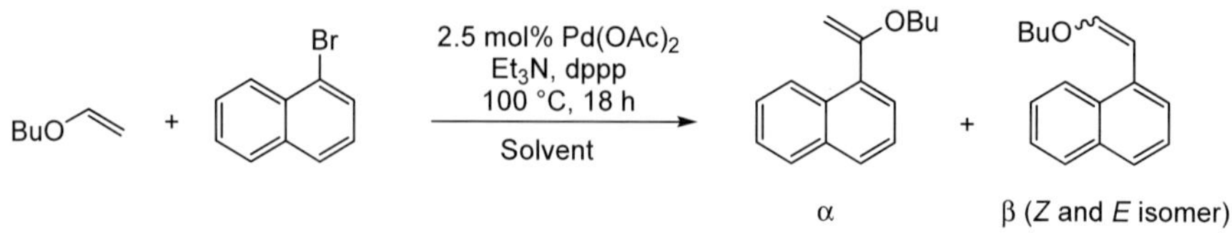

	conversion	α/β
toluene	23%	46/54
acetonitrile	36%	69/31
DMF	100%	60/40
DMSO	100%	75/25
$[C_4C_1im][BF_4]$	50%	>99/1

Scheme 6.8: Effect of solvent on the regioselectivity of the Heck coupling between butyl vinylether and 1-bromonaphthalene

As carbene-formation from imidazolium ionic liquids has been demonstrated in several instances (vide supra) it was reasoned that a chiral imidazolium ionic liquid might lead to formation of a chiral palladium complex and thus chiral induction in the synthesis of the product. The effect of chiral ionic liquids on enantioselective Heck oxyarylation reactions was therefore explored, but only very low *ee* (below 5%) was observed in the reaction between 2-iodophenol and 7-benzyloxychromene with either $PdCl_2$ or $Pd(OAc)_2$ as catalyst, as shown in Scheme 6.9.[13] When PPh_3 was added as ligand, no enantiomeric excess was observed at all. It is not unambiguously clear whether the chiral induction is due to formation of a carbene complex or due to a solvent template effect. As stated earlier, the presence of a halide source might be important for the formation of palladium carbene complexes, and it is therefore possible, that hardly any Pd-carbene compound was generated under the given reaction conditions. In addition, silver carbonate can react with imidazole and silver-carbene complexes have been prepared in a comparable manner.[89]

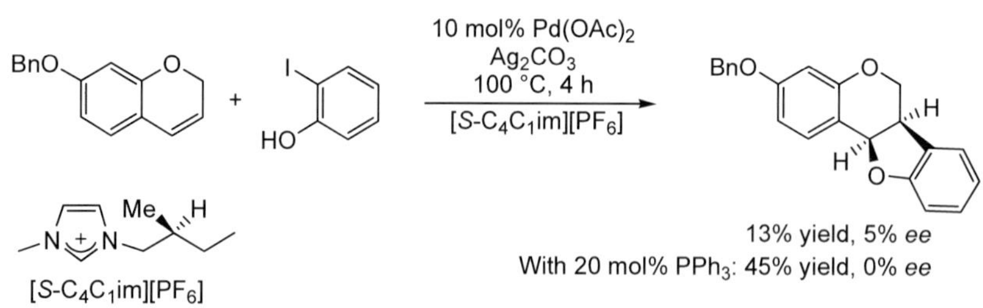

Scheme 6.9: Heck oxyarylation in a chiral ionic liquid

6.3.3 Microwave and Ultrasound Accelerated Reactions

Irradiation of the reaction solution with either microwaves or ultrasound can lead to dramatically reduced reaction times and both techniques have been applied successfully to the Heck reaction in ionic liquids. Ultrasound has been found to greatly accelerate Heck coupling reactions in solutions of either $[C_4C_1im]Br$ or $[C_4C_1im][BF_4]$, affording good conversion in less than three hours at 30°C.[66] On the basis of ^{1}H-NMR and mass spectroscopic data, formation of a palladium carbene species $PdX_2(C_4C_1imy)_2$ appears to take place, which was proposed to be a precursor to zero-valent nanoparticle clusters. TEM analysis reveals monodispersed grains of about 20 nm diameter, composed of particles of approximately 1 nm size.

Microwave-assisted Heck coupling reactions have been reported in several instances.[75,81,88] Using microwave dielectric heating, reaction times as short as only a few minutes can be realised and some data from such experiments are listed in Table 6.3. A wide selection of arylhalides were reacted with olefins with Pd/C immobilised in $[C_8C_1im][BF_4]$ under microwave irradiation.[88]

Table 6.3: Microwave-assisted olefination of arylhalides

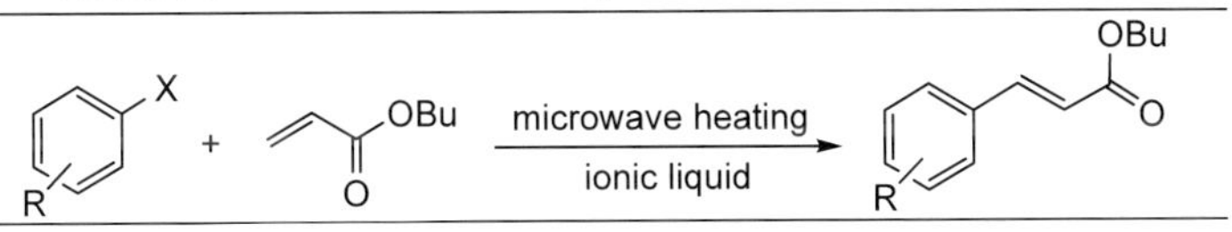

4 mol% $PdCl_2$; 8 mol% $P(o\text{-}Tol)_3$; 1.5 eq. NEt_3; $[C_4C_1im][PF_6]$

R	X	time [min]	temp [°C]	yield [%]
H	Br	20	220	87
p-MeO	Br	20	220	61
p-CF_3	Br	20	220	90

3 mol% Pd/C; 2 eq. NBu_3; $[C_8C_1im][BF_4]$

R	X	time [min]	P [W]	yield [%]
H	Cl	1.5	375	0
p-NO_2	Cl	2.0	375	42
H	Br	1.5	225	80
p-MeO	Br	1.5	225	30
H	I	1.5	225	86

6.3.4 Heck Reactions with Heterogeneous Catalysts

Of the heterogeneous catalysts employed in conjunction with ionic liquids, palladium on activated charcoal is most frequently used.[60,75,86,88]

Other examples include palladium species immobilised on other solid supports[81,82] as well as isolated (naked) Pd-nanoparticles[31,83]

So far there is only one example where a supported ionic liquid phase catalyst has been used in the Heck reaction.[87] Stirring a suspension of amorphous silica in a mixture of THF, $[C_4C_1im][PF_6]$ and $Pd(OAc)_2$ afforded a powdery immobilised catalyst. The coupling reaction was performed with the catalyst suspended in *n*-dodecane and following filtration it could be reused several times, with only a minor decrease in activity. The decrease in activity might be due to blockage of the silica pores with the ammonium salt formed during the reaction and washing the solid catalyst with aqueous sodium hydroxide indeed restored the original activity almost completely. From ICP-OES analysis it was established that less than 0.24% of the palladium leached into the organic layer.

In an other example where palladium was also immobilised on silica, but the reaction carried out in $[C_4C_1im][PF_6]$ as solvent, considerable leaching of the palladium(0) and palladium(II) catalyst from the support into the ionic liquid as well as into the organic phase was observed.[85]

It has been suggested that active sites in heterogeneous Heck reactions are actually dissolved palladium species.[90] Such a hypothesis is in agreement with a mechanism where palladium nanoparticles suspended in an ionic liquid are suggested to act as reservoir to a molecular catalyst, which is formed upon oxidative addition, as shown in Figure 6.4.[31]

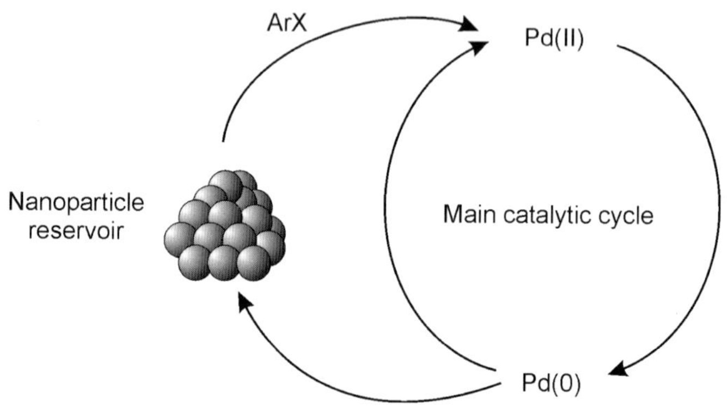

Figure 6.4: Proposed pathway for nanoparticle-promoted Heck reactions in ionic liquids

Palladium nanoparticles were further immobilised on chitosan flakes and from TEM it was established that they have a core-shell structure in which the metallic palladium is surrounded by a stabilising shell composed of a $[(C_4)_4N]^+$ monolayer that is surrounded by Br^- and $[PdCl_4]^{2-}$ species.[82] High catalytic activity for arylhalides including activated arylchlorides was observed at 130°C in $[(C_4)_4N]Br$ using tetrabutylammonium acetate as base.

Nonetheless, chlorobenzene was essentially unreactive under the conditions employed.

A charged support, double layered hydroxide, that can be described by the general formula, $Mg_{1-x}Al_x(OH)_2Cl \cdot zH_2O$, was used to immobilise $[PdCl_4]^{2-}$.[81] While results with this catalyst were only poor in *N*-methyl pyrrolidinone, good conversion and selectivity was obtained in $[(C_4)_4N]Br$ with arylchlorides as substrates. By employing microwave irradiation coupling between chlorobenzene and styrene afforded the desired product in 95% yield within 30 minutes. The products were isolated by distillation and the catalytic activity remained essentially stable for at least 5 runs.

Although copper salts are well-known additives for C-C coupling reactions,[91] only few examples on copper-catalysed Heck reactions exist.[92-94] The coupling of arylhalides with butyl acrylate has been achieved in $[(C_4)_4N]Br$ with 3 mol% copper bronze as pre-catalyst.[84] When the reactions were recycled, the rates increase markedly and it was suggested that the arylhalide reacted with the copper bronze to yield highly active nanoparticles. The average size of these colloids was determined by TEM to be 4.2 nm, which appear to be homogeneously dispersed in the ionic liquid.

6.4 Other Palladium Catalysed Cross-Coupling Reactions

While Heck coupling reactions have been studied in some detail in ionic liquids, other important C-C bond forming reactions have received less attention. Overall, the data available indicate that ionic liquids have considerable potential to not only facilitate product extraction and catalyst recovery, but also to improve reaction rates and catalyst lifetimes.

Many of the following reactions, e.g. Suzuki, Stille, Negishi and Sonogashira couplings, follow the same basic reaction mechanism shown in Scheme 6.10, but more complicated pathways can also be involved. Initially, the *cis*-adduct is formed during oxidative addition, which subsequently isomerises to the *trans*-complex. Trans-metallation with, for example, an organo-boron, -tin, -zinc or -copper reagent affords $PdRR'L_2$, which undergoes reductive elimination to give the desired product.

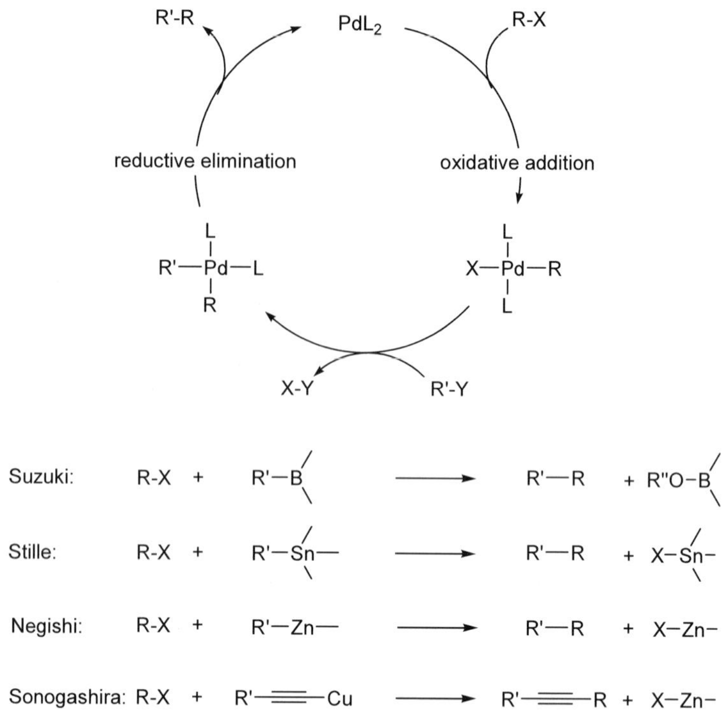

Scheme 6.10: Proposed general mechanism for some palladium catalysed cross-coupling reactions other than the Heck reaction

6.4.1 Suzuki Coupling Reactions

The Suzuki reaction comprises the coupling of an arylhalide with a boronic acid and it is perceived to be one of the most environmentally benign methods for the formation of new C-C bonds.[95] Numerous reviews are available and only some of the more recent shall be referenced here.[96-99] The mechanism, as depicted in Scheme 6.10,[100] shares close similarities to that of the Heck reaction although other intermediates have also been suggested.[101] Addition of a base is usually required to account for the low nucleophilicity of the borane reagent. In polar solvents such as THF or water, NaOH or NaOMe are commonly employed, while Na_2CO_3 or K_3PO_4 are frequently used in DMF. Likewise, rather weak bases such as Et_3N, NaOAc or Na_2CO_3 are usually used in ionic liquids. In conventional solvents such as toluene and dioxane extensive catalyst decomposition is often observed as is evident from the formation of black deposits.

A number of studies concerning Suzuki coupling reactions in ionic liquids have been undertaken and they are summarised in Table 6.4.

Table 6.4: Summary of palladium-catalysed Suzuki-coupling reactions in ionic liquids

Ionic liquid	Catalyst precursor	Base	Comments	Ref
$[C_2C_1im][BF_4]$	$Pd(PPh_3)_4$	Na_2CO_3	110 °C. Solid-phase reaction; coupling of 4-iodophenol immobilised on a polystyrene-Wang resin with arylboronic acids; DMF as co-solvent; acceleration in the presence of the ionic liquid; catalyst recycling not practical.	[102]
$[(C_4)_4N]Br$	$Pd(PPh_3)_4$, $Pd(OAc)_2$ $PdCl_2(PPh_3)_2$	K_2CO_3	30-120 °C. High reaction rates with bromides and chlorides; products extracted with Et_2O; catalyst recycling possible.	[103]
$[C_4C_1im][PF_6]$	Carbapalladacyle **27**	NaOAc	130 °C. Very poor conversion, probably due to base-induced catalyst decomposition.	[104]
$[C_4C_1im]$[various] $[C_4C_1C_1im][BF_4]$	$Pd(PPh_3)_4$, $Pd(OAc)_2$ and many others	Na_2CO_3	110 °C. Broad study used to identify the active catalyst species; both imidazoylidenes (carbenes derived from the imidazolium cation) and halide ligands are implicated in the catalytic cycle; products extracted with Et_2O; catalyst stable for at least 4 runs.	[105]
$[C_4C_1im]$[various] $[C_4C_1C_1im][BF_4]$ $[C_4C_1py][Tf_2N]$	$[C_4C_1im]_2[PdCl_4]$ $PdCl_2(CH_3CN)_2$	Na_2CO_3	110 °C. Study on the *in situ* formation of complexes bearing *N*-donor ligands derived from imidazoles and other nitrogen bases; ionic liquid anion and cation effect investigated; products extracted with hexane; catalyst stable for at least 5 runs.	[106]
$[C_4py][Tf_2N]$ $[C_3CNpy][Tf_2N]$	$PdCl_2$ and related nitrile complexes	Na_2CO_3	110 °C. Effect of coordinating ionic liquid (*via* nitrile functionality) studied; catalyst activity is maintained for at least 9 runs when $[C_3CNpy][Tf_2N]$ is used as solvent; reaction rate only moderate; products extracted with Et_2O.	[15]
$[(C_6)_3C_{14}P]Cl$	$Pd_2(dba)_3$	Et_3N, K_2PO_4 and others	50-70 °C. Activated chlorides can be coupled at 70 °C if additional phosphine ligand is present; products extracted with hexane; catalyst stable for at least 5 runs.	[10-12]
$[C_4C_1im][BF_4]$	$Pd(PPh_3)_4$	Na_2CO_3	110 °C. Coupling of arylbromides with arylboronic acids; high reaction rates; products extracted with Et_2O; catalyst activity maintained for at least 5 runs.	[107]
$[C_4C_1im][BF_4]$	$Pd(OAc)_2$	NaOAc	30 °C. Ultrasound irradiation; coupling of arylhalides including chlorobenzenes with phenylboronic acid; products extracted with Et_2O; catalyst active for at least 4 runs.	[108]

In keeping with Heck reactions, both palladium(0) and palladium(II) catalyst precursors have been successfully used and there seems to be little to distinguish the two in terms to activity, stability and reuse. Relatively mild reaction conditions have been realised by means of ultrasound irradiation or in phosphonium-based ionic liquids.

Rather long reaction times tend to be required in imidazolium-based ionic liquids when the established protocol[109] for the Suzuki coupling reaction is applied. However, catalyst activity increases markedly if an initiation procedure is put into operation, i.e. prior to the addition of the boronic acid reagent, $Pd(PPh_3)_4$ is stirred for one hour under inert conditions at 110°C in $[C_4C_1im][BF_4]$ in the presence of an arylhalide and a halide ion source.[105,107] Only then is an aqueous solution of Na_2CO_3 and the boronic acid added. After such an induction period handling of the catalyst solution in air was possible without any observed catalyst decomposition. In this manner markedly increased reaction rates were achieved for aryliodides and bromides, whereas arylchlorides remained unreactive. It was subsequently found that mixed ligand phosphine-imidazolylidene palladium species, $[PdX(PPh_3)_2(C_4C_1imy)]^+$ (X = Br or Cl), are formed under the conditions employed, as outlined in Scheme 6.11.[19] The isolated chloride derivative $[PdCl(PPh_3)_2(C_4C_1imy)][BF_4]$ was found to be highly active and reaction rates as high as 930 $mol{\cdot}mol^{-1}{\cdot}h^{-1}$ were observed in the coupling of bromobenzene with phenylboronic acid. A similar process also takes place when other palladium sources, *e.g.* $PdCl_2$, $Pd_2(dba)_3$ or $Pd(OAc)_2$, are used together with a phosphine ligand. It thus appears that the formation of a mixed phosphine-carbene palladium complex is crucial in imidazolium ionic liquids in order to achieve high reaction rates. Accordingly, much lower activity and extensive catalyst decomposition was observed with $[C_4C_1C_1im][BF_4]$ as in this solvent formation of a carbene complex is not possible due to the presence of the methyl substituent in the 2-position. Anion effects were also observed and inferior results were obtained with $[PF_6]^-$ or $[SbF_6]^-$, while no reaction took place in $[C_4C_1im]Cl$.

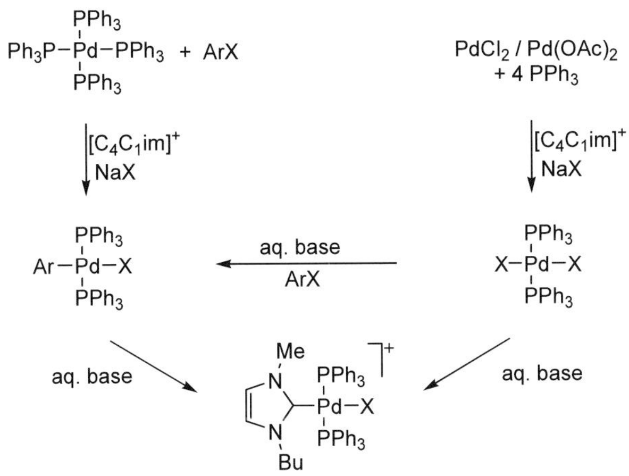

Scheme 6.11: Proposed mechanism for the formation of palladium-imidazolylidene species under the conditions employed in Suzuki reactions

Only few palladium salts readily dissolve in $[C_4C_1im][BF_4]$, one exception being $[C_4C_1im]_2[PdCl_4]$. It was found that this salt can form *N*-donor palladium(II) complexes when an aqueous base is present[106] although water alone may already be sufficient (Scheme 6.4).[29] Such species were deliberately prepared *in situ* from $[C_4C_1im]_2[PdCl_4]$ and four equivalents of methylimidazole and the resulting complex exhibits good activity in the Suzuki reaction with arylbromides. However, reaction rates were inferior to those in molecular solvents (Table 6.5). Nonetheless, no visible catalyst decomposition takes place in the ionic liquid and recycling was possible. Ionic liquids which cannot form palladium carbene complexes afforded activities similar to molecular solvents, but at the expense of catalyst decomposition. Activity also depended on the nitrogen ligand employed and the most encouraging results were found with 1-methylbenzimidazole.

Table 6.5: Reaction of bromobenzene and tolylboronic acid with catalyst formed *in situ* from $[C_4C_1im]_2[PdCl_4]$ and 4 equivalents of 1-methylimidazole

Solvent	yield (%)	TOF
dioxane	96[a)]	236
toluene	90[a)]	225
$[C_4C_1im]Cl$	0	0
$[C_4C_1im]Br$	25	63
$[C_4C_1im][BF_4]$	42	106
$[C_4C_1im][PF_6]$	60	150
$[C_4C_1C_1im][BF_4]$	85[a)]	213
$[C_4C_1py][Tf_2N]$	83[a)]	209

Conditions: 110°C, 20 minutes; [a)] Extensive catalyst decomposition and leaching.

The coupling between arylhalides and phenylboronic acid proceeds under very mild conditions (30°C) and in a short time in $[C_4C_4im][BF_4]$ when the reaction mixture is sonicated. Methanol was required as co-solvent to dissolve the phenylboronic acid, affording a homogeneous solution. In air, homo-coupling of the boronic acid was significant, but under inert conditions, no such products were observed. A stronger base such as sodium methoxide was required when arylchlorides are used as substrates. Formation of the palladium bis-carbene complex $[Pd(C_4C_4imy)_2]^{2+}$ occurs during the reaction, and palladium colloids are also formed, preventing efficient recycling of the catalyst. If, however, the preformed bis-carbene complex is employed as the catalyst with methanol as the only solvent present, the formation of zerovalent palladium species was not observed and reuse was possible, though with a steady decrease in activity.

Rather mild reaction conditions, 50-70°C were also realised when the reaction was performed in a phosphonium ionic liquid and even aryl-chlorides prove to be suitable substrates in this solvent. A range of arylhalides were coupled with the palladium(0) precursor $Pd_2(dba)_3$ in $[C_{14}(C_6)_3P]Cl$ in the presence of potassium phosphate base.[11,12] High yields, typically greater than 85% were observed after 1–3 hours for arylbromides and iodides at 50°C, see Table 6.6. A slightly higher temperature is required with arylchlorides together with triphenylphosphine as a co-ligand, but the reaction was very slow when an electron rich substrate such as 4-methoxychlorobenzene was used. The active species has not been identified, but it has been suggested[105] that the phosphonium ionic liquid might act as a source for alkylphosphines and such ligands have previously been found to be successful in Suzuki reactions of arylchlorides.[110-112] Water and toluene were added in small quantities to facilitate salt and substrate solubility, and in many cases the biaryl product crystallised from the reaction as it formed. At the end of the reaction, water and hexane was added, resulting in the formation of a triphasic system with the ionic liquid as the middle phase. The resulting biaryl is thus extracted into the upper hexane phase while inorganic salts dissolve in the aqueous phase. In this manner recycling of the ionic liquid phase was possible for at least five runs with essentially stable activity.

Table 6.6: Coupling between phenylboronic acid and *p*-haloacetophenone in $[C_{14}(C_6)_3P]Cl$ a)

Halide	Ligand	Temp. [°C]	Time [h]	Yield [%]
I	None	50	1	100
Br	PPh_3	50	1	99
Cl	PPh_3	70	30	84

a) 1 mol% $Pd_2(dba)_3$; 3.3 equiv. of K_3PO_4

Molten tetrabutylammonium bromide was shown to be another suitable solvent for the coupling of arylhalides with arylboronic acids.[103] With 1.2 mol% of $Pd(PPh_3)_4$ and two equivalents of aqueous K_2CO_3 high yields were observed after 10 minutes at 120°C. Both the presence of water and base was necessary for the reaction to proceed. Other palladium sources such as $Pd(OAc)_2$ or palladium black were also active, albeit with lower reaction rates. Notably, arylchlorides also react and chlorobenzene reacted with phenylboronic acid to afford biphenyl in 64% yield after 30 minutes. Recycling of the catalyst was possible and a higher activity was observed in the second and third cycles.

The Suzuki-coupling reaction between iodobenzene and phenylboronic acid was further investigated in the nitrile-functionalised pyridinium ionic liquid $[C_3CNpy][Tf_2N]$, shown in Figure 6.5. The reaction was performed at 110°C with 1.2 mol% of either of the palladium complexes **28**–**32**, derived from $[C_3CNpy]^+$ and $PdCl_2$, immobilised in $[C_3CNpy][Tf_2N]$ or in the non-functionalised analogue $[C_4py][Tf_2N]$. After 12 hours at 110°C yields between 81-88% were observed in either $[C_3CNpy][Tf_2N]$ or $[C_4py][Tf_2N]$, bromobenzene as substrate afforded only poor conversion, ca. 35% yield, while less than 1% yield was detected with chlorobenzene as substrate.

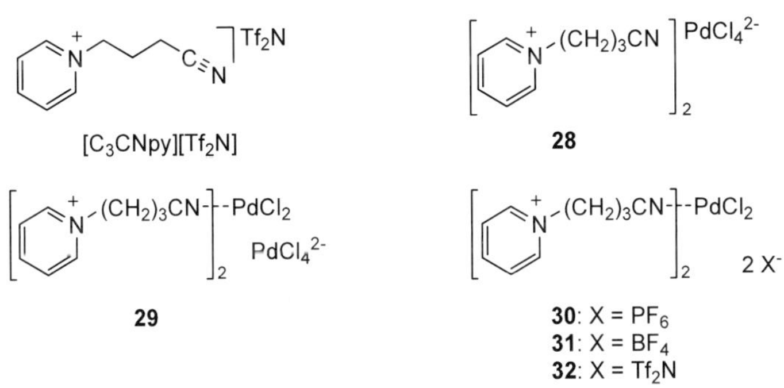

Figure 6.5: Nitrile-functionalised pyridinium ionic liquid and palladium derivatives

The difference between the reactions conducted in the nitrile-functionalised ionic liquid and $[C_4py][Tf_2N]$ became apparent upon recycling of the catalyst. After nine cycles with complex **32** as catalyst precursor in the functionalised ionic liquid, there was no significant decrease in activity, but if **32** was immobilised in $[C_4py][Tf_2N]$, the catalyst solution rapidly lost its activity to become completely inactive after the fifth cycle. ICP-OES analysis shows that less than 5 ppm of palladium was lost from $[C_3CNpy][Tf_2N]$ into the organic phase, whereas typically 28 ppm of palladium was leached from $[C_4py][Tf_2N]$. It is conceivable to attribute the superior retention of the palladium complex in $[C_3CNpy][Tf_2N]$ to coordination of the palladium to the ionic liquid via the nitrile moiety, resulting in higher capacity for immobilisation of the palladium complex. Formation of carbene complexes from pyridinium-based ionic liquids is highly unlikely and attempts to identify the nature of the catalyst failed, which prevents delineation of the role of the nitrile group in the Suzuki reaction.

The cross-coupling of allyl alcohols with aryl- and vinylboronic acids has been accomplished in $[C_4C_1im][BF_4]$ and $[C_4C_1im][PF_6]$ with rhodium catalysts.[113] While no reaction takes place in polar solvents such as water or DMF, good results were obtained in $[C_4C_1im][PF_6]$, see Scheme 6.12, and rhodium(I) as well as rhodium(III) compounds worked well in the latter solvent. The presence of copper salts as well as acids led to an increase of the reaction rate.

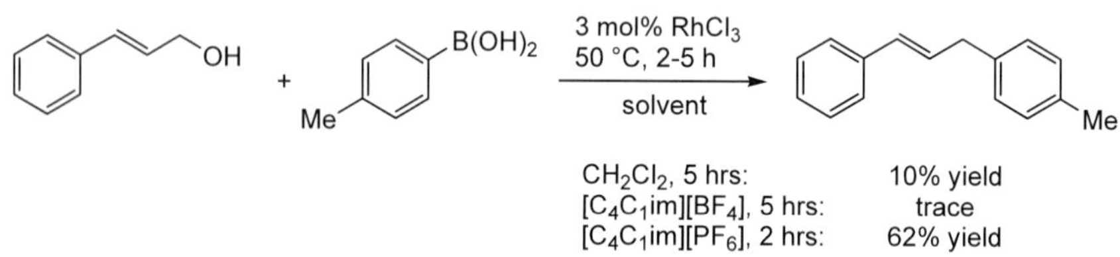

Scheme 6.12: Rhodium-mediated coupling of cinnamyl alcohol and *p*-tolylboronic acid

6.4.2 Stille Coupling Reactions

The Stille reaction involves the coupling of organic electrophiles, typically aryl triflates or halides, with organostannane reagents such as vinyltributyltin, and base is not required for the reaction to proceed.[114] Copper(I) salts are frequently added as co-catalyst and their role is presumably to capture free ligand liberated from the metal complex prior to the reductive elimination of the coupling product.[115] Depending on the solvent (and other factors), different intermediates have been proposed.[116,117] With coordinating solvents such as water or THF an open transition state appears to be formed, weakly coordinating solvents (such as most ionic liquids) seem to lead to the formation of a cyclic intermediate, as shown in Figure 6.6.

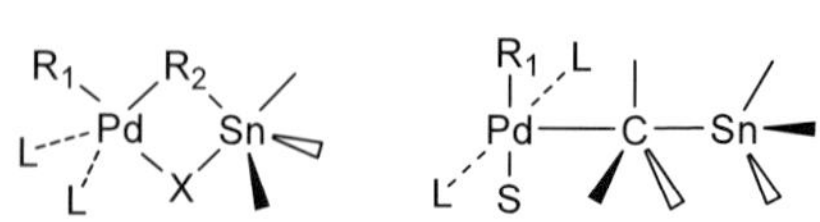

Figure 6.6: Cyclic (left) and open (right) transition states proposed in the Stille coupling reaction: S = solvent (or ligand)

The coupling between organic halides and stannanes is very general with respect to the employable halides and stannanes and many functional groups are tolerated. Arylbromides and iodides were reacted in $[C_4C_1im][BF_4]$ with either tributylphenyltin or tributylvinyltin.[118] In the reaction with α-iodoenones reaction rates were inferior in the ionic liquid relative to *N*-methylpyrolidone (NMP) as solvent. The performance of a given catalyst depends on the organo halide used. $PdCl_2(PhCN)_2$ afforded the best results when iodides were used, while $Pd(PPh_3)_4$ worked best for bromides, as shown in Scheme 6.13. The product may be extracted with diethyl ether and reuse of the catalyst is possible for 2-5 more runs, depending on the substrate, with only a slight decrease in activity.

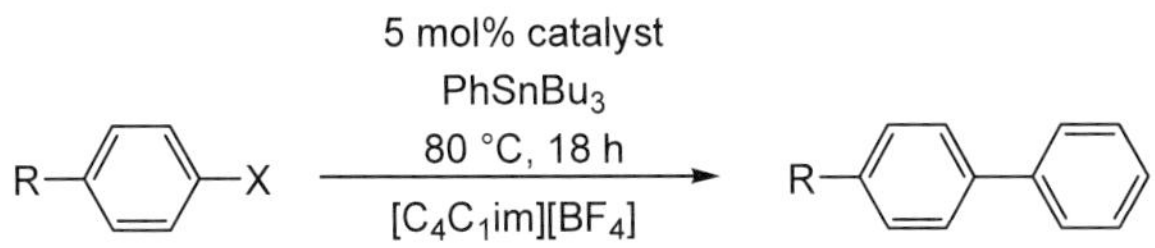

R	X	catalyst system	yield
H	Br	$PdCl_2(PhCN)_2$, CuI, $AsPh_3$	35%
H	Br	$Pd(PPh_3)_4$	90%
OMe	I	$PdCl_2(PhCN)_2$, CuI, $AsPh_3$	82%
OMe	I	$Pd(PPh_3)_4$	30%

Scheme 6.13: Stille-coupling of arylhalides

An array of ionic liquids, mostly based on the imidazolium cation, but also including pyridinium and pyrrolidinium salts, have been screened in the Stille-coupling of iodobenzene with tributylvinyltin in the presence (or absence) of $AsPh_3$ and the results are shown in Table 6.7.[119] $Pd_2(dba)_3$ was used in the presence of $AsPh_3$ while $Pd(OAc)_2$ was used as catalyst in the ligandless reaction. Reaction rates were always lower when no arsine ligand was present, independent of the nature of the ionic liquid, and it took up to 24 hours (relative to 4 hours with $Pd_2(dba)_3/AsPh_3$) to achieve complete conversion. Ionic liquids with nucleophilic anions, namely with Br^- or $[OTs]^-$ generally afforded inferior yields, while $[OTf]^-$ and $[C_8H_{17}SO_4]^-$ gave the best results. Catalytic activity was lost in most cases after the first run, presumably due to the precipitation of Pd(0) during the product extraction step. The reaction between iodobenzene and tetraalkyltin proceeded only sluggishly and in the best case only ca. 60% of the desired product was observed after 48 hours at 80°C in $[C_4C_1im][Tf_2N]$.

Table 6.7: Conversion in the Stille-coupling reaction between iodobenzene and tributylvinyltin with respect to different ionic liquids at a catalyst loading of 4 mol%

Solvent	$Pd_2(dba)_3/AsPh_3$ 1 hour at 80°C	$Pd(OAc)_2$ 4 hours at 80°C
$[C_4C_1im]Br$	41	10
$[C_4C_1im][BF_4]$	50	49
$[C_4C_1im][PF_6]$	49	30
$[C_4C_1im][Tf_2N]$	73	75
$[C_4C_1im][C_8H_{17}SO_4]$	99	55
$[C_6py][TF_2N]$	85	84
$[C_4C_1pyrr][TF_2N]$	74	85

Copper-free coupling between 4-iodotoluene and tributylphenyltin in $[C_4C_1im][Tf_2N]$ with 2 mol% of the palladium carbene complex $PdCl_2(PPh_3)(C_4C_1imy)$ is of only modest activity, ca. 39% yield after 3 hours at 100°C.[120] Addition of 10 mol% $AsPh_3$ increased the yield to 75% and recycling was possible, giving 68% yield upon reuse of the catalyst solution.

The effect of nitrile-functionalised ionic liquids on the recycling potential was investigated in the coupling between iodobenzene and tributylphenyltin.[15] Several, closely related palladium-precursors, based on the ionic liquid cation $[C_3CNpy]^+$ and $PdCl_2$ (Figure 6.5), were tested in the reaction, but the nature of the pre-catalyst did not have a significant influence on the rate. Independent from the ionic liquid used, rather low catalytic activity was observed and with 5 mol% catalyst, yields ranged between 43-65% after 12 hours at 80°C.

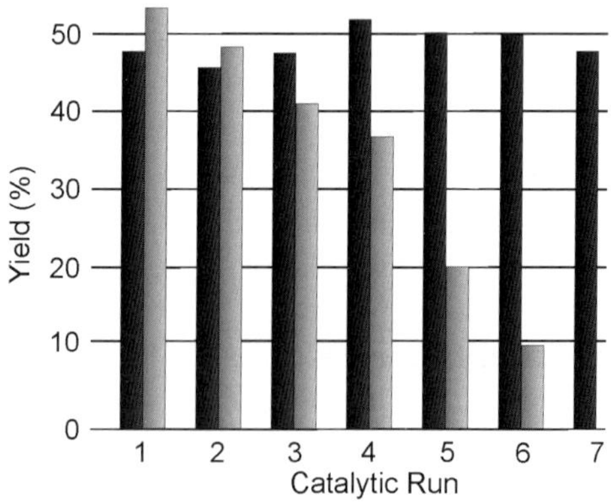

Figure 6.7: Recyclability of the pre-catalyst **32** in $[C_3CNpy][Tf_2N]$ (dark) and $[C_4py][Tf_2N]$ (light) in the Stille reaction between phenyltributylstannane and iodobenzene; 5 mol% catalyst, 80°C, 12 hours

While the activity remains essentially constant in the nitrile functionalised ionic liquid, activity decreased rapidly in $[C_4py][Tf_2N]$, as is clear from Figure 6.7. The catalyst in the reaction was shown to be colloidal and isolated nanoparticles from the first run afforded similar conversion to those formed *in situ*. A much lower palladium content was found in the organic phase with $[C_3CNpy][Tf_2N]$ relative to $[C_4py][Tf_2N]$, ca. 7 versus 46 ppm, indicating that the nitrile-functionalised ionic liquid improves catalyst retention and stability. Transmission electron microscopy images of the isolated palladium nanoparticles are shown in Figure 6.8.

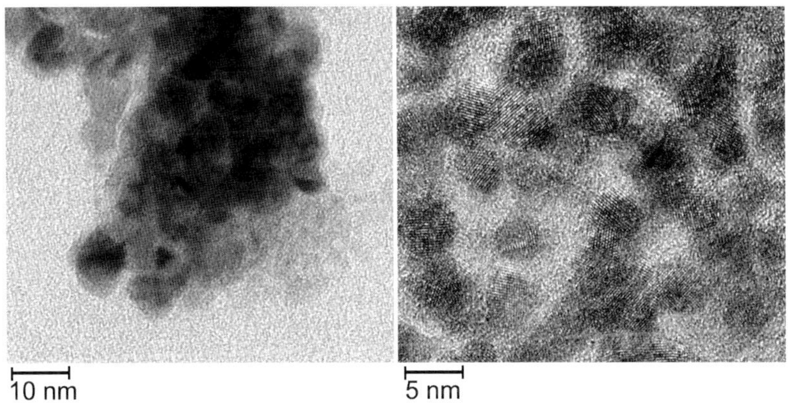

Figure 6.8: Comparison of TEM images of the palladium nanoparticles isolated after the Stille reaction from $[C_4py][Tf_2N]$ (left) and $[C_3CNpy][Tf_2N]$ (right)

The nanoparticles obtained from $[C_4py][Tf_2N]$ and $[C_3CNpy][Tf_2N]$ both have a diameter of ca. 5 nm, but they exhibit different states. Those isolated from $[C_4py][Tf_2N]$ are aggregated whereas the nanoparticles generated in the nitrile-functionalised ionic liquid were evenly distributed and the surface lattice can be clearly observed. The TEM images support the hypothesis that the nitrile-functionalised ionic liquid stabilises the palladium nanoparticles. In either case, agglomeration is prevented and ultimately catalyst deactivation suppressed. The former hypothesis has been suggested previously for a different system.[121]

6.4.3 Negishi Coupling Reactions

In the Negishi cross-coupling reaction an aryl or benzyl zinchalide is reacted with an arylhalide to yield the corresponding biaryl. The reaction of a range of aryl zinc compounds with mostly aryliodides proceeds well in a biphasic system comprised of $[C_4C_1C_1im][BF_4]$ and toluene (Scheme 6.14).[122] With 2 mol% $Pd(dba)_2$ as catalyst precursor and 4 mol% of the charged phosphine ligand DPP-$[C_4C_1im]^+$, fast conversion was observed at room temperature with aryliodides. Arylbromides and triflates are also suitable substrates but require longer reaction times (3-5 hours). The catalyst could be reused after decantation of the organic phase, but a significant decrease in activity was observed by the third run.

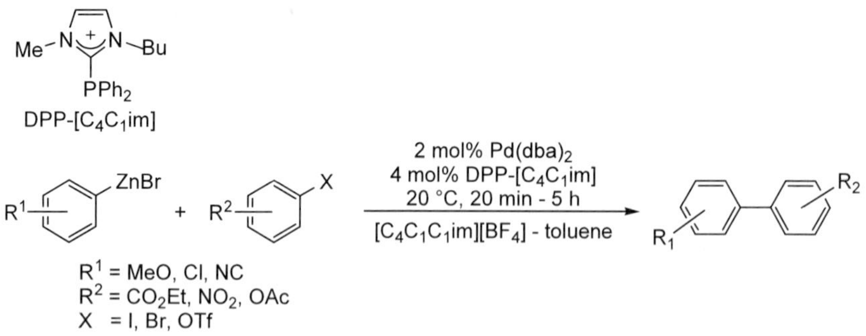

Scheme 6.14: Negishi-coupling between arylzinc bromides and arylhalides

6.4.4 Sonogashira Coupling Reactions

The coupling between an alkyne and an arylhalide is referred to as Sonogashira coupling. Highly active solid supported palladium bis(pyrimidine) complexes have been shown to catalyse the reaction in the absence of copper(I) co-catalysts.[123,124] With $PdCl_2(PPh_3)_2$ as catalyst and diisopropylamine or piperidine as base, the coupling reaction proceeds efficiently in $[C_4C_1im][PF_6]$ in the absence of a copper(I) co-catalyst.[125] Slightly lower yields were obtained in either $[C_4C_1im][BF_4]$ or $[C_2C_1im][BF_4]$. The product could be extracted with hexane and the ionic liquid washed with water to remove the ammonium salt formed during the reaction. Following this procedure, recycling of the catalyst was possible, albeit with gradual loss of activity (Scheme 6.15).

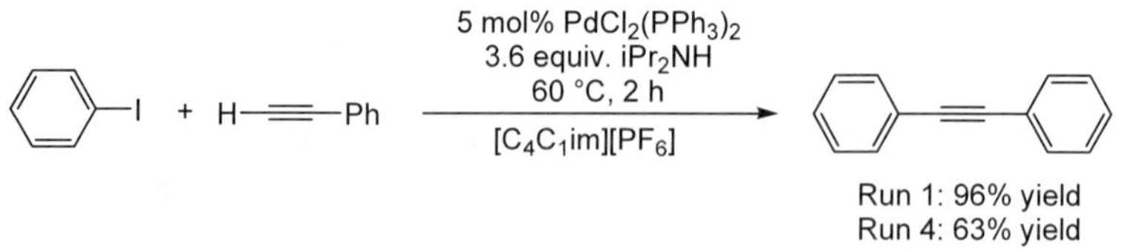

Scheme 6.15: Sonogashira-coupling of phenyliodide and phenylacetylene

6.4.5 Homocoupling Reactions

The homocoupling of aryliodides has been investigated in $[C_4C_1im][PF_6]$[126] as well as in $[C_4C_1im][BF_4]$ and $[C_8C_1im][BF_4]$.[127] The reaction proceeds well with the cyclopalladated complex **33**, as illustrated in Scheme 6.16, but rather high reaction temperatures of 160°C were required. Relative to DMA as solvent, catalytic activity decreased when the reaction was performed in either $[C_4C_1im][BF_4]$ or $[C_8C_1im][BF_4]$, but the isolation of the product was simplified. In contrast to some of the coupling reactions described previously, no reaction took place in $[C_4C_1im]Br$.

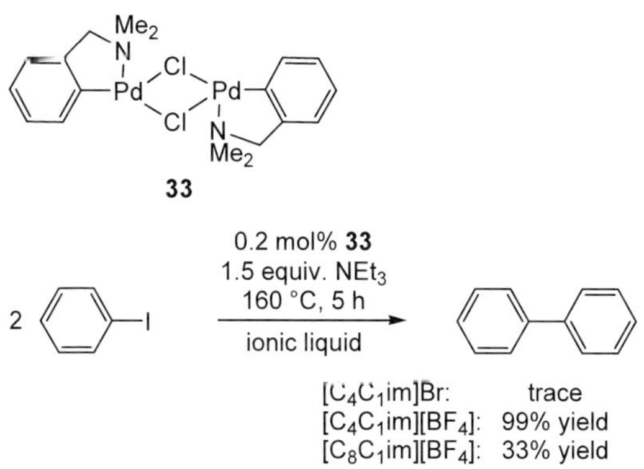

Scheme 6.16: Homocoupling of iodobenzene with a cyclopalladated complex

A series of palladium salts and complexes was further screened for the coupling of arylbromides in $[C_4C_1im][PF_6]$. A stronger base such as tetrakis(dimethylamino)ethylene was required for electron-poor arenes and good conversion was observed independent from the catalyst precursor employed. However, bromoanisole remained unreactive under the conditions used. Catalyst recycling was possible after extraction of the product with diethyl ether but activity decreased considerably with each run.

6.5 Allylic Alkylation

The palladium(0) catalysed allylic alkylation of soft carbon nucleophiles represents a very useful tool for organic synthesis. The reaction conditions often involve heating a mixture of stabilised carbanions together with the substrate and the catalyst mixture in THF.

The biphasic allylic alkylation of ethylcinnamyl carbonate with ethyl acetoacetate has been studied in $[C_4C_1im]Cl$–methylcyclohexane with $PdCl_2$/TPPTS as catalyst.[128] Relative to the same reaction under aqueous biphasic conditions, much higher reaction rates were observed, probably due to the improved solubility of the substrates in the imidazolium chloride. In addition, formation of cinnamyl alcohol was entirely suppressed in the ionic liquid.

A series of different nucleophiles was tested in the reaction with 3-acetoxy-1,3-diphenylprop-1-ene in $[C_4C_1im][BF_4]$ with $Pd(OAc)_2$ as catalyst precursor.[129] As with molecular solvents, the presence of an excess of phosphine ligand was found to be necessary for the reaction to take place. The effect of employing different phosphines has been studied and it was found that the reaction rate is greatly enhanced with electron-rich or σ-donating phosphines.[130] With π-accepting phosphines, little or no reaction took place. As can be seen from Table 6.8, the reaction proceeds faster in the ionic liquid than in THF and a wider range of phosphines can be applied.

Table 6.8: Ligand effects in the allylic alkylation of 3-acetoxy-1,3-diphenylprop-1-ene with dimethyl malonate

	$[C_4C_1im][BF_4]$		THF	
Ligand	time [h]	conversion [%]	time [h]	conversion [%]
$P(nBu)_3$	1	100	1	100
PCy_3	1	100	20	0
PPh_3	6	100	6	33
$P(tBu)_3$	22	75	22	0
$P(OPh)_3$	18	0	18	0
$P(p\text{-}C_6H_4CF_3)_3$	20	0	20	0

It was reasoned, that in the ionic liquid the electronic effects appear to be the dominating factor while in THF both steric and electronic effects are significant. As the organic compounds dissolve in the ionic liquid, a product extraction step was required, and unless charged phosphine ligands like TPPTS were used, palladium leaching was considerable. DBU gave better results than K_2CO_3 and this is probably a reflection of their different solubilities in the ionic liquid. While the latter is only sparingly soluble, DBU readily dissolves.

The effect of hydrogen bonding on allylic alkylation was studied in the base-free reaction of phenylallyl carbonate with dimethyl malonate.[131,132] While the reaction proceeds rapidly in THF in the presence of four equivalents of PPh_3, it is very sluggish in $[C_4C_1im][BF_4]$. The reaction in THF was significantly inhibited when small quantities of the ionic liquid were added. Only with excess external base did the reaction proceed in a comparable rate in the ionic liquid. It was shown that the oxidative addition of allylic acetate to Pd(0) is reversible and in THF the resulting acetate anion and Pd-allyl cation can form tight ion pairs rather than free ions. The free carbonate anion hydrogen bonds to the acidic 2-proton of the imidazolium cation and methoxide formation, which is required to deprotonate the nucleophile, can therefore not function effectively, see Scheme 6.17. It was argued that the slow conversion in the ionic liquid was due to insufficient availability of nucleophiles.

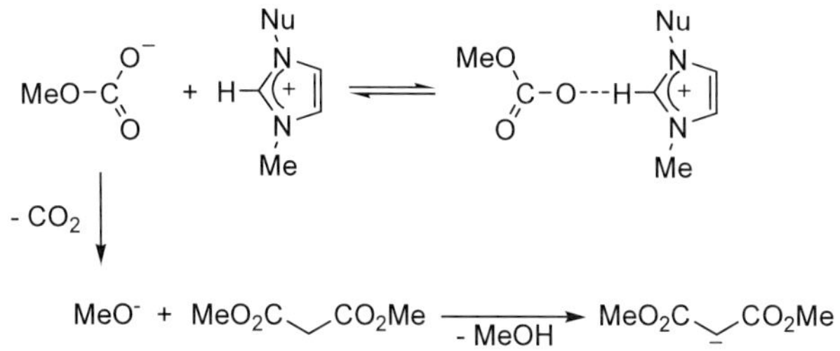

Scheme 6.17: Role of the imidazolium cation as H-bond donor, thus inhibiting nucleophilic attack from Nu^- in allylic alkylation reactions

The same reaction conditions as above (2 mol% Pd, P:Pd 1:4) were employed in the enantioselective allylation of 3-acetoxy-1,3-diphenylprop-1-ene in $[C_4C_1im][PF_6]$, using ferrocenylphosphine ligands.[133] Relative to the reaction in THF, the *ee* increased from 40% to 68% in the ionic liquid. After extraction of the product with toluene, reuse of the catalyst was possible, however both yield and selectivity decreased.

A series of chiral chelate phosphine ligands, mainly based on the ferrocene backbone (Figure 6.9), have been tested in $[C_4C_1im][PF_6]$ as catalysts for enantioselective allylic alkylation reactions and results are listed in Table 6.9.[134]

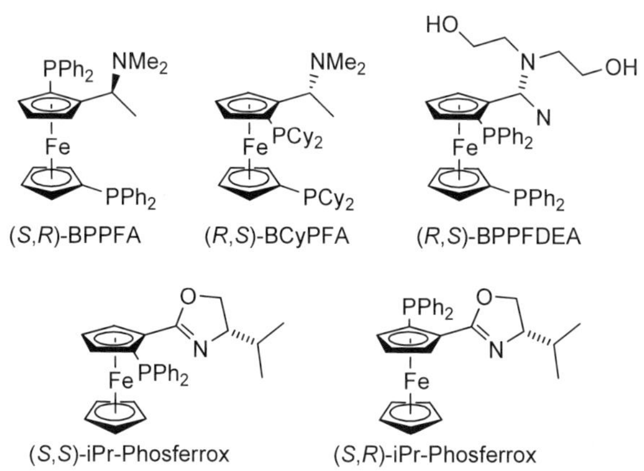

Figure 6.9: Chiral ferrocene-based ligands tested in allylic alkylation

Table 6.9: Allylic alkylation of 3-acetoxy-1,3-diphenylprop-1-ene with dimethylmalonate in $[C_4C_1im][PF_6]$ at 20°C with 2 mol% $Pd_2(dba)_3$ and 4 mol% chiral ligand

Ligand	Base	Cycle	Yield [%]	*ee* [%]
BPPFDEA	K_2CO_3	1	81	74
*i*Pr-Phosferrox	K_2CO_3	1	34	86
*i*Pr-Phosferrox	K_2CO_3	2	27	84
*i*Pr-Phosferrox	BSA/AcOK	1	55	68
*i*Pr-Phosferrox	BSA/AcOK	2	47	66
BCyPFA	BSA/AcOK	1	69	56
BCyPFA	BSA/AcOK	2	25	46

The product was extracted with toluene and the remaining ionic liquid phase washed with water to remove residual base. Bis(trimethylsilyl)acetamide afforded higher yields than K_2CO_3. While the selectivity decreased only slightly upon recycling, yields dropped markedly. This behaviour was ascribed to a 10-20% loss of ionic liquid during the workup. Reaction conditions were optimised for the (*RS*,*S*)-*i*Pr-Phosferrox ligand and almost quantitative yield (98%) and good selectivity (88%) was obtained after 15 hours at 60°C. In contrast to higher reaction temperatures the selectivity decreased significantly during the reaction at room temperature.

6.6 References

[1] Welton, T.; Smith, P. J. *Adv. Organomet. Chem.* **2004**, *51*, 251-284.

[2] Gruber, A. S.; Pozebon, D.; Monteiro, A. L.; Dupont, J. *Tetrahedron Lett.* **2001**, *42*, 7345-7348.

[3] Wolf, C.; Lerebours, R. *Org. Biomol. Chem.* **2004**, *2*, 2161-2164.

[4] Deveau, A. M.; Macdonald, T. L. *Tetrahedron Lett.* **2004**, *45*, 803-807.

[5] Mukhopadhyay, S.; Yaghmur, A.; Baidossi, M.; Kundu, B.; Sasson, Y. *Org. Process. Res. Dev.* **2003**, *7*, 641-643.

[6] Badone, D.; Baroni, M.; Cardamone, R.; Ielmini, A.; Guzzi, U. *J. Org. Chem.* **1997**, *62*, 7170-7173.

[7] Kaufmann, D. E.; Nouroozian, M.; Henze, H. *Synlett* **1996**, 1091-1092.

[8] Moreno-Manas, M.; Pleixats, R. *Acc. Chem. Res.* **2003**, *36*, 638-643.

[9] Le Bras, J.; Mukherjee, D. K.; Gonzalez, S.; Tristany, M.; Ganchegui, B.; Moreno-Manas, M.; Pleixats, R.; Henin, F.; Muzart, J. *New J. Chem.* **2004**, *28*, 1550-1553.

[10] Gerritsma, D. A.; Robertson, A.; McNulty, J.; Capretta, A. *Tetrahedron Lett.* **2004**, *45*, 7629-7631.

[11] McNulty, J.; Capretta, A.; Cheekoori, S.; Clyburne, J. A. C.; Robertson, A. J. *Chim. Oggi* **2004**, *22*, 13-16.

[12] McNulty, J.; Capretta, A.; Wilson, J.; Dyck, J.; Adjabeng, G.; Robertson, A. *Chem. Commun.* **2002**, 1986-1987.

[13] Kiss, L.; Kurtan, T.; Antus, S.; Brunner, H. *Arkivoc* **2003**, 69-76.
[14] Xiao, J. C.; Twamley, B.; Shreeve, J. M. *Org. Lett.* **2004**, *6*, 3845-3847.
[15] Zhao, D. B.; Fei, Z. F.; Geldbach, T. J.; Scopelliti, R.; Dyson, P. J. *J. Am. Chem. Soc.* **2004**, *126*, 15876-15882.
[16] Dupont, J.; Spencer, J. *Angew. Chem. Int. Ed.*. **2004**, *43*, 5296-5297.
[17] Cavell, K. J.; McGuinness, D. S. *Coord. Chem. Rev.* **2004**, *248*, 671-681.
[18] Husan, M.; Kozhevnikov, I. V.; Siddiqui, M. R. H.; Steiner, A.; Winterton, N. *J. Chem. Res. S* **2000**, 392-393.
[19] Mathews, C. J.; Smith, P. J.; Welton, T.; White, A. J. P.; Williams, D. J. *Organometallics* **2001**, *20*, 3848-3850.
[20] Xu, L. J.; Chen, W. P.; Xiao, J. L. *Organometallics* **2000**, *19*, 1123-1127.
[21] Handy, S. T.; Okello, M. *Tetrahedron Lett.* **2003**, *44*, 8395-8397.
[22] Fehlhammer, W. P.; Bliss, T.; Kernbach, U.; Brüdgam, I. *J. Organomet. Chem.* **1995**, *490*, 149-153.
[23] Herrmann, W. A.; Schwarz, J.; Gardiner, M. G.; Spiegler, M. *J. Organomet. Chem.* **1999**, *575*, 80-86.
[24] Lebel, H.; Janes, M. K.; Charette, A. B.; Nolan, S. P. *J. Am. Chem. Soc.* **2004**, *126*, 5046-5047.
[25] Clement, N. D.; Cavell, K. J.; Jones, C.; Elsevier, C. J. *Angew. Chem. Int. Ed.* **2004**, *43*, 1277-1279.
[26] McGuinness, D. S.; Cavell, K. J.; Yates, B. F. *Chem. Commun.* **2001**, 355-356.
[27] McGuinness, D. S.; Cavell, K. J.; Yates, B. F.; Skelton, B. W.; White, A. H. *J. Am. Chem. Soc.* **2001**, *123*, 8317-8328.
[28] Clement, N. D.; Cavell, K. J. *Angew. Chem. Int. Ed.* **2004**, *43*, 3845-3847.
[29] Dullius, J. E. L.; Suarez, P. A. Z.; Einloft, S.; de Souza, R. F.; Dupont, J.; Fischer, J.; De Cian, A. *Organometallics* **1998**, *17*, 815-819.
[30] Zhao, D. B.; Fei, Z. F.; Geldbach, T. J.; Scopelliti, R.; Laurenczy, G.; Dyson, P. J. *Helv. Chim. Acta* **2005**, *88*, 665-675.
[31] Cassol, C. C.; Umpierre, A. P.; Machado, G.; Wolke, S. I.; Dupont, J. *J. Am. Chem. Soc.* **2005**, *127*, 3298-3299.
[32] Mizoroki, T.; Mori, K.; Ozaki, A. *Bull. Chem. Soc. Jpn.* **1971**, *44*, 581.
[33] Heck, R. F.; Nolley, J. P. *J. Org. Chem.* **1972**, *37*, 2320-2322.
[34] Farina, V. *Adv. Synth. Catal.* **2004**, *346*, 1553-1582.
[35] Peris, E.; Crabtree, R. H. *Coord. Chem. Rev.* **2004**, *248*, 2239-2246.
[36] Biffis, A.; Zecca, M.; Basato, M. *J. Mol. Catal. A* **2001**, *173*, 249-274.
[37] Bräse, S.; de Meijere, A. In *Metal-Catalyzed Cross-Coupling Reactions*; de Meijere, A., Diederich, F., Eds.; Wiley-VCH: Weinheim, **2004**; Vol. 1, pp 217-315.
[38] Crisp, G. T. *Chem. Soc. Rev.* **1998**, *27*, 427-436.
[39] Amatore, C.; Jutand, A. *J. Organomet. Chem.* **1999**, *576*, 254-278.
[40] Amatore, C.; Jutand, A. *Acc. Chem. Res.* **2000**, *33*, 314-321.
[41] de Meijere, A.; Diederich, F., Eds. *Metal-Catalyzed Cross-Coupling Reactions*; 2 ed.; Wiley-VCH: Weinheim, **2004**.
[42] Tsuji, J. *Palladium Reagents and Catalysts*; Wiley: Chichester, **1995**.
[43] Herrmann, W. A.; Ofele, K.; Von Preysing, D.; Schneider, S. K. *J. Organomet. Chem.* **2003**, *687*, 229-248.
[44] Whitcombe, N. J.; Hii, K. K.; Gibson, S. E. *Tetrahedron* **2001**, *57*, 7449-7476.
[45] Miyaura, N., Ed. *Cross-Coupling Reactions - A Practical Guide*; Springer: Berlin, **2002**.
[46] Jeffrey, T. *J. Chem. Soc., Chem. Commun.* **1984**, 1287-1289.
[47] Jeffrey, T. *Tetrahedron Lett.* **1985**, *26*, 2667-2669.

[48] Mehnert, C. P.; Weaver, D. W.; Ying, J. Y. *J. Am. Chem. Soc.* **1998**, *1120*, 12289-12296.
[49] Benaglia, M.; Puglisi, A.; Cozzi, F. *Chem. Rev.* **2003**, *103*, 3401-3429.
[50] Zhao, S. F.; Zhou, R. X.; Zheng, X. M. *J. Mol. Catal. A* **2004**, *211*, 139-142.
[51] Fujita, S.; Yuzawa, K.; Bhanage, B. M.; Ikushima, Y.; Arai, M. *J. Mol. Catal. A* **2002**, *180*, 35-42.
[52] Yeung, L. K.; Lee, C. T.; Johnston, K. P.; Crooks, R. M. *Chem. Commun.* **2001**, 2290-2291.
[53] Kayaki, Y.; Noguchi, Y.; Ikariya, T. *Chem. Commun.* **2000**, 2245-2246.
[54] Fukuyama, T.; Arai, M.; Matsubara, H.; Ryu, L. *J. Org. Chem.* **2004**, *69*, 8105-8107.
[55] Curran, D. P.; Fischer, K.; Moura-Letts, G. *Synlett* **2004**, 1379-1382.
[56] Nakamura, Y.; Takeuchi, S.; Ohgo, Y. *J. Fluorine Chem.* **2003**, *120*, 121-129.
[57] Reetz, M. T.; Westermann, E. *Angew. Chem. Int. Ed.*. **2000**, *39*, 165-168.
[58] Carmichael, A. J.; Earle, M. J.; Holbrey, J. D.; McCormac, P. B.; Seddon, K. R. *Org. Lett.* **1999**, *1*, 997-1000.
[59] Hamill, N. A.; Hardacre, C.; McMath, S. E. J. *Green Chem.* **2002**, *4*, 139-142.
[60] Perosa, A.; Tundo, P.; Selva, M.; Zinovyev, S.; Testa, A. *Org. Biomol. Chem.* **2004**, *2*, 2249-2252.
[61] Böhm, V. P. W.; Herrmann, W. A. *Chem. Eur. J.* **2000**, *6*, 1017-1025.
[62] Herrmann, W. A.; Böhm, V. P. W. *J. Organomet. Chem.* **1999**, *572*, 141-145.
[63] Bouquillon, S.; Ganchegui, B.; Estrine, B.; Henin, F.; Muzart, J. *J. Organomet. Chem.* **2001**, *634*, 153-156.
[64] Zou, G.; Wang, A. E.; Zhu, J.; Tang, J.; He, M. Y. *J. Mol. Catal. A* **2003**, *206*, 193-198.
[65] Xu, L. J.; Chen, W. P.; Ross, J.; Xiao, J. L. *Org. Lett.* **2001**, *3*, 295-297.
[66] Deshmukh, R. R.; Rajagopal, R.; Srinivasan, K. V. *Chem. Commun.* **2001**, 1544-1545.
[67] Selvakumar, K.; Zapf, A.; Beller, M. *Org. Lett.* **2002**, *4*, 3031-3033.
[68] Calo, V.; Nacci, A.; Lopez, L.; Mannarini, N. *Tetrahedron Lett.* **2000**, *41*, 8973-8976.
[69] Calo, V.; Nacci, A. *Z. Naturforsch. A* **2001**, *56*, 702-706.
[70] Calo, V.; Nacci, A.; Monopoli, A.; Lopez, L.; di Cosmo, A. *Tetrahedron* **2001**, *57*, 6071-6077.
[71] Calo, V.; Nacci, A.; Lopez, L.; Napola, A. *Tetrahedron Lett.* **2001**, *42*, 4701-4703.
[72] Calo, V.; Nacci, A.; Monopoli, A.; Laera, S.; Cioffi, N. *J. Org. Chem.* **2003**, *68*, 2929-2933.
[73] Calo, V.; Nacci, A.; Monopoli, A. *J. Mol. Catal. A* **2004**, *214*, 45-56.
[74] Calo, V.; Nacci, A.; Monopoli, A.; Spinelli, M. *Eur. J. Org. Chem.* **2003**, 1382-1385.
[75] Vallin, K. S. A.; Emilsson, P.; Larhed, M.; Hallberg, A. *J. Org. Chem.* **2002**, *67*, 6243-6246.
[76] Kabalka, G. W.; Dong, G.; Venkataiah, B. *Tetrahedron Lett.* **2004**, *45*, 2775-2777.
[77] Liu, S. F.; Fukuyama, T.; Sato, M.; Ryu, I. *Org. Process. Res. Dev.* **2004**, *8*, 477-481.
[78] Park, S. B.; Alper, H. *Org. Lett.* **2003**, *5*, 3209-3212.
[79] Handy, S. T.; Okello, M.; Dickenson, G. *Org. Lett.* **2003**, *5*, 2513-2515.
[80] Mo, J.; Xu, L. J.; Xiao, J. L. *J. Am. Chem. Soc.* **2005**, *127*, 751-760.
[81] Choudary, B. M.; Madhi, S.; Chowdari, N. S.; Kantam, M. L.; Sreedhar, B. *J. Am. Chem. Soc.* **2002**, *124*, 14127-14136.
[82] Calo, V.; Nacci, A.; Monopoli, A.; Fornaro, A.; Sabbatini, L.; Cioffi, N.; Ditaranto, N. *Organometallics* **2004**, *23*, 5154-5158.
[83] Calo, V.; Nacci, A.; Monopoli, A.; Detomaso, A.; Iliade, P. *Organometallics* **2003**, *22*, 4193-4197.
[84] Calo, V.; Nacci, A.; Monopoli, A.; Ieva, E.; Cioffi, N. *Org. Lett.* **2005**, *7*, 617-620.

[85] Okubo, K.; Shirai, M.; Yokoyama, C. *Tetrahedron Lett.* **2002**, *43*, 7115-7118.
[86] Hagiwara, H.; Shimizu, Y.; Hoshi, T.; Suzuki, T.; Ando, M.; Ohkubo, K.; Yokoyama, C. *Tetrahedron Lett.* **2001**, *42*, 4349-4351.
[87] Hagiwara, H.; Sugawara, Y.; Isobe, K.; Hoshi, T.; Suzuki, T. *Org. Lett.* **2004**, *6*, 2325-2328.
[88] Xie, X. G.; Lu, J. P.; Chen, B.; Han, J. J.; She, X. G.; Pan, X. F. *Tetrahedron Lett.* **2004**, *45*, 809-811.
[89] Csabai, P.; Joó, F. *Organometallics* **2004**, *23*, 5640-5643.
[90] Biffis, A.; Zecca, M.; Basato, M. *J. Mol. Catal. A* **2001**, *173*, 249-274.
[91] Beletskaya, I. P.; Cheprakov, A. V. *Coord. Chem. Rev.* **2004**, *248*, 2337-2364.
[92] Deng, W.; Liu, L.; Guo, Q. X. *Chin. J. Org. Chem.* **2004**, *24*, 150-165.
[93] Iyer, S.; Thakur, V. V. *J. Mol. Catal. A* **2000**, *157*, 275-278.
[94] Yang, Y.; Zhou, R.; Zhao, S.; Li, Q.; Zheng, X. *J. Mol. Catal. A* **2003**, *192*, 303-306.
[95] Suzuki, A. *J. Organomet. Chem.* **1999**, *576*, 147-168.
[96] Bellina, F.; Carpita, A.; Rossi, R. *Synthesis* **2004**, 2419-2440.
[97] Kotha, S.; Lahiri, K.; Kashinath, D. *Tetrahedron* **2002**, *58*, 9633-9695.
[98] Miyaura, N. *Top. Curr. Chem.* **2002**, *219*, 11-59.
[99] Littke, A. F.; Fu, G. C. *Angew. Chem. Int. Ed.* **2002**, *41*, 4176-4211.
[100] Aliprantis, A. O.; Canary, J. W. *J. Am. Chem. Soc.* **1994**, *116*, 6985-6986.
[101] Miyaura, N.; Yamada, K.; Suginome, H.; Suzuki, A. *J. Am. Chem. Soc.* **1985**, *107*, 972-980.
[102] Revell, J. D.; Ganesan, A. *Org. Lett.* **2002**, *4*, 3071-3073.
[103] Zou, Y.; Wang, Q. R.; Tao, F. G.; Ding, Z. B. *Chin. J. Chem.* **2004**, *22*, 215-218.
[104] Corma, A.; Garcia, H.; Leyva, A. *Tetrahedron* **2004**, *60*, 8553-8560.
[105] McLachlan, F.; Mathews, C. J.; Smith, P. J.; Welton, T. *Organometallics* **2003**, *22*, 5350-5357.
[106] Mathews, C. J.; Smith, P. J.; Welton, T. *J. Mol. Catal. A* **2004**, *214*, 27-32.
[107] Mathews, C. J.; Smith, P. J.; Welton, T. *Chem. Commun.* **2000**, 1249-1250.
[108] Rajagopal, R.; Jarikote, D. V.; Srinivasan, K. V. *Chem. Commun.* **2002**, 616-617.
[109] Miyaura, N.; Yanagi, T.; Suzuki, A. *Synth. Commun.* **1981**, *11*, 513-519.
[110] Wolfe, J. P.; Singer, R. A.; Yang, B. H.; Buchwald, S. L. *J. Am. Chem. Soc.* **1999**, *121*, 9550-9561.
[111] Kirchhoff, J. H.; Dai, C.; Fu, G. C. *Angew. Chem. Int. Ed.* **2002**, *41*, 1945-1947.
[112] Littke, A. F.; Fu, G. C. *J. Org. Chem.* **1999**, *64*, 10-11.
[113] Kabalka, G. W.; Dong, G.; Venkataiah, B. *Org. Lett.* **2003**, *5*, 893-895.
[114] Stille, J. K. *Angew. Chem. Int. Ed.* **1986**, *25*, 508-523.
[115] Casado, A. L.; Espinet, P. *Organometallics* **2003**, *22*, 1305-1309.
[116] Casado, A. L.; Espinet, P. *Organometallics* **1998**, *17*, 954-959.
[117] Casado, A. L.; Espinet, P. *J. Am. Chem. Soc.* **1998**, *120*, 8978-8958.
[118] Handy, S. T.; Zhang, X. L. *Org. Lett.* **2001**, *3*, 233-236.
[119] Chiappe, C.; Imperato, G.; Napolitano, E.; Pieraccini, D. *Green Chem.* **2004**, *6*, 33-36.
[120] Liu, S. F.; Fukuyama, T.; Sato, M.; Ryu, L. *Synlett* **2004**, 1814-1816.
[121] Bucher, S.; Hormes, J.; Modrow, H.; Brinckmann, R.; Waldöfer, N.; Bönnemann, H.; Beuermann, L.; Krischok, S.; Maus-Friedrichs, W.; Kempter, V. *Surf. Sci.* **2002**, *497*, 321-332.
[122] Sirieix, J.; Ossberger, M.; Betzemeier, B.; Knochel, P. *Synlett* **2000**, 1613-1615.
[123] Buchmeiser, M. R.; Schareina, T.; Kempe, R.; Wurst, K. *J. Organomet. Chem.* **2001**, *634*, 39-46.
[124] Hillerich, J.; Plenio, H. *Chem. Commun.* **2003**, 3024-3025.

[125] Fukuyama, T.; Shinmen, M.; Nishitani, S.; Sato, M.; Ryu, I. *Org. Lett.* **2002**, *4*, 1691-1694.
[126] Park, S. B.; Alper, H. *Tetrahedron Lett.* **2004**, *45*, 5515-5517.
[127] Li, Q.; Nie, J.; Yang, F.; Zheng, R.; Zou, G.; Tang, J. *Chin. J. Chem.* **2004**, *22*, 419-421.
[128] de Bellefon, C.; Pollet, E.; Grenouillet, P. *J. Mol. Catal. A* **1999**, *145*, 121-126.
[129] Chen, W. P.; Xu, L. J.; Chatterton, C.; Xiao, J. L. *J. Chem. Soc., Chem. Commun.* **1999**, 1247-1248.
[130] Ross, J.; Chen, W. P.; Xu, L. J.; Xiao, J. L. *Organometallics* **2001**, *20*, 138-142.
[131] Ross, J.; Xiao, J. L. *Chem. Eur. J.* **2003**, *9*, 4900-4906.
[132] Ross, J.; Xiao, J. L. In *Ionic Liquids as Green Solvents: Progress and Prospects*, ACS Symposium Series 856, **2003**, 314-322.
[133] Toma, S.; Gotov, B.; Kmentova, I.; Solcaniova, E. *Green Chem.* **2000**, *2*, 149-151.
[134] Kmentova, I.; Gotov, B.; Solcaniova, E.; Toma, S. *Green Chem.* **2002**, *4*, 103-106.

Chapter 7

OLEFIN METATHESIS

7.1 Introduction

Catalytic olefin metathesis represents one of the most important methods for C-C bond formation. Specifically, ring-closing metathesis (RCM) can often provide an elegant solution in the synthesis of complex organic molecules.[1] To date, numerous well-defined homogenous catalysts have been used and many reviews are available on the topic.[2-5] Among the most widely used homogeneous catalysts are molybdenum and tungsten alkylidenes of the general formula M(OR')$_2$(NAr)(=CHR) **34**, and the more air- and water-stable ruthenium complexes RuCl$_2$(PCy$_3$)(L)(=CHR) **35**, and [RuCl(η^6-*p*-cymene)(PCy$_3$)(=C=C=CRR')]$^+$ **36**.

One drawback of homogeneously catalysed metathesis reactions lies in the high catalyst loadings that are required for acceptable reaction rates. Catalyst ratios of 0.5 - 10 mol% are commonly employed and contribute notably to the reaction costs. Furthermore, removal of ruthenium-based catalysts from the product, especially from polymers, is not always facile.[6,7] Accordingly, significant effort has been directed towards the immobilisation and recovery of the catalyst, by attaching the complex onto a solid support,[8-17] use of *sc*CO$_2$ as reaction medium[18,19] or *via* biphasic (aqueous) catalysis.[20,21] Within this context, ionic liquids have only very recently been recognised as a potential solution. With the exception of the pioneering work by Chauvin *et al.*, in 1995 on the cross-metathesis of linear olefins with W(OAr)$_2$Cl$_4$ as catalyst in a chloroaluminate ionic liquid,[22] the remaining literature stems from 2000 onwards and is focused almost exclusively on ring-closing metathesis with ruthenium catalysts.

The widely accepted mechanism for olefin metathesis[23] is shown in Scheme 7.1, exemplified by the ring closing metathesis of diallyltosylamide to tosyldihydropyrrole, using either of the ruthenium complexes **35** or **36**. Formation of the active catalyst proceeds via dissociation of a neutral ligand, e.g. tricyclohexyl phosphine or cymene, respectively, a step that is accelerated by the presence of Lewis and Brønsted acids such as CuCl or HCl. The actual catalytic cycle consists of a series of cycloaddition and cycloreversion steps, with metal-carbenes and metalacyclobutanes as intermediates. In principle, all steps are reversible, but as the catalyst is sensitive to the substitution pattern of the olefins, the equilibrium usually lies well to one side. It should be noted that the charge of the catalyst remains unchanged during the entire catalytic cycle, which may have implications for immobilisation in an ionic liquid.

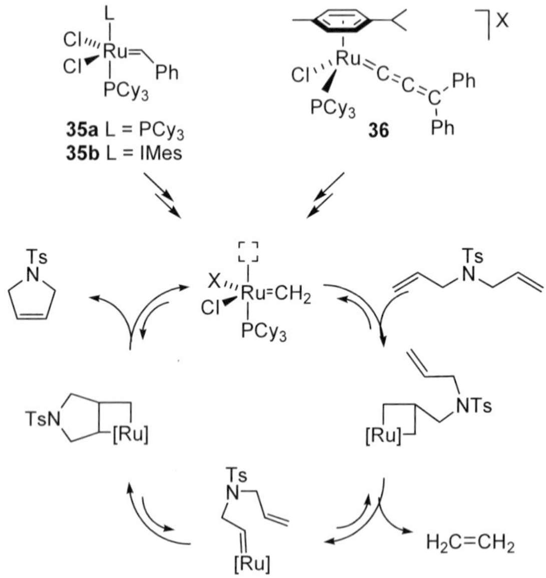

Scheme 7.1: Mechanism in ruthenium-catalysed ring-closing metathesis

7. 2 General Considerations

A summary of metathesis reactions in ionic liquids is presented in Table 7.1. From the data available it appears that these reactions can easily be performed in neat ionic liquids and increased reaction rates are sometimes observed relative to molecular solvents. Co-solvents are used in some cases, mainly to obtain a biphasic system and thereby facilitate product isolation. So far, only imidazolium-type solvents have been employed with variations in the alkyl substitution pattern and the nature of the anion. Apart from the common perfluorinated anions, chloroaluminates have also been used, specifically in cases where early transition metals were employed as catalysts.[22,24,25] When choosing the solvent one should remember that the imidazolium cation may alter or deactivate the catalyst via formation of stable carbene complexes. Indeed, it was shown that the presence of a relatively weak base, such as Cs_2CO_3, is sufficient to promote this step.[26-28] Alternatively, silver(I) oxide can be used as base and halide abstracting agent.[29] Several ruthenium carbene catalysts or catalyst precursors have been prepared in this way, though not tested with an ionic liquid as solvent, and Scheme 7.2 shows some recent examples. In some instances, the carbene complex is prepared *in situ* and directly used in catalysis. Other ruthenium sources than $[RuCl_2(p\text{-cymene})]_2$ have also been employed and these include $Ru(methallyl)_2(Ph_2P(CH_2)_nPPh_2)$, $RuCl_2(DMSO)_4$, $[RuCl_2(cod)]_n$ and $[RuCl_2(nbd)]_n$.[28,30]

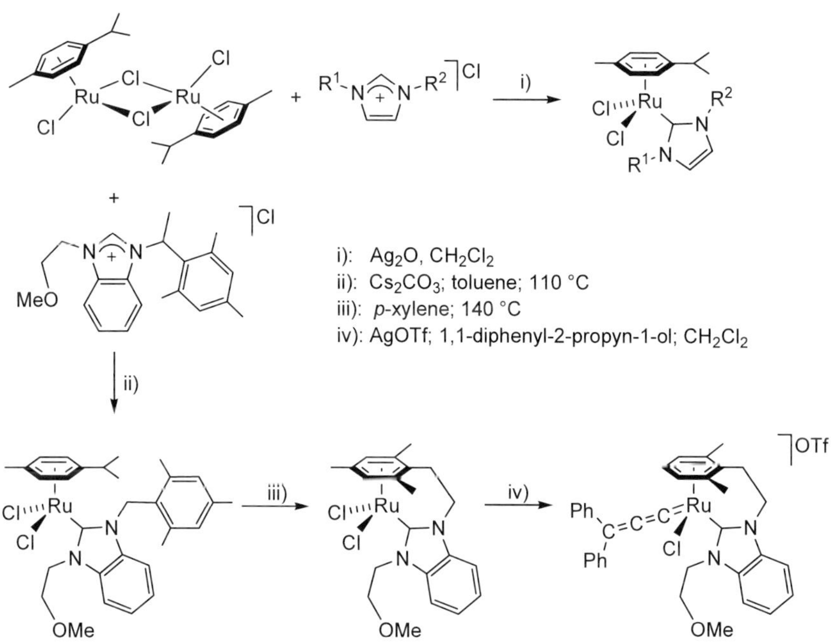

Scheme 7.2: Formation of ruthenium carbene complexes from imidazolium salts

Carbene formation can be suppressed by substituting the acidic proton in the 2-position by, for example, a methyl group, although this has the effect of increasing the melting point and the viscosity of the ionic liquid. Alternatively, a protic ionic liquid such as $[HC_4im][OTf]$ may be used.[31] Here, the presence of an acidic NH^+-functionality prevents formation of a carbene complex and further accelerates the formation of the catalytically active species.

The reaction products are usually obtained in good to excellent yield and in most cases can be separated by decantation or extraction of the reaction mixture with a suitable solvent. With neutral Grubbs-type ruthenium complexes, a significant amount of catalyst is lost during the product-isolation step as these are partially soluble in the organic phase and reaction rates tend to decrease drastically after the first run. To achieve good retention of the catalyst in the ionic liquid phase, the presence of charged moieties within the coordination sphere are therefore crucial. Retention in the ionic phase seems to be somewhat better when the charged allylidene complex **36** is employed. However, with this compound catalyst-stability is an issue and ionic liquid solutions containing **36** remain reasonably active for only about 36 hours at ambient temperature.[32] The nature and charge of the catalytically active species arising from allylidene complexes derived from **36** is not unambiguously resolved. Recent results from ring-closing metathesis with a η^6-polystyrene analogue of **36** suggest that complexation of the arene is not entirely lost during catalysis.[12] With complex **36**, the nature of the anion of both the ionic liquid and the catalyst are important factors influencing both rate and selectivity:[26,32,33] For ring-closing metathesis, best results are obtained with $[OTf]^-$ whereas in ring-opening polymerisation, $[PF_6]^-$ gives the highest reaction rates. It is highly unlikely that this is an effect of ion-pairing,[34,35] but rather a reflection of the fact that on the one hand anions such as $[BF_4]^-$ and $[PF_6]^-$ are susceptible towards hydrolysis[36] and on the other hand $[OTf]^-$ is a reasonably good ligand which can coordinate to the metal centre.

Table 7.1: Transition-metal catalysed metathesis reactions in ionic liquids

Catalyst	Ionic Liquid	Comments	Ref
		Metathesis with metals other than ruthenium	
$W(OAr)_2Cl_4$	$[C_4C_1im]Cl$-$AlCl_3$-Et_2AlCl_2	**Cross-metathesis** of linear olefins, e.g. conversion of 1-pentene to form ethylene and 4-octene; no reaction details given; system active for several runs	[22]
WCl_6 $Na[ReO_4]$	$[C_4C_1im][BF_4]$ $[C_4C_1im]Cl$-$AlCl_3$	**Cross-metathesis** of linear olefins, e.g. conversion of 1-hexene to 4-octene and 5-decene; rather low conversions (<15-30%); addition of 10-30% $SnBu_4$ significantly increases the yield and the selectivity for 4-octene (> 95%).	[25]
$Mo(OR')_2(NAr)(=CHR)$	$[C_2C_1im]Cl$-$AlCl_3$	**Ring-opening cross-metathesis** of norbornene with allyltrimethylsilylsilane; product extracted with hexane; conversion after one hour at RT ca. 80%, selectivity for the disilylated product ca. 75%.	[24]
		Ruthenium-catalysed metathesis reactions	
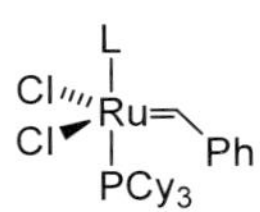 **35a** R = PCy_3 **35b** R = IMes	$[C_4C_1im][various]$ $[C_8C_1im][various]$	**RCM** of diallyltosylamide and other dienes; product extracted with diethyl ether - significant decrease in activity after second cycle; yield depends on nature of ionic liquid, best results with $[C_4C_1im][PF_6]$.	[37]
	$[C_4C_1im][BF_4]$	**RCM** of diallyltosylamide and related dienes; full conversion within seconds with microwave irradiation; no data on recycling given.	[38]
	$[C_4C_1im][PF_6]$	**RCM** of diallyltosylamide and related dienes in a CH_2Cl_2/ionic liquid solution; full conversion within seconds with microwave irradiation; no data on recycling provided.	[39]
	$[C_4C_1im][PF_6]$	**RCM** of polyallyl arenes at 80 °C with **2a**.	[40]
	$[C_2C_1im]Cl$-$AlCl_3$	**RCM** of N-carboxylmethyl-3-amino-1,7-octadiene; aqueous work-up.	[24]
	$[C_4C_1C_1im][PF_6]$	**ROMP** of norbornene at 40 °C in an ionic liquid-toluene 1:4 biphasic system; product obtained via decantation; with **35a** significant decrease in activity after first run, **35b** remains highly active for three cycles then decreases significantly.	[33]

Catalyst	Ionic Liquid	Comments	Ref
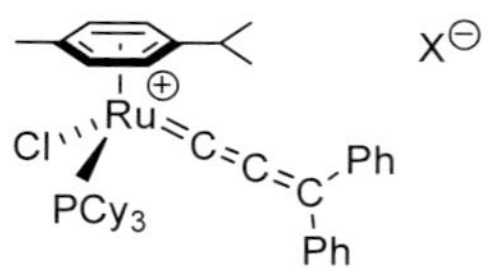	$[C_4C_1im][various]$	**RCM** of diallyltosylamide and other dienes; products extracted with toluene; yield and rate strongly depend on the anions (catalyst and ionic liquid), $[OTf]^-$ gives the best results; slow decomposition of the catalyst, recyclability limited to two cycles.	[32] [41]
	$[HC_4im][BF_4]$ $[HC_4im][OTf]$ $[C_4C_1im][BF_4]$	**RCM** of diallyltosylamide; product extracted with Et_2O; reaction proceeds significantly faster in the protic solvent based on $[HC_4im]^+$; reaction rate is anion-dependent, with $[OTf]^-$ giving the best results, no data on catalyst reuse given.	[31]
	$[C_4C_1im]Cl\text{-}AlCl_3$	**RCM** of 1,7-Octadiene; product extracted with methyl-*tert*-butyl ether, yield unchanged in second cycle (32%).	[24]
	$[C_4C_1im][BF_4]$ $[C_4C_1im][PF_6]$	**RCM** of diallyltosylamide and related dienes in a CH_2Cl_2 solution doped with ionic liquid with microwave irradiation; low conversion, probably due to catalyst decomposition.	[39]
	$[C_4C_1C_1im][PF_6]$	**ROMP** of norbornene in at 40 °C in an ionic liquid/toluene 1:4 biphasic system; product obtained via decantation; activity remains high for four cycles, then decreases significantly.	[33]
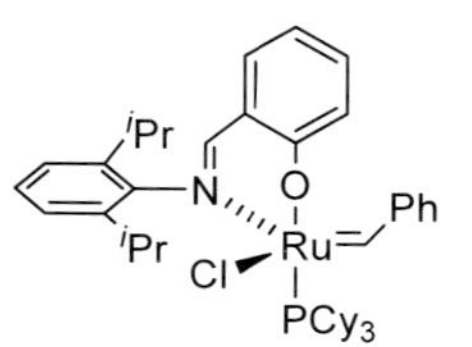	$[C_4C_1im][PF_6]$ $[C_2C_1im]Cl\text{-}AlCl_3$ $[C_4py]Cl\text{-}AlCl_3$	**RCM** of α,ω-dienes with various heteroatom-substituents in neat ionic liquid and ionic liquid-co-solvent mixtures; products isolated via decantation/extraction (pentane, hexane, CH_2Cl_2, toluene) or aqueous work-up; significant decrease of catalyst activity after the first run.	[24]
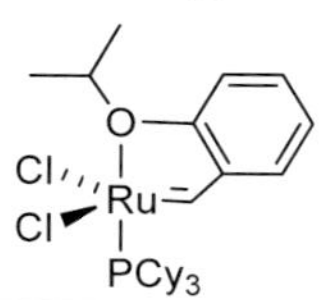	$[C_4C_1im][PF_6]$	**RCM** of diallyltosylamide and related dienes; product extraction with either Et_2O or toluene; subsequent reuse of catalyst results in drastic decrease of catalytic activity due to leaching.	[42] [43]

Catalyst	Ionic Liquid	Comments	Ref
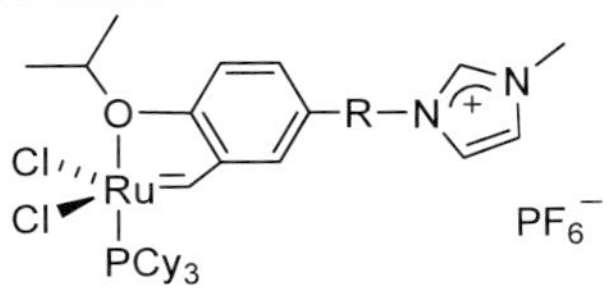 **38a** R = —$(CH_2)_3$— **38b** R = —O—$(CH_2)_4$—	$[C_4C_1im][PF_6]$	**RCM** of diallyltosylamide and other dienes with **38a** at 60 °C; high catalyst stability and very good conversion even after 10 cycles; catalyst is stable in the ionic liquid for months.	[42]
	$[C_4C_1im][PF_6]$	**RCM** of diallyltosylamide and other dienes with **38b** in an ionic liquid-CH_2Cl_2 1:9 mixture at 50 °C; high catalyst stability and very good conversion even after 10 cycles.	[43]
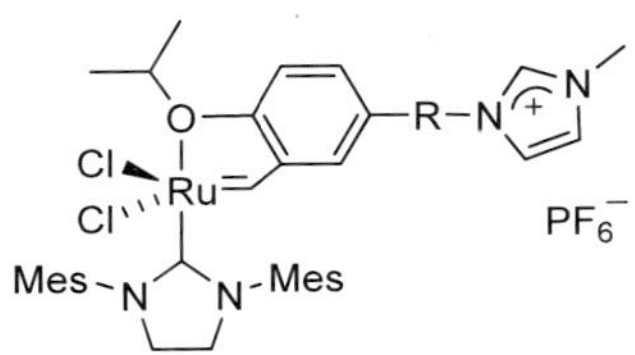 **39** R = —$(CH_2)_3$—	$[C_4C_1im][PF_6]$ $[C_4C_1im][Tf_2N]$	**RCM** of diallyltosylamide and other dienes; toluene as co-solvent; high recyclability at 25 °C; ruthenium leaching ca. one order of magnitude lower than previously reported; tetra-substituted dienes could not be cyclised.	[44]

7. 3 Task-Specific Catalysts

Consideration of the widely used second generation metathesis pre-catalyst **35b**, reveals four different sites where an ionic tag could be attached to improve catalyst retention in an ionic liquid phase: (i) the phosphine ligand, (ii) the *N*-heterocyclic carbene, (iii) the alkylidene and (iv) exchange of the chloride for a strongly electron withdrawing group like e.g. a functionalised perfluoro carboxylate. In principle, the lessons learnt from solid-supported metathesis catalysts[16,17] can be applied to the synthesis of novel ionic liquid phase metathesis catalysts, simply substituting the support for an appropriate charged moiety. Accordingly, two separate research groups have described almost identical solutions to enhance both catalyst lifetime and retention in ionic liquids.[42-44] Both adopted the (boomerang) concept, first introduced with complex **37** by Hoveyda and co-workers.[45] It is proposed that, after consumption of the substrate, the catalytically active species encounters the initially released 2-isopropoxystyrene to regenerate complex **37**, hence prolonging its lifetime, as shown in Scheme 7.3.

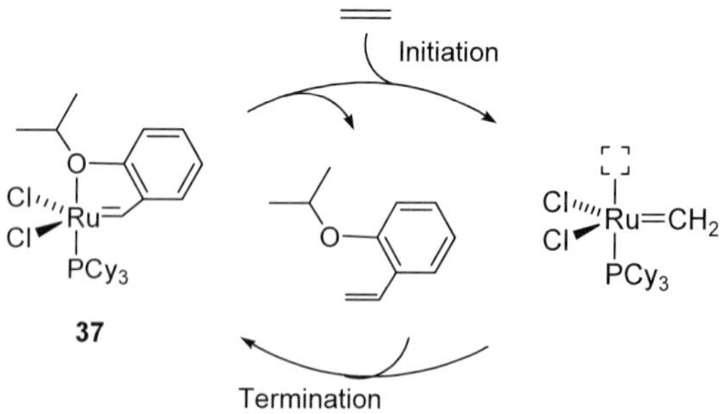

Scheme 7.3: Boomerang-concept to increase the lifetime a metathesis catalyst

Various polymeric and solid supports, such as polyethylene glycol (PEG), can be used to immobilise these catalysts.[46-48] Exchanging the polymer support on the styrene moiety for charged ionic liquid tags affords complexes **38** and **39,** which are retained to a significantly higher degree in the ionic liquid phase (Scheme 7.4).

The combined features of charge and catalyst stabilisation find their expression in markedly enhanced recyclability. Using catalyst loadings of 2.5 - 5 mol%, the activity decreased by only a few percent after 10 cycles and remained above 90% of the initial value. Furthermore, the catalyst solution could be kept for months without measurable loss in activity. By substituting the tricyclohexylphosphine ligand for a *N*-heterocyclic carbene, catalyst stability was further improved which finds its expression in very low residual ruthenium levels in the isolated product.[44] As these are only the first examples of metathesis catalysts specifically designed to operate in ionic liquids, the potential for further optimisation is probably considerable.

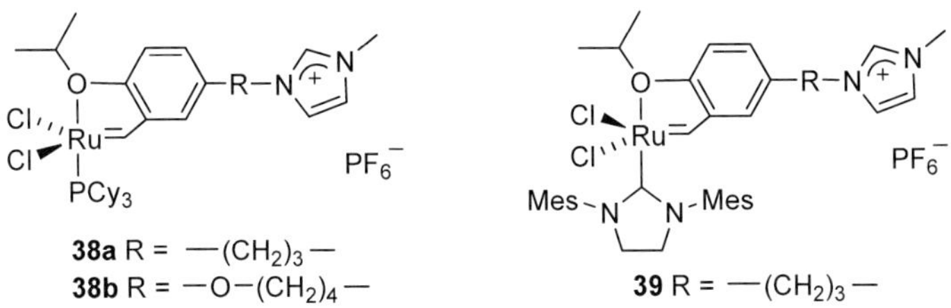

Scheme 7.4: Task-specific complexes for ring-closing metathesis

Many ionic liquids appear to be sufficiently transparent to microwave irradiation.[38,39] By using microwave dielectric heating, diallylaminotosylate was cyclised quantitatively in 15 seconds in $[C_4C_1im][BF_4]$ using complex **35b** as the catalyst. Conversions were considerably inferior with **36** in a mixture of CH_2Cl_2 and $[C_4C_1im][PF_6]$, which is attributed to catalyst degradation under the conditions employed. No attempts were made to reuse the catalyst solution, but it is conceivable that such short reaction times may increase the recyclability and thus the turnover number of a given catalyst.

7.4 Ring-Opening Polymerisation

By the end of 2004 ring-opening polymerisation in ionic liquids has been reported in only one paper, using complexes **35a**, **35b** or **36** as catalysts.[33] A biphasic system comprised of $[C_4C_1C_1im][PF_6]$ and toluene in a 1:4 ratio was used to minimise catalyst loss during product isolation and the polymer could be obtained almost quantitatively by simple decantation. The activity of ionic liquid solutions of **35a** drop markedly after the first cycle and become inactive in the third cycle, whereas **35b** and **36** could be used for three and five cycles, respectively, with very good conversions, as shown in Scheme 7.5.

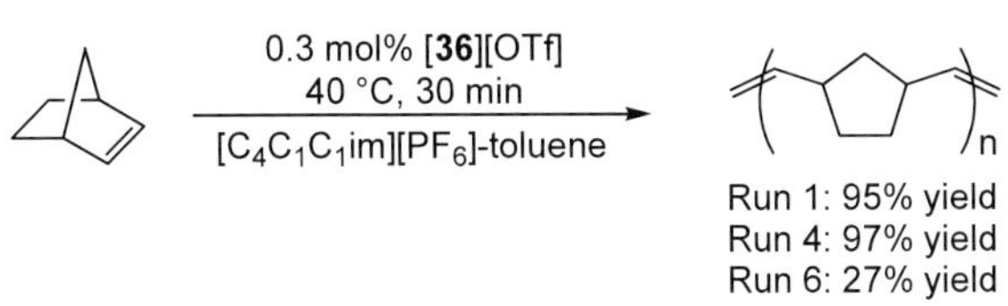

Scheme 7.5: Ring-opening polymerisation metathesis of norbornene

The molecular weight of the polymers obtained increased with the number of catalytic runs, ranging between 79 and 303 kDa, indicating a gradual loss of active catalyst able to initiate a polymer chain. The polydispersity as well as the *cis*/*trans*-ratio remained almost constant during the recycling.

7. 5 Concluding Remarks

As in many other areas of catalysis in ionic liquids, the research of the past years in ionic liquid supported olefin metathesis has been dominated by demonstrating the general feasibility in this reaction medium. With the development of the task-specific complexes new prospects have been opened and the improved recyclability of these compounds relative to the common metathesis catalysts is impressive. Apart from more active and stable catalysts, the development of enantioselective catalysts and of continuous processes are likely to be the next goals.

It would appear that some more fundamental improvements are required to make metathesis in ionic liquids sufficiently attractive: rather than extracting the product, which commonly requires a large amount of conventional solvent, biphasic systems that allow decantation of the product ought to be envisaged. As metathesis reactions are often performed at elevated temperatures, ionic liquids with higher melting points could also be considered to optimise and facilitate product separation.

7.6 References

[1] Nicolaou, K. C.; King, N. P.; He, Y. In *Alkene Metathesis in Organic Synthesis*; Fürstner, A., Ed.; Springer: Berlin Heidelberg, 1998, pp 73-104.
[2] Fürstner, A. *Angew. Chem. Int. Ed.* **2000**, *39*, 3013-3043.
[3] Fürstner, A., Ed. *Alkene Metathesis in Organic Synthesis*; Springer: Berlin Heidelberg, 1998; Vol. 1.
[4] Buchmeiser, M. R. *Chem. Rev.* **2000**, *100*, 1565-1604.
[5] Trnka, T. M.; Grubbs, R. H. *Acc. Chem. Res.* **2001**, *34*, 18-29.
[6] Maynard, H. D.; Grubbs, R. H. *Tetrahedron Lett.* **1999**, *40*, 4137-4140.
[7] Cho, J.-Y.; Kim, B. M. *Org. Lett.* **2003**, *5*, 531-533.
[8] Buchmeiser, M. R.; Lubbad, S.; Mayr, M.; Wurst, K. *Inorg. Chim. Acta* **2003**, *345*, 145-153.
[9] Krause, J. O.; Lubbad, S.; Nuyken, O.; Buchmeiser, M. R. *Adv. Synth. Catal.* **2003**, *345*, 996-1004.
[10] Varray, S.; Lazaro, R.; Martinez, J.; Lamaty, F. *Organometallics* **2003**, *22*, 2426-2435.

[11] Kobayashi, S.; Akiyama, R. *Chem. Commun.* **2003**, 449-460.

[12] Akiyama, R.; Kobayashi, S. *Angew. Chem. Int. Ed.* **2002**, *41*, 2602-2604.

[13] Connon, S. J.; Blechert, S. *Bioorg. Med. Chem. Lett.* **2002**, *12*, 1873-1876.

[14] Hultzsch, K. C.; Jernelius, J. A.; Hoveyda, A. H.; Schrock, R. R. *Angew. Chem. Int. Ed.* **2002**, *41*, 589-593.

[15] Chabanas, M.; Coperet, C.; Basset, J. M. *Chem. Eur. J.* **2003**, *9*, 971-975.

[16] Buchmeiser, M. R. *New J. Chem.* **2004**, *28*, 549-557.

[17] Kingsbury, J. S.; Hoveyda, A. H. In *Polymeric Materials in Organic Synthesis and Catalysis*; Buchmeiser, M. R., Ed.; Wiley-VCH: Weinheim, **2003**, pp 467.

[18] Fürstner, A.; Ackermann, L.; Beck, K.; Hori, H.; Koch, D.; Langemann, K.; Liebl, M.; Six, C.; Leitner, W. **2001**, *123*, 9000-9006.

[19] Leitner, W. *C. R. Acad. Sci. Ser., II C* **2000**, *3*, 595-600.

[20] Grubbs, R. H.; Lynn, D. M. In *Aqueous-Phase Organometallic Chemistry*; Cornils, B., Herrmann, W. A., Eds.; Wiley-VCH: Weinheim, **1998**.

[21] Lynn, D. M.; Mohr, B.; Grubbs, R. H.; Henling, L. M.; Day, M. W. *J. Am. Chem. Soc.* **2000**, *122*, 6601-6609.

[22] Chauvin, Y.; Olivierbourbigou, H. *Chemtech* **1995**, *25*, 26-30.

[23] Hérisson, J.-L.; Chauvin, Y. *Makromol. Chem.* **1970**, *141*, 161.

[24] Gürtler, C.; Jautelat, M. European Patent EP 1035093A2, **2000**.

[25] Vasnev, A. V.; Greish, A. A.; Kustov, L. M. *Mendeleev Commun.* **2004**, 59-61.

[26] Sémeril, D.; Bruneau, C.; Dixneuf, P. H. *Adv. Synth. Catal.* **2002**, *344*, 585-595.

[27] Cetinkaya, B.; Demir, S.; Ozdemir, I.; Toupet, L.; Semeril, D.; Bruneau, C.; Dixneuf, P. H. *Chem. Eur. J.* **2003**, *9*, 2323-2330.

[28] Castarlenas, R.; Fischmeister, C.; Dixneuf, P. H. *New J. Chem.* **2003**, *27*, 215-217.

[29] Csabai, P.; Joó, F. *Organometallics* **2004**, *23*, 5640-5643.

[30] Castarlenas, R.; Alaoui-Abdallaoui, I.; Sémeril, D.; Mernari, B.; Guesmi, S.; Dixneuf, P. H. *New J. Chem.* **2003**, *27*, 6-8.

[31] Picquet, M.; Tkatchenko, I.; Tommasi, I.; Wasserscheid, P.; Zimmermann, J. *Adv. Synth. Catal.* **2003**, *345*, 959-962.

[32] Sémeril, D.; Olivier-Bourbigou, H.; Bruneau, C.; Dixneuf, P. H. *Chem. Commun.* **2002**, 146-147.

[33] Csihony, S.; Fischmeister, C.; Bruneau, C.; Horvath, I. T.; Dixneuf, P. H. *New J. Chem.* **2002**, *26*, 1667-1670.

[34] Valentini, M.; Rüegger, H.; Pregosin, P. S. *Helv. Chim. Acta* **2001**, *84*, 2833-2853.

[35] Drago, D.; Pregosin, P. S.; Pfaltz, A. *Chem. Commun.* **2002**, 286-287.

[36] Swatloski, R. P.; Holbrey, J. D.; Rogers, R. D. *Chem. Commun.* **2003**, 361-363.

[37] Buijsman, R. C.; van Vuuren, E.; Sterrenburg, J. G. *Org. Lett.* **2001**, *3*, 3785-3787.

[38] Mayo, K. G.; Nearhoof, E. H.; Kiddle, J. J. *Org. Lett.* **2002**, *4*, 1567-1570.

[39] Garbacia, S.; Desai, B.; Lavastre, O.; Kappe, O. C. *J. Org. Chem.* **2003**, *68*, 9136-9139.

[40] Martinez, V.; Blais, J. C.; Bravic, G.; Astruc, D. *Organometallics* **2004**, *23*, 861-874.

[41] Castarlenas, R.; Fischmeister, C.; Bruneau, C.; Dixneuf, P. H. *J. Mol. Catal. A* **2004**, *213*, 31-37.

[42] Audic, N.; Clavier, H.; Mauduit, M.; Guillemin, J. C. *J. Am. Chem. Soc.* **2003**, *125*, 9248-9249.

[43] Yao, Q.; Zhang, Y. L. *Angew. Chem. Int. Ed.* **2003**, *42*, 3395-3398.

[44] Clavier, H.; Audic, N.; Mauduit, M.; Guillemin, J. C. *Chem. Commun.* **2004**, 2282-2283.

[45] Kingsbury, J. S.; Harrity, J. P. A.; Bonitatebus, P. J.; Hoveyda, A. H. *J. Am. Chem. Soc.* **1999**, *121*, 791-799.

[46] Garber, S. B.; Kingsbury, J. S.; Gray, B. L.; Hoveyda, A. H. *J. Am. Chem. Soc.* **2000**, *122*, 8168-8179.

[47] Kingsbury, J. S.; Garber, S. B.; Giftos, J. M.; Gray, B. L.; Okamoto, M. M.; Farrer, R. A.; Fourkas, J. T.; Hoveyda, A. H. *Angew. Chem. Int. Ed.* **2001**, *40*, 4251-4256.

[48] Yao, Q. *Angew. Chem. Int. Ed.* **2000**, *39*, 3896-3898.

Chapter 8

DIMERISATION - POLYMERISATION

8.1 Dimerisation and Oligomerisation

8.1.1 Introduction

Dimerisation of olefins is a major industrial process, and is carried out on a multi million ton scale annually.[1] One of the most important methods is represented by the Shell Higher Olefin Process (SHOP), which can even be run under biphasic conditions. In the oligomerisation of ethylene, the catalyst is generated *in situ* in 1,4-butanediol from a nickel salt, $Na[BH_4]$ and a chelating ligand. The olefins formed in the reaction are immiscible with the polar solvent and are isolated by phase separation and subsequent distillation.[2]

The oligomerisation of propenes and butenes is frequently catalysed by trialkylaluminium or trialkylaluminium/nickel species, the latter being applied in the IFP-Dimersol® process. Here, the reaction is commonly performed in neat substrate under mild conditions. With $[NiBr(\eta^3\text{-}C_3H_5)(PCy_3)]\text{-}AlEtCl_2$ as catalyst, turnover frequencies as high as 800,000 $mol{\cdot}mol^{-1}{\cdot}h^{-1}$ have been observed in the oligomerisation of propene.[1]

Selectivity is of major concern in dimerisation reactions. Even if no oligomers are formed, the product can comprise a mixture of vinylidenes and linear or branched olefin dimers (of which various isomers can be present). The possible pathways in the dimerisation of terminal olefins are illustrated in Scheme 8.1, not taking oligomerisation products into account. As solubility properties of ionic liquids can be tuned over a wide range these solvents should facilitate the synthesis of well-defined oligomers with a narrow weight distribution. Once the oligomer chain exceeds a certain length, it no longer dissolves in the ionic liquid and the chain growth should come to a halt. Such beneficial effects due to the formation of a separate product phase are demonstrated in some of the following examples.

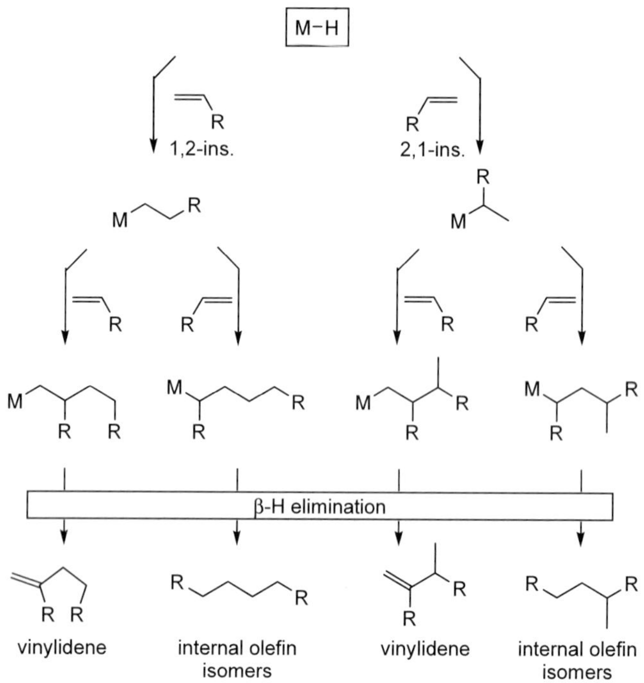

Scheme 8.1: Reaction pathways in olefin dimerisation, excluding higher oligomerisation products

Summaries on dimerisation and oligomerisation reactions in ionic liquids have been published previously.[3,4] Predominantly complexes of either palladium or nickel have been employed in ionic liquids for dimerisation/oligomerisation reactions, but there are also examples where iron,[5] tungsten[6] and rhodium[7] have been used. Apart from metal catalysed dimerisation/oligomerisation reactions in ionic liquids, examples of electrochemical dimerisation have been reported, which include arylhalides[8] and 3-(4-fluorophenyl) thiophene[9] as substrates.

8.1.2 Nickel-Catalysed Reactions

The first example for biphasic oligomerisation of olefins in ionic liquids was published in 1990, reporting on the dimerisation of propene by nickel(II) catalysts in chloroaluminate ionic liquids of the general formula [cation]Cl_x-$(AlCl_3)_y$ with either $[C_4C_1im]^+$, $[C_4py]^+$ or $[(C_4)_4P]^+$ as cation.[10] It was found that in basic ionic liquids, $y < 0.5$, no catalysis took place. Excess chloride, which is present in such basic chloroaluminates, poisons the catalyst and it was shown that nickel compounds of the type $NiCl_2(PR_3)_2$

were transformed into inactive, paramagnetic complexes, as described in Equation 1.[11]

$$[NiCl_2(PR_3)_2] + 2\,Cl^- \rightleftharpoons [NiCl_4]^{2-} + 2\,PR_3 \quad (1)$$

In acidic ionic liquids, where the molar fraction of $AlCl_3$ is greater than 0.5, unwanted formation of trimers and higher oligomers was observed during catalysis. However, such acid-induced side reactions are successfully suppressed if small amounts of ethylaluminium dichloride, which acts as a proton scavenger, are present. Thus, when the pre-catalyst $[Ni(MeCN)_6][BF_4]_2$ was immobilised in a mixture of $[C_4C_1im]Cl$-$AlCl_3$-$AlEtCl_2$ (0.41:0.50:0.04), C_8-products with a selectivity of 92-98% were obtained from the dimerisation of *n*-butenes.[12] While addition of PPh_3, PCy_3 or PBu_3 did not affect the reaction rates, turnover frequencies were almost one order of magnitude higher in the presence of PCy_3CS_2. The composition of the dimer products was independent from the phosphine additive used with an observed branch-index (BI) of 1.3, which is similar to that obtained in the homogeneous Dimersol® system. Higher TOFs at comparable selectivity were further achieved when $[Ni(MeCN)_6][AlCl_4]_2$ was employed as catalyst instead of the analogous tetrafluoroborate salt.[13] The product was isolated by decantation and in that manner the catalyst was reused for six runs without change in activity or product composition. It was proposed that the difference in activity between the two nickel salts was due to decomposition of the $[BF_4]^-$ anion upon contact with alkyl aluminium chloride, affording a species that ultimately gave rise to a less active Ni-compound.

The above protocol for the dimerisation of butenes has also been tested on a pilot-plant scale and the reaction run continuously for 5,500 hours, after which it was stopped deliberately. By immobilising the catalyst in the chloroaluminate ionic liquid, nickel consumption was lowered by a factor of 10 relative to the process in neat substrate. The process is meanwhile available for licensing from IFP under the name of Difasol® but has not yet been commercialised. The technical implications of the process are described in some detail elsewhere.[3,14]

Square planar nickel complexes bearing *O,O'*-chelating ligands are known to afford the linear dimerisation product with good selectivity.[15] In toluene the dimerisation of 1-butene with the Ni(hfacac) complex **40** (shown in Scheme 8.2) yields linear octene with 75% selectivity. Yet, high ligand cost and low activity, 500 $mol{\cdot}mol^{-1}{\cdot}h^{-1}$ at 90°C, together with difficult catalyst recovery are factors that have so far prevented the industrial application of this otherwise useful complex. When this catalyst is immobilised in a slightly acidic chloroaluminate melt buffered with small amounts of weak organic bases, markedly higher reaction rates are obtained with only a slightly inferior selectivity.[16,17] For example, with $[C_4C_1py]Cl$-$AlCl_3$-chinoline (0.43:0.53:0.04) as solvent system, linear octenes were formed with 64% selectivity and a turnover frequency of 1,240 $mol \cdot mol^{-1} \cdot h^{-1}$ at 25°C.

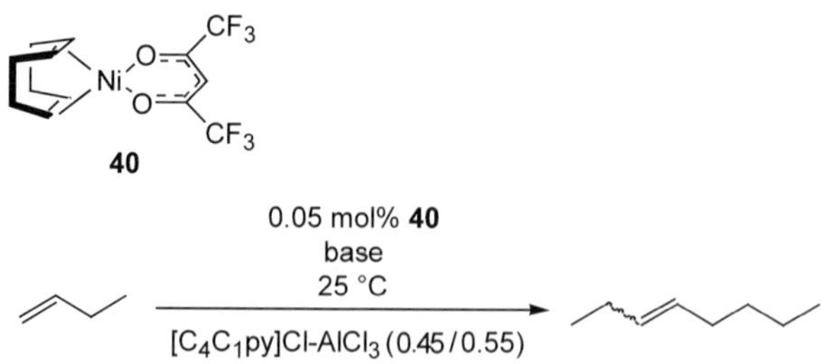

	TOF	C_8-Dimer [%]	Linear [%]
Pyrrole	1,350	86	56
Chinoline	1,240	98	64
Pyridine	550	78	33
1,6-Lutidine	2,480	55	68

Scheme 8.2: Dimerisation of 1-butene with Ni(hfacac)(cod)

The role of the base is to trap free acidic species generated during the reaction and thereby to suppresses cationic side reactions, as outlined in Equation 2. The relatively high selectivity of complex **40** for the formation of linear dienes is thus maintained in the ionic liquid.

$$\begin{aligned} Al_2Cl_7^- + \text{Base} &\rightleftharpoons \text{Base-}AlCl_3 + AlCl_4^- \\ \text{Base-}AlCl_3 + HCl &\rightleftharpoons \text{HBase}^+ + 2AlCl_4^- \end{aligned} \qquad (2)$$

Turnover frequencies could be further increased (reaction rates as high as 7,500 $mol \cdot mol^{-1} \cdot h^{-1}$) if LiCl was added instead of an organic base, however at a pronounced cost to the selectivity. While a temperature of 50°C is required in toluene to activate the catalyst, complex **40** exhibits activity already at –10°C in the ionic liquid. This indicates that the *in situ* generation of the catalyst, which is believed to require the formation of a Ni-hydride complex, proceeds more efficiently in the ionic liquid. On the other hand, the use of aluminiumalkyles as the proton scavenger led to poor results and the catalyst decomposed rapidly at ambient temperature. The catalyst stability was sufficient at low temperature, –10°C, but the linear product was formed with only 12% selectivity under these conditions. The biphasic nature of the system allows for easy product separation and catalyst recycling. Accordingly, the performance was also tested in a continuous mode and catalytic activity was maintained for at least three hours.[17] After that time, an overall turnover number greater than 18,000 was realised. Linear selectivity decreased somewhat with time, which might be indicative of slow abstraction of the hfacac-ligand from the nickel catalyst.

Several (bis)carbene nickel complexes were tested as catalysts for the dimerisation of propene or 1-butene (Scheme 8.3).[18] While the activity of these complexes was very poor in the dimerisation of 1-butene when toluene was used as solvent, turnover frequencies as high as 7,000 mol·mol^{-1}·h^{-1} were observed at ambient temperature with [C_4C_1im]Cl-$AlCl_3$-*N*-methylpyrrole (0.45:0.55:0.10) as solvent. With propene as substrate, TOFs of 75,000 mol·mol^{-1}·h^{-1} were achieved. Compared to $NiCl_2(PCy_3)_2$, the activity of the carbene complexes is considerably higher, but selectivity towards the desired, highly branched propene dimer is low.

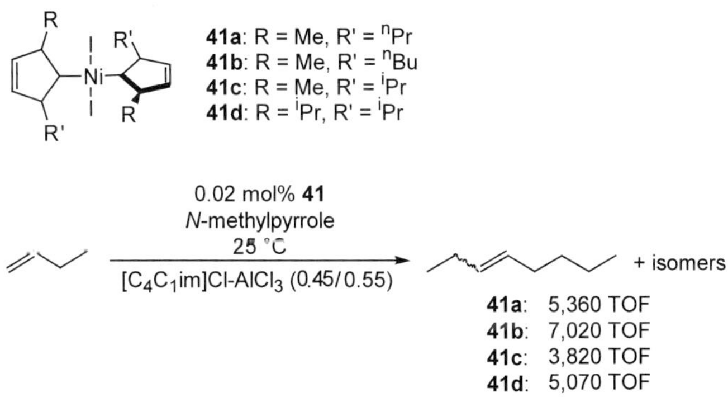

Scheme 8.3: Dimerisation with Nickel bis-carbene complexes

The oligomerisation of ethylene has been investigated in a series of hydrophobic ionic liquids with the nickel complex **42** as catalyst (Scheme 8.4).[19,20] Use of high quality ionic liquids, *i.e.* those in which chloride and water impurities are absent, was mandatory for good activity. With [C_4C_1im][PF_6] as solvent and under an ethylene pressure of 50 bar, turnover frequencies as high as 12,000 mol·mol^{-1}·h^{-1} were observed. The product formed a separate phase, which allowed for facile recycling of the catalyst, although accompanied by a slight decrease in catalyst activity. When 1,4-butanediol was used as solvent instead of the ionic liquid, hardly any ethylene conversion was detected. Ionic liquids with longer alkyl chains, C_6-C_{10}, also led to markedly lower reaction rates, possibly a consequence of their greater lipophilicity. Relative to dichloromethane, narrower oligomer distributions were observed under biphasic conditions.

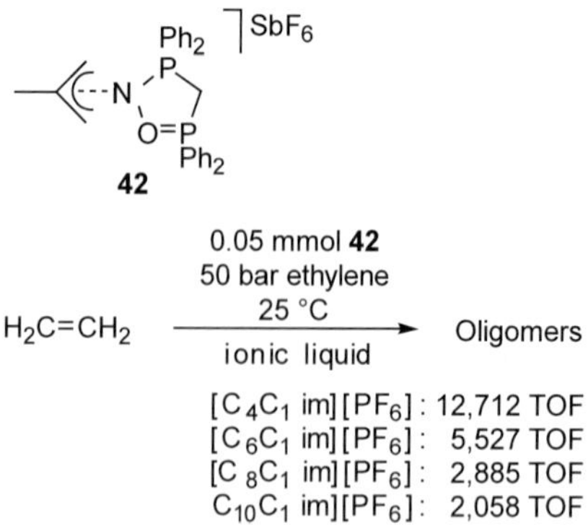

Scheme 8.4: Ethylene oligomerisation in different ionic liquids

8.1.3 Palladium-Catalysed Reactions

The dimerisation of methyl acrylate is an important reaction since the product, dimethyl dihydromuconate, represents a useful precursor for speciality chemicals such as cyclopentenones, as well as for the production of nylon-6,6. The synthetic challenge is to avoid formation of head-to-tail dimerisation products. The tail-to-tail dimerisation of methyl acrylate with palladium catalysts has been evaluated in protic and common imidazolium ionic liquids.[7,21-23]

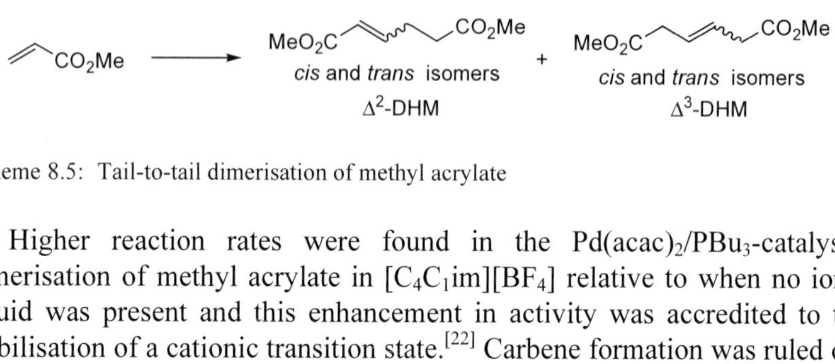

Scheme 8.5: Tail-to-tail dimerisation of methyl acrylate

Higher reaction rates were found in the $Pd(acac)_2/PBu_3$-catalysed dimerisation of methyl acrylate in $[C_4C_1im][BF_4]$ relative to when no ionic liquid was present and this enhancement in activity was accredited to the stabilisation of a cationic transition state.[22] Carbene formation was ruled out by testing a separately prepared palladium-carbene complex, which showed no catalytic activity. At *ca.* 80% conversion, the reaction is quenched, indicating that the product inhibits the reaction. Accordingly, the reaction was run under biphasic conditions with toluene as a co-solvent to induce the formation of two phases in the otherwise monophasic reaction. While the ligand PBu_3 is readily extracted from the ionic liquid, activity remains essentially stable for at least four runs when an ammoniumphosphine ligand, 1-dibutylphosphino-2-(dimethylamino)ethane, is used instead. Not only can this ligand stabilise the active catalytic species by means of chelation, it can also reoxidise the Pd(0) species and assist in the opening of the metallacycle via σ-bond metathesis involving the acidic N–H bond, as shown in Scheme 8.6. The biphasic system was also tested in a continuous-flow reactor and after 50 hours a total turnover number above 4,000 was achieved.

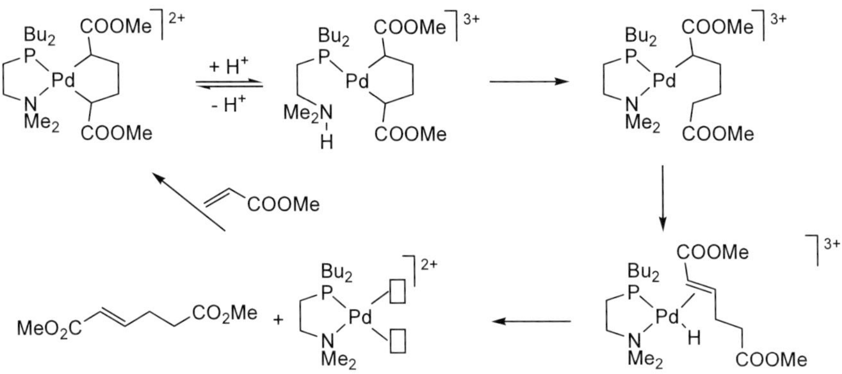

Scheme 8.6: Catalyst stabilisation with a chelating P-N ligand

An acceleration of the reaction rate in the palladium-catalysed dimerisation of methyl acrylate was observed with a protic ionic liquid, $[HC_4im][BF_4]$, affording a turnover frequency of 220 mol·mol^{-1}·h^{-1}, relative to 72-100 mol·mol^{-1}·h^{-1} in $[C_4C_1im][BF_4]$.[7] The reason for the improved activity is probably twofold. First, as this liquid is commonly made from the reaction of *N*-butylimidazole with HBF_4, no halide impurities, which were found to inhibit the reaction, are present. Second, $Pd(acac)_2$ was found to react with $[HC_4im][BF_4]$ to yield a mixture of $[Pd(acac)(C_4im)_2][BF_4]$ and $[Pd(C_4im)_4][BF_4]_2$. Thus, the ionic liquid acts as proton reservoir for the protonation of the acetyl acetonate ligand in $Pd(acac)_2$, facilitating the generation of the catalytically active species. However, if $[HC_1im][BF_4]$ is used as solvent instead, formation of a biphasic system takes place, which, under the same conditions, led to lower turnover frequencies, 25–30 mol·mol^{-1}·h^{-1}, probably due to mass-transfer limitations.

Biphasic reaction conditions, and thereby easy isolation of the product, are also obtained with a *sc*CO_2-ionic liquid mixture.[21] Methyl acrylate is highly soluble in CO_2 over a wide density range while a density higher than ca. 0.4 g·cm^{-3} is required to dissolve the resulting product dimethyl-dihydromuconate. Therefore reactions were run at 80°C and at a CO_2 pressure greater than 150 bar, allowing the product to be extracted into the CO_2-rich phase. Under these conditions the selectivity, TOF and TON values were comparable to those achieved under monophasic conditions.

The linear dimerisation of butadiene with palladium(II) catalyst precursors has been investigated in $[C_4C_1im]^+$ with a variety of different anions.[24] Observed turnover frequencies, which range from 37-49 mol·mol^{-1}·h^{-1}, are affected only slightly by the nature of the ionic liquid or catalyst precursor. Best activities were obtained with four equivalents of triphenylphosphine per palladium at a reaction temperature of 70°C. Contrary to the reaction in THF, no formation of metallic palladium was observed and reuse of the catalyst solution was possible. Pressurising the reaction mixture with 5-10 bar of carbon dioxide led to a decrease in reaction rates, which was explained by decreased substrate solubility in the CO_2-expanded ionic liquid.

Dimerisation, rather than the expected Heck coupling, took place in the reaction between styrene and arenediazonium salts in $[C_4C_1im][PF_6]$ to afford 1,3-diphenyl-1-butene.[25] Only catalytic amounts of the diazonium salt were required to trigger the reaction and their role in the reaction remain so far unclear.

The hydrodimerisation of butadiene was achieved in a phosphine-free system affording the telomer octa-2,7-dien-1-ol with a selectivity of 84-95%, as shown in Scheme 8.7.[26] At ambient temperature the reaction mixture, comprising of substrate, the palladium catalyst, water and $[C_4C_1im][BF_4]$, forms a homogeneous phase, but the products are insoluble in the ionic liquid-water solution below 5°C and could be separated by decantation. Higher TOFs were observed in CO_2-expanded ionic liquids, but at the cost of decreased selectivity. Whereas formation of metallic palladium took place in the ionic liquids with catalyst precursors such as $[(\eta^3\text{-}C_4H_7)Pd\text{-}\mu\text{-}Cl]_2$ or $Pd(OAc)_2$, hardly any decomposition was observed when $[C_4C_1im]_2[PdCl_4]$ was used as the catalyst.

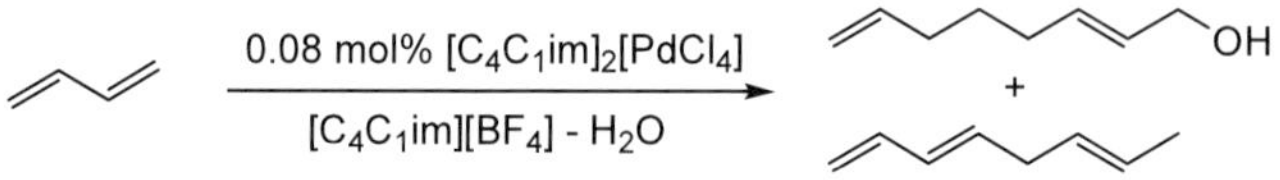

Scheme 8.7: Hydrodimerisation of butadiene

The number of alkyl substituents attached to the imidazolium cation was found to be of great importance in the telomerisation of butadiene and methanol with a palladium/phosphine catalyst system, see Scheme 8.8.[27] In the presence of imidazolium ionic liquids with an acidic proton in the 2-position, rapid deactivation of the catalyst took place and it was proposed that formation of stable and inactive palladium-carbene species occurred. In contrast, both a pyridinium-based ionic liquid as well as imidazolium ionic liquids bearing a methyl substituent in the 2-position led to active systems.

However, with the former liquid, formation of palladium black was observed, together with increased yields of linear dimerisation and oligomerisation products. It is well known that alkyl-pyridinium species can transfer an alkyl group to a phosphine ligand,[28,29] giving rise to phosphonium salts which in turn are no longer capable of stabilising the Pd(0) species. Thus, $[C_4C_1C_1im]^+$ with either $[BF_4]^-$ or $[Tf_2N]^-$ as the anion proved to be the better choice of solvent and the desired telomeres were formed with around 83% selectivity. After the reaction was complete, both phases are yellow indicating that immobilisation of the palladium catalyst is not yet very efficient. When the ionic phosphine TPPMS was used as ligand instead, a single phase was observed, suggesting that the sulfonated ligand also acts as surfactant. Accordingly, heptane was added as co-solvent to induce phase separation, but some 1-4% palladium was nevertheless present in the organic phase. Despite the loss of palladium catalyst recycling was possible several times with only a slight deterioration of activity.

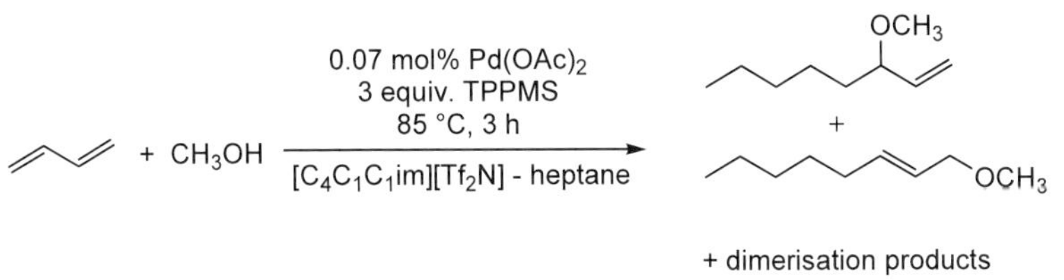

Scheme 8.8: Palladium-catalysed telomerisation of butadiene with methanol

8.1.4 Dimerisation Reactions with other Metal-Catalysts

Ethylene was dimerised and oligomerised in acidic $[C_4C_1im]Cl\text{-}AlCl_3$ with the tungsten catalyst $WCl_2(=NPh)(PMe_3)_3$, affording a mixture of butenes and hexenes.[6] The products formed a separate phase and could be isolated by simple decantation, allowing reuse of the ionic liquid phase. The same reaction also proceeded with $WCl_4(=NAr)$ as catalyst, but addition of $AlEtCl_2$ was necessary to activate the complex. Furthermore, the organic phase was coloured after the reaction, indicating that some tungsten species had been extracted.

Cyclodimerisation of 1,3-butadiene was performed with $[Fe(NO)_2Cl]_2$ as catalyst precursor, together with a reducing agent such as zinc or $AlEt_2Cl$ in $[C_4C_1im][BF_4]$ or $[C_4C_1im][PF_6]$, as shown in Scheme 8.9.[5] At 10°C, higher turnover frequencies were observed in the hydrophilic ionic liquid $[C_4C_1im][BF_4]$ while at 50°C the activity was greater in the hydrophobic ionic liquid. The addition of phosphine ligands led to lower reaction rates. Highest activities were obtained with ca. 0.5 mol% zinc at 50°C, affording turnover frequencies above 1,400 mol·mol^{-1}·h^{-1} with 100% selectivity towards 4-vinyl-1-cyclohexene. With isoprene as substrate, four isomeric products were observed. After the reaction conditions were tuned by addition of zinc and phosphine ligands, 1,4-dimethyl-4-vinylcyclohexene was obtained with up to 70% selectivity, although at relatively low reaction rates.

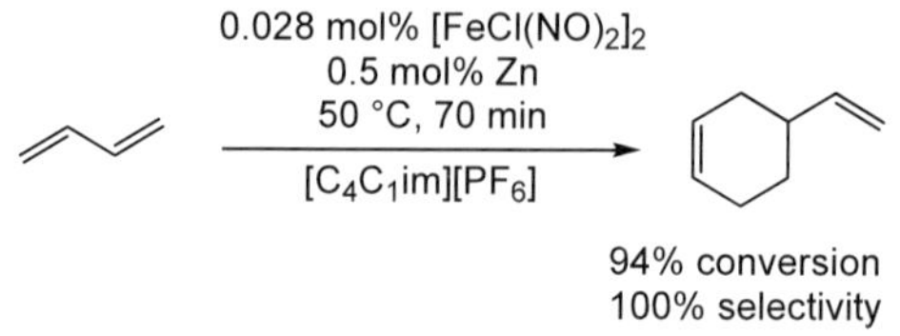

Scheme 8.9: Cyclodimerisation of 1,4-butadiene

Tail-to-tail dimerisation of methyl acrylate has been investigated with a rhodium catalyst in protic and neutral imidazolium ionic liquids.[7] While acidic ionic liquids showed some advantage with palladium catalysts (*vide supra*), reaction rates with $Rh(Cp^*)(C_2H_4)_2$ as catalyst were much lower in $[HC_4im][BF_4]$ and $[HC_1im][BF_4]$ relative to those in $[C_4C_1im][BF_4]$. It was suggested that this was due to coordination of imidazole to the rhodium centre.

In the classical oxidative alkyne-dimerisation reaction large quantities of solvent are often required which makes alternative procedures attractive. With 20 mol% CuCl and 20 mol% TMEDA, the dimerisation product of phenylacetylene, *viz.* 1,4-diphenyl-1,3-butadiene, was obtained at ambient temperature in $[C_4C_1im][PF_6]$ in 95% yield.[32] Products were isolated by extraction of the ionic liquid phase with toluene and reuse of the catalyst was possible, but addition of further catalyst was necessary every two runs to keep reaction rates more or less constant.

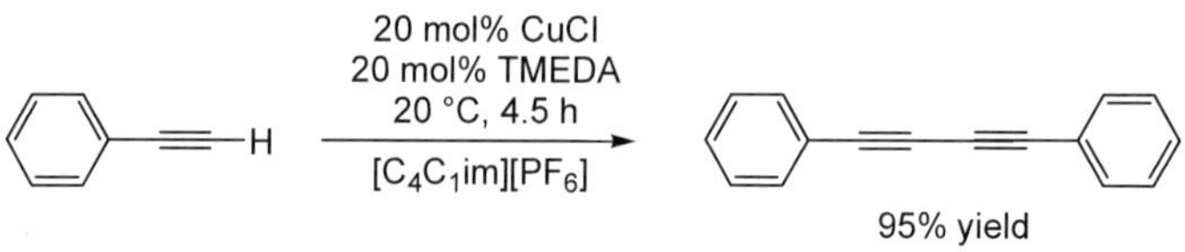

Scheme 8.10: Oxidative alkyne dimerisation

8. 2 Polymerisation

8.2.1 Introduction

A summary of polymerisation reactions in ionic liquids, which also covers non-metal catalysed reactions, is available[33] and ring-opening polymerisation (ROMP) was described in Chapter 6.

Polymerisation reactions in ionic liquids have so far focused on processes that do not involve a transition metal catalyst. Examples include acid-catalysed,[34-36] free-radical,[37-50] electrochemical[9,51-55] and laser[56] induced polymerisation reactions and a review is available on the topic.[57]

As polymers are not volatile, separation of the ionic liquid from the polymer will almost inevitably involve extraction and/or washing with conventional organic solvent. The proposed environmental advantage of using ionic liquids will therefore be at least partially lost. On the other hand the often low solubility of polymerisation catalysts in common reaction media, as well as arduous separation of catalyst residue from the resulting polymer may provide a justification for employing ionic liquids as solvents in polymerisation reactions. The prospect of controlling the molecular weight of the polymer using task-specific ionic liquids represents another motivation for studying polymerisation reactions in ionic liquids. As they can be easily tuned with respect to their solubility properties ionic liquids appear to be promising solvents for the synthesis of block co-polymers. From the few examples available it seems that, compared to polymerisation in neat substrate, higher rates of propagation with lower rates for termination are observed in ionic liquids.

8.2.2 Atom Transfer Radical Polymerisation (ATRP)

The general mechanism in atom transfer radical polymerisation is depicted in Scheme 8.11. The main difference to conventional radical polymerisation is in the presence of a metal complex. Free radicals are generated from reaction between the initiator (such as an organic halide) and the metal species which further controls the reaction by reversibly transforming the free radicals into a dormant species.[58] However, it ought to be pointed out that in ATRP contrary to, for example, Ziegler-Natta-type catalysts, the polymerisation does not take place at the metal centre.

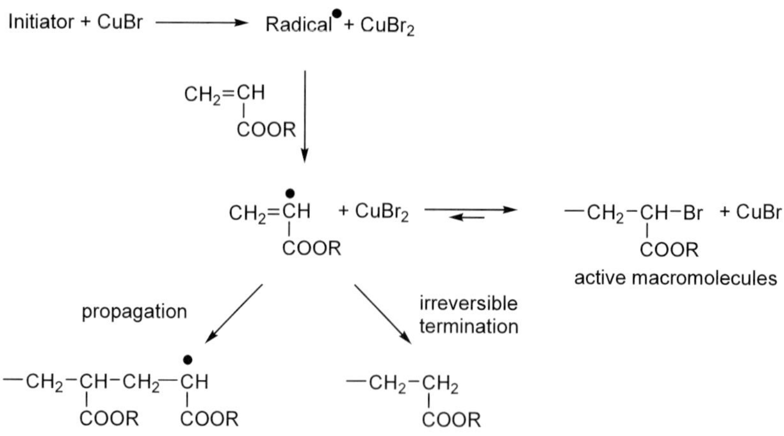

Scheme 8.11: Propagation and termination steps in ATRP with CuBr as catalyst

In the classic ATRP system, the activity of the catalyst depends dramatically on the nature of the preferably polydentate N-containing ligands. Mainly copper(I) salts have been used for radical polymerisation in ionic liquids,[59-62] but an example involving an iron catalyst has also been reported.[63] Apart from ionic liquid immobilised catalysts, solid-supported catalysts have been widely used to avoid polymer contamination.[64] A major drawback of copper-mediated radical polymerisation reactions lies in the high catalyst concentrations that are required. Equimolar amounts of metal relative to the initiator are frequently used in order to achieve acceptable reaction rates. Accordingly, protocols for catalyst reuse and facile product separation are highly desirable.

Most of the metal-catalysed atom transfer radical polymerisation reactions so far conducted in an ionic liquid focus on methyl methacrylate (MMA) or closely related olefins as substrate. The first example was published in 2000, reporting on the polymerisation of methyl methacrylate in $[C_4C_1im][PF_6]$ with a CuBr/alkylpyridylmethanimine catalyst system using ethyl-2-bromoisobutyrate as the initiator.[59] As the substrate is soluble in the ionic liquid the polymerisation proceeded under homogeneous conditions. After 90 minutes at 70°C, a conversion of 87% was reached which is substantially faster than in non-polar organic solvents or in bulk substrate. A further acceleration in the reaction rate can be achieved by using two equivalents of ligand per copper ion. Depending on the reaction temperature and ligand concentration, polydispersity varied between 1.30 and 1.43. Since the polymer is insoluble in the ionic liquid it was removed by washing the reaction mixture with toluene. Due to the good solubility of the catalyst in the ionic liquid contamination of the polymer product with copper was markedly reduced.

In another study, the possible interaction of different anions in a series of imidazolium-based ionic liquids with the transition metal catalyst was investigated.[63] The anions included halides, carbonate, dodecylsulfate and dibutyl phosphonate, and it was found that ATRP in these ionic liquids proceeded with rather low initiation efficiency. This was ascribed to the relatively low concentration of the copper catalyst in the organic medium and the high metal concentration in the ionic liquid phase. Yet, after a short non-stationary state polymerisation proceeded in a controlled manner and molecular weights evolved linearly with conversion, resulting in polymers with low polydispersity. A high ionic liquid/initiator ratio, 10:1, led to higher reaction rates and lower initiation activity while ionic liquid/initiator ratios of 5:1 afforded higher initiation activity at lower reaction rates, however, with the benefit of lower polydispersity.

No organic ligand was needed for the iron-catalysed atom transfer radical polymerisation of MMA in the ionic liquids used. In contrast, controlled ATPR of methyl methacrylate was possible, for the first time, in the absence of an organic ligand with a copper catalyst using $[C_4C_1im][PO_2(OBu)_2]$ as solvent.[63] However, when copper(I) bromide was used as a catalyst, an organic ligand was required in those ionic liquids that contained a halide or carbonate anion. After the removal of the polymer and the unreacted dimer the catalyst was successfully regenerated.

The effect of substrate solubility on polymerisation was studied in $[C_4C_1im][PF_6]$ with a CuBr/alkyltriamine catalyst system employing ethyl-2-bromopropionate as initiator.[37] Whereas methyl acrylate is miscible with $[C_4C_1im][PF_6]$, only traces of decyl acrylate are soluble in the ionic liquid and the solubility of the substrate also affects that of the initiator, as shown in Table 8.1. In the polymerisation of butyl acrylate it was possible to isolate the product essentially free of catalyst residues by a simple phase separation. Independent of the nature of the substrate, more than 99.9% of catalyst was retained in the ionic liquid phase in all cases. Under homogeneous reaction conditions, fairly narrow molecular weight distributions were observed, which correlated well with the calculated values according to the ATRP mechanism. Significant deviations from controlled polymerisation conditions were observed for acrylates that were only sparingly soluble, namely with hexyl and dodecyl substituents. This may be taken as an indication that a different mechanism operates, although this result could also be due to mass transfer effects.

Table 8.1: Solubility of acrylates and ethyl 2-bromopropionate in $[C_4C_1im][PF_6]$. Monomer/ionic liquid/initiator 1:1:0.1 by weight

Monomer	Monomer dissolved in IL phase [wt.-%]	Initiator dissolved in IL phase [wt.-%]
Methyl acrylate	100	100
Butyl acrylate	38	41
Hexyl acrylate	10	25
Dodecyl acrylate	trace	< 10

An important benefit of performing the polymerisation of acrylates in $[C_4C_1im][PF_6]$ is that the number of undesirable side-reactions is considerably reduced. This observation correlates with the fraction of active macromolecules found in the reaction mixture, which was much higher in the ionic liquid relative to the solvent-free reaction, see Table 8.2.[43] Especially towards the end of a polymerisation reaction, when the monomer concentration is low, side-reactions such as irreversible chain termination become increasingly important.

Table 8.2: Fraction of potentially active macromolecules in ATRP of butylacrylate

Monomer conversion	Active macromolecules in bulk	Active macromolecules in ionic liquid solution
23	88	90
95	44	84

Irreversible termination of growing macromolecules during the final stages of ATRP are particularly disadvantageous if the synthesis of block co-polymers by sequential polymerisation is attempted. Due to different solubilities of catalyst, monomer and polymer in the ionic liquid phase, a larger amount of active molecules may be observed in the presence of an ionic liquid.

Block co-polymers have been synthesised in $[C_4C_1im][PF_6]$ by ATRP of butylacrylate and acrylate monomer.[62] The outcome of the reaction depends significantly on the order of substrate addition. If, for example, methyl acrylate was added to a two-phase system of poly-butylacrylate and ionic liquid, the resulting copolymer has a narrow polydispersity and is essentially free of homopolymer. A markedly higher amount of homopolymer was formed when butyl acrylate was added to a solution of poly-methyl acrylate and the degree depended on the stage of the MA polymerisation. Below 70% conversion, copolymer without homopolymer was formed, while above 90% conversion, practically no co-polymer was produced.

By using an imidazolium ionic liquid with a chiral alkyl chain, *R*-(+)-2-methyl butane, induction of isotactic sequences in the poly(methyl acrylate)

was attempted.[43] It was found that fractions of isotactic sequences increased with increasing ratio of chiral ionic liquid to monomer, but the effect was relatively small.

Transition metal compounds in their higher oxidation states are used in reverse ATRP, and initiators such as 2,2'-azabisisobutyronitrile (AIBN) or benzoyl peroxide may be used in place of toxic organic halides. In conventional solvents, the $CuCl_2$/bipyridine catalysed reverse ATRP reaction proceeds in an uncontrolled manner, probably due to poor solubility of the catalyst. In contrast, the reverse atom transfer radical polymerisation of methyl methacrylate in $[C_4C_1im][PF_6]$, with 2,2-azobisisobutyronitrile as the initiator affords polymethyl methacrylate with M_W/M_n values below 1.2.[60] However, the polymerisation also proceeded in an uncontrolled manner in the ionic liquid if $CuCl_2$ was absent or not present in sufficient quantities. Excess $CuCl_2$, on the other hand, not only slows the polymerisation rate significantly, but it could also contaminate the resulting polymer. The ionic liquid phase could be recycled, with the reaction again proceeding in a controlled fashion, after addition of further aliquots of AIBN, initiator and substrate.

A well controlled reverse ATPR of MMA was further achieved in $[C_{12}C_1im][BF_4]$ with AIBN/$CuCl_2$/bipy as the initiating system.[61] The ionic liquid could be reused, but more copper(II) chloride had to be added in order to sustain catalytic activity.

8.2.3 Non-Radical Polymerisation Reactions

Classic Ziegler-Natta catalysts of the general formula MCp_2Cl_2 (M = Ti, Zr, Hf) have been evaluated in the polymerisation of ethylene in $[C_2C_1im]Cl$-$AlCl_3$.[65] No catalysis occurred in basic chloroaluminate ionic liquid and from electrochemical experiments it appears that $[TiCp_2Cl_3]^-$ is formed in such media. In acidic melts, with $Al_2Me_3Cl_3$ as co-catalyst, only $TiCp_2Cl_2$ is active, giving 0.02-0.15 (g PE) min^{-1} (mmol Ti)$^{-1}$ bar^{-1}. It is likely that the lack of catalytic activity with zirconium and hafnium analogues is due to the stronger metal-chloride bond in these complexes, which precludes formation of an active catalyst species, believed to be $TiCp_2R^+$ in the active system.

Ethylene may also be polymerised in $[C_4C_1im]Cl$-$AlCl_3$-$AlCl_2Et$ (1:1:0.32) in the presence of the nickel(II) bisimine complex, **43**, and toluene as co-solvent.[66] After the reaction, the upper toluene layer contains the product while the catalyst remains in the ionic liquid. After decantation, the second run was initiated by introducing a new batch of ethylene-saturated co-solvent and an aliquot of trimethyl aluminium, in order to compensate for the extraction of the free alkylaluminium compound into the organic phase. While in the first run only 5 kg·mol^{-1}·h^{-1} polyethylene was produced at –10°C, productivity increased to 96 kg·mol^{-1}·h^{-1} in the second run, and to 198 kg·mol^{-1}·h^{-1} in the third run, and raising the temperature by 20°C led to a further increase in activity. It is important to note, that the polymer properties change upon recycling of the catalyst and this is a reflection of the changing content and nature of organoaluminates in the ionic liquid phase, giving rise to different active species.

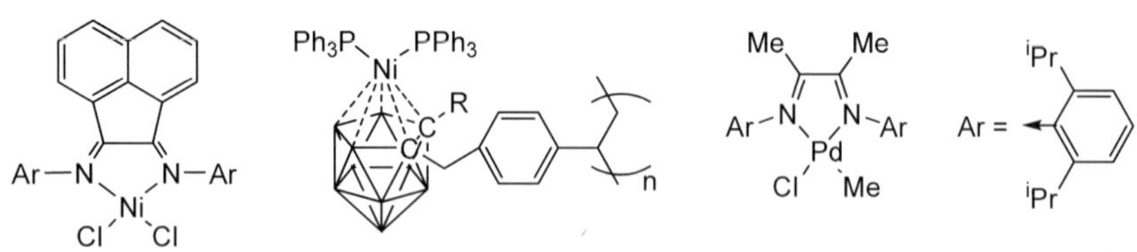

43 **44a**: R = Me **44b**: R = Ph **45**

Scheme 8.12: Nickel and palladium catalysts used in biphasic olefin polymerisation

Polystyrene supported nickel(II) carborane complexes **44a** and **44b** have been tested in the polymerisation of ethylene, styrene and vinyl chloride in [C_4C_1im][BF_4]. The compounds were active for all three substrates, even in the absence of a co-catalyst. Increased catalytic activity was observed relative to THF, affording polymers with higher molecular weight and, in most cases, lower polydispersity.[67]

Polymerisation of ethylene was also attempted in [C_4C_1im][PF_6] or [C_6py][Tf_2N], employing the palladium compound **45** together with either Ag[SbF_6] or Na[BARF] to activate the complex.[68] However, the best activity obtained in these ionic liquids was one order of magnitude lower compared to using dichloromethane as solvent.

Phenylacetylene has been polymerised in ionic liquids using rhodium(I) catalysts.[69,70] The polymerisation of phenylacetylene, using the rhodium complexes Rh(acac)(diene) and [RhCl(diene)]$_2$ (diene = norbornadiene or 1,5-cyclooctadiene), was tested in [C_4C_1im][BF_4] and [C_4py][BF_4]. Those complexes with norbornadiene generally afforded higher yields. It was found that halide impurities present in the ionic liquid suppress the catalytic activity and the presence of NEt_3 was therefore be essential in order to trigger the catalytic cycle. Indeed, phenylacetylene was polymerised smoothly with Rh(acac)(nbd) at 25°C within one minute to give 91% isolated yield when halide-free [C_4C_1im][BF_4] was used.[69]

A series of ionic liquids have been tested for the polymerisation of phenylacetylene with the pyrazolylborate rhodium complexes **46** and **47**, shown in Scheme 8.13.[70] Complex **46** afforded higher molecular weights and lower polydispersity at comparable activity in the ionic liquid relative to dichloromethane. The addition of small amounts of methanol to the ionic liquid was found to have a positive effect on the catalytic activity. With complex **47** best results were obtained in [C_4C_1im]Cl, giving 64% yield after 2 hours at 65°C. Molecular weights were usually lower with **46** than those obtained with either **47a** or **47b** as catalyst.

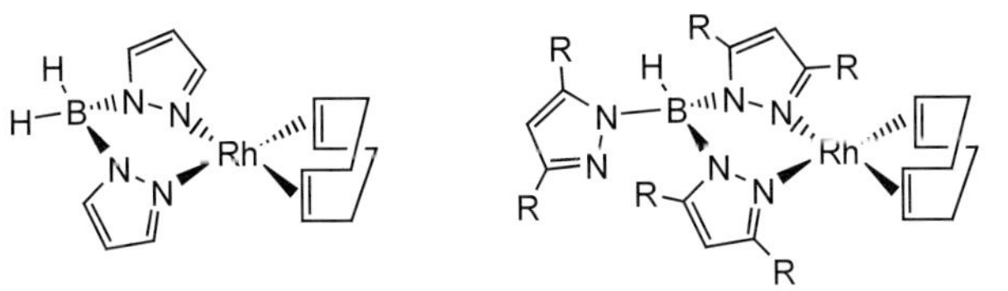

46

47a: R = H
47b: R = Me

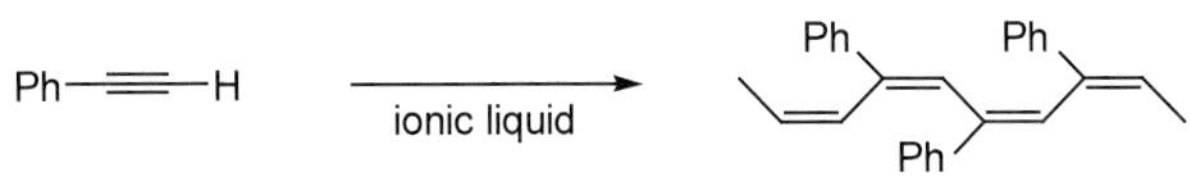

	46 (65°C)	**47a** (65 °C)	**47b** (20 °C)
CH_2Cl_2: M_W =		21,800	
[C_4C_1im][BF_4]: M_W =	50,800	55,000	61,600
[C_8C_1im][BF_4]: M_W =	67,800	60,000	94,800

Scheme 8.13: Rhodium pyrazolylborate complexes for polymerization of phenyl acetylene

Alternating co-polymerisation of styrene and carbon monoxide has been tested with palladium complexes in a range of ionic liquids (see Scheme 8.14).[68,71] With [Pd(bipy)$_2$][PF_6]$_2$ as catalyst, yields were found to decrease in the order [Tf_2N]$^-$ > [PF_6]$^-$ > [BF_4]$^-$ for both imidazolium and pyridinium ionic liquids.[71] The length of the alkyl substituent also plays a significant role in terms of reactivity and recyclability. For [$C_n$py][Tf_2N], yields increased as the alkyl chain was increased from n = 4 to n = 10. However, with long alkyl chains, the isolated polymer was grey in colour, indicating that inclusion of palladium had occurred. Turnover numbers in [C_6py][Tf_2N] were much higher than in alcohols. Raising the temperature from 20°C to 100°C resulted in marked increase in activity while increasing the CO pressure only had a positive effect up to ca. 10 bar. Further increasing the pressure led to polymers with lower molecular weight. Recycling of the catalyst phase was possible for at least four runs without decrease in activity or a change in the polymer quality.

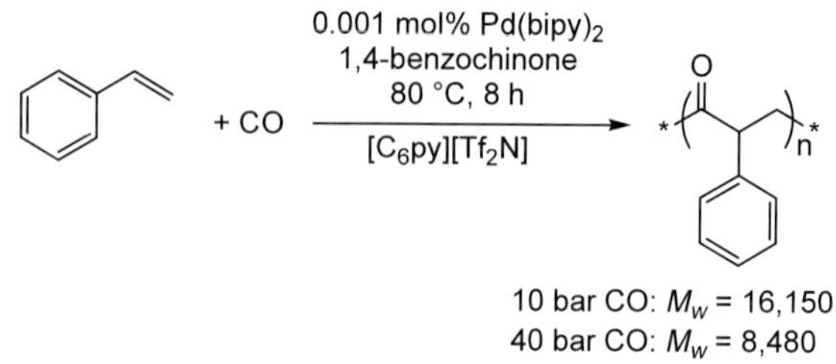

Scheme 8.14: Alternating co-polymerisation of styrene and CO

With $Pd(OAc)_2(bipy)$ as the catalyst, methanol was necessary as a co-solvent in $[C_6py][Tf_2N]$ in order to accomplish co-polymerisation.[68] In the absence of methanol polystyrene was the only product observed. At an ionic liquid-methanol ratio of 10:1, 70°C and 40 bar CO, 2.73 kg co-polymer per gram palladium was obtained although residual halide inhibits catalytic activity. Ionic liquids with non-coordinating anions are important as the inactivity in $[C_4C_1im][CF_3CO_2]$ suggests that CO cannot compete with trifluoroacetic acid for the open site on the palladium catalyst. About three times as much co-polymer was produced in $[C_6py][Tf_2N]$ at 50°C relative to dichloromethane as solvent, using $PdCl(CH_3)(bipy)$ and $Ag[SbF_6]$ as the catalyst system.

8.3 References

[1] *Ullmann's Encylopedia of Industrial Chemistry*; 6 ed.; Wiley-VCH, **2003**; Vol. 17.

[2] Freitas, E. R.; Gum, C. R. *Chem. Eng. Prog.* **1979**, 73-76.

[3] Olivier-Bourbigou, H.; Forestière, A. In *Ionic Liquids in Synthesis*; Wasserscheid, P., Welton, T., Eds.; Wiley-VCH: Weinheim, **2003**, pp 271-275.

[4] Wasserscheid, P. In *Ionic Liquids in Synthesis*; Wasserscheid, P., Welton, T., Eds.; Wiley-VCH: Weinheim, **2003**, pp 244-257.

[5] Ligabue, R. A.; Dupont, J.; de Souza, R. F. *J. Mol. Catal. A* **2001**, *169*, 11-17.

[6] Olivier, H.; Laurent-Gérot, P. *J. Mol. Catal. A* **1999**, *148*, 43-48.

[7] Picquet, M.; Tkatchenko, I.; Tommasi, I.; Wasserscheid, P.; Zimmermann, J. *Adv. Synth. Catal.* **2003**, *345*, 959-962.

[8] Mellah, M.; Gmouh, S.; Vaultier, M.; Jouikov, V. *Electrochem. Commun.* **2003**, *5*, 591-593.

[9] Naudin, E.; Ho, H. A.; Branchaud, S.; Breau, L.; Belanger, D. *J. Phys. Chem. B* **2002**, *106*, 10585-10593.

[10] Chauvin, Y.; Gilbert, B.; Guibard, I. *Chem. Commun.* **1990**, 1715-1716.

[11] Chauvin, Y.; Einloft, S.; Olivier, H. *Ind. Eng. Chem. Res.* **1995**, *34*, 1149-1155.

[12] Chauvin, Y.; Olivier, H.; Wyrvalski, C. N.; Simon, L. C.; de Souza, R. F. *J. Catal.* **1997**, *165*, 275-278.

[13] Simon, L. C.; Dupont, J.; de Souza, R. F. *Appl. Catal. A* **1998**, *175*, 215-220.

[14] Olivier-Bourbigou, H.; Hugues, F. In *Green Industrial Applications of Ionic Liquids*; Volkov, S., Ed.; Kluwer Academic Publishers: Dordrecht, **2002**, pp 67-84.

[15] Keim, W.; Hoffmann, B.; Lodewick, R.; Peukert, M.; Schmitt, G.; Fleischhauer, J.; Meier, U. *J. Mol. Catal.* **1979**, *6*, 79-97.

[16] Ellis, B.; Keim, W.; Wasserscheid, P. *Chem. Commun.* **1999**, 1049-1049.

[17] Wasserscheid, P.; Eichmann, M. *Catal. Today* **2001**, *66*, 309-316.

[18] McGuinness, D. S.; Mueller, W.; Wasserscheid, P.; Cavell, K. J.; Skelton, B. W.; White, A. H.; Englert, U. *Organometallics* **2002**, *21*, 175-181.

[19] Wasserscheid, P.; Gordon, C. M.; Hilgers, C.; Muldoon, M. J.; Dunkin, I. R. *Chem. Commun.* **2001**, 1186-1187.

[20] Wasserscheid, P.; Hilgers, C.; Keim, W. *J. Mol. Catal. A* **2004**, *214*, 83-90.

[21] Ballivet-Tkatchenko, D.; Picquet, M.; Solinas, M.; Francio, G.; Wasserscheid, P.; Leitner, W. *Green Chem.* **2003**, *5*, 232-235.

[22] Zimmermann, J.; Wasserscheid, P.; Tkatchenko, I.; Stutzmann, S. *Chem. Commun.* **2002**, 760-761.

[23] Picquet, M.; Stutzmann, S.; Tkatchenko, I.; Tommasi, I.; Zimmermann, J.; Wasserscheid, P. *Green Chem.* **2003**, *5*, 153-162.

[24] Silva, S. M.; Suarez, P. A. Z.; de Souza, R. F.; Dupont, J. *Pol. Bull.* **1998**, *40*, 401-405.

[25] Kabalka, G. W.; Dong, G.; Venkataiah, B. *Tetrahedron Lett.* **2004**, *45*, 2775-2777.

[26] Dullius, J. E. L.; Suarez, P. A. Z.; Einloft, S.; de Souza, R. F.; Dupont, J.; Fischer, J.; De Cian, A. *Organometallics* **1998**, *17*, 815-819.

[27] Magna, L.; Chauvin, Y.; Niccolai, G. P.; Basset, J. M. *Organometallics* **2003**, *22*, 4418-4425.

[28] Kutney, J. P.; Greenhouse, R. *Synth. Commun.* **1975**, *5*, 119-124.

[29] Deady, L. W.; Finlayson, W. L.; Korytky, O. L. *Austr. J. Chem.* **1979**, *32*, 1735-1742.

[30] Small, B. L.; Schmidt, R. *Chem. Eur. J.* **2004**, *10*, 1014-1020.

[31] Small, B. L.; Marcucci, A. J. *Organometallics* **2001**, *20*, 5738-5744.

[32] Yadav, J. S.; Reddy, B. V. S.; Reddy, K. B.; Gayathri, K. U.; Prasad, A. R. *Tetrahedron Lett.* **2003**, *44*, 6493-6496.

[33] Carmichael, A. J.; Haddleton, D. M. In *Ionic Liquids in Synthesis*; Wasserscheid, P., Welton, T., Eds.; Wiley-VCH: Weinheim, **2003**, pp 319-335.

[34] Biedron, T.; Kubisa, P. *J. Pol. Sci. A* **2004**, *42*, 3230 3235.

[35] Biedron, T.; Bednarek, M.; Kubisa, P. *Macromol. Rap. Commun.* **2004**, *25*, 878-881.

[36] Vijayaraghavan, R.; MacFarlane, D. R. *Chem. Commun.* **2004**, 700-701.

[37] Biedron, T.; Kubisa, P. *Macromol. Rap. Commun.* **2001**, *22*, 1237-1242.

[38] Benton, M. G.; Brazel, C. S. In *Ionic Liquids, Industrial Applications for Green Chemistry*, ACS Symposium Series 818, **2002**, 125-133.

[39] Harrisson, S.; Mackenzie, S. R.; Haddleton, D. M. *Chem. Commun.* **2002**, 2850-2851.

[40] Hong, H. L.; Zhang, H. W.; Mays, J. W.; Visser, A. E.; Brazel, C. S.; Holbrey, J. D.; Reichert, W. M.; Rogers, R. D. *Chem. Commun.***2002**, 1368-1369.

[41] Perrier, S.; Davis, T. P.; Carmichael, A. J.; Haddleton, D. M. *Chem. Commun.* **2002**, 2226-2227.

[42] Scott, M. P.; Brazel, C. S.; Benton, M. G.; Mays, J. W.; Holbrey, J. D.; Rogers, R. D. *Chem. Commun.* **2002**, 1370-1371.

[43] Biedron, T.; Kubisa, P. *Poly. Int.* **2003**, *52*, 1584-1588.

[44] Perrier, S.; Davis, T. P.; Carmichael, A. J.; Haddleton, D. M. *Eur. Poly. J.* **2003**, *39*, 417-422.

[45] Scott, M. P.; Rahman, M.; Brazel, C. S. *Eur. Poly. J.* **2003**, *39*, 1947-1953.

[46] Zhang, H. W.; Hong, K. L.; Jablonsky, M.; Mays, J. W. *Chem. Commun.* **2003**, 1356-1357.
[47] Zhao, Y. L.; Chen, C. F.; Xi, F. *J. Poly. Sci. A* **2003**, *41*, 2156-2165.
[48] Benton, M. G.; Brazel, C. S. *Poly. Int.* **2004**, *53*, 1113-1117.
[49] Ryan, J.; Aldabbagh, F.; Zetterlund, P. B.; Yamada, B. *Macromol. Rap. Commun.* **2004**, *25*, 930-934.
[50] Zhang, H. W.; Hong, K.; Mays, J. W. *Poly. Bull.* **2004**, *52*, 9-16.
[51] Goldenberg, L. M.; Osteryoung, R. A. *Synth. Met.* **1994**, *64*, 63-68.
[52] Arnautov, S. A. *Synth. Met.* **1997**, *84*, 295-296.
[53] Sekiguchi, K.; Atobe, M.; Fuchigami, T. *Electrochem. Commun.* **2002**, *4*, 881-885.
[54] Sekiguchi, K.; Atobe, M.; Fuchigami, T. *J. Electroanal. Chem.* **2003**, *557*, 1-7.
[55] El Abedin, S. Z.; Borissenko, N.; Endres, F. *Electrochem. Commun.* **2004**, *6*, 422-426.
[56] Harrisson, S.; Mackenzie, S. R.; Haddleton, D. M. *Macromol.* **2003**, *36*, 5072-5075.
[57] Kubisa, P. *Prog. Poly. Sci.* **2004**, *29*, 3-12.
[58] Matyjaszewski, K. *Macromol. Symp.* **2002**, *182*, 209-224.
[59] Carmichael, A. J.; Haddleton, D. M.; Bon, S. A. F.; Seddon, K. R. *Chem. Commun.* **2000**, 1237-1238.
[60] Ma, H. Y.; Wan, X. H.; Chen, X. F.; Zhou, Q. F. *J. Poly. Sci. A* **2003**, *41*, 143-151.
[61] Ma, H. Y.; Wan, X. H.; Chen, X. F.; Zhou, Q. F. *Polymer* **2003**, *44*, 5311-5316.
[62] Biedron, T.; Kubisa, P. *J. Poly. Sci. A* **2002**, *40*, 2799-2809.
[63] Sarbu, T.; Matyjaszewski, K. *Macromol. Chem. Phys.* **2001**, *202*, 3379-3391.
[64] Shen, Y. Q.; Tang, H. D.; Ding, S. J. *Prog. Poly. Sci.* **2004**, *29*, 1053-1078.
[65] Carlin, R. T.; Wilkes, J. S. *J. Mol. Catal.* **1990**, *63*, 125-129.
[66] Pinheiro, M. F.; Mauler, R. S.; de Souza, R. F. *Macromol. Rap. Commun.* **2001**, *22*, 425-428.
[67] Yinghuai, Z.; Parthiban, A.; Ben Hedi Kooli, F. *Catal. Today* **2004**, *96*, 143-146.
[68] Shaughnessy, K. H.; Klingshirn, M. A.; P'Pool, S. J.; Holbrey, J. D.; Rogers, R. D. In *Ionic Liquids as Green Solvents: Progress and Prospects,* ACS Symposium Series 856, **2003**, pp 300-313.
[69] Mastrorilli, P.; Nobile, C. F.; Gallo, V.; Suranna, G. P.; Giannandrea, R. *Sci. Tech. Cataly.* **2003**, *145*, 535-536.
[70] Trzeciak, A. M.; Ziolkowski, J. J. *Appl. Organomet. Chem.* **2004**, *18*, 124-129.
[71] Hardacre, C.; Holbrey, J. D.; Katdare, S. P.; Seddon, K. R. *Green Chem.* **2002**, *4*, 143-146.

Chapter 9

MISCELLANEOUS REACTIONS

Over the past ten years, ionic liquids have been applied to a wide array of organic transformations and within the confines of this book it is not possible to cover all the different reactions. The following pages provide a selection of catalysed reactions in ionic liquids that have not been covered in the previous chapters. In many cases, there are only a few examples of a certain reaction available, making it difficult to deduce any clear trends with respect to the possible potential of ionic liquids as reaction media. As the following pages will show, rare-earth metal salts are often used with remarkable success to catalyse reactions in ionic liquids. This is probably due to the fact that this class of solvents offers an environment in which these salts exhibit good solubility but that is less reactive than water, allowing for the presence of more sensitive functional groups in the substrate. It is surprising that in many examples hydrophobic ionic liquids give much better results with these strong Lewis-acid catalysts although their solubility is much higher in hydrophilic ionic liquids, and some cases are certainly worth studying further in order to rationalise these solvent effects.

9.1 Halogenation and Dehalogenation

There are numerous examples in which ionic liquids have been used as reaction media for the halogenation of organic substrates in the absence of a metal catalyst and only some of the more recent references are listed.[1-9] In this context, it is worth mentioning task-specific ionic liquids such as pentylpyridinium tribromide, a non-volatile ionic liquid analogue of bromine, which acts as both solvent and reagent, as shown in Scheme 9.1.[10]

In a similar manner, easy-to-handle solvents/reagents for electrophilic iodochlorination and iodobromination have been demonstrated.[11]

Scheme 9.1: Synthetic pathway to a vapour-pressure free ionic liquid analogue of bromine

Not all ionic liquids are sufficiently inert towards the often highly reactive reagents used in halogenation reactions. As an example, decomposition of imidazolium-based ionic liquids was observed in nucleophilic fluorination reactions employing alkali metal fluorides as the fluorinating reagent.[12] Analysis of the volatile decomposition products of the solvent by GC-MS revealed the formation of several species including alkyl imidazoles, fluoroalkanes and alkenes. Similar observations were also made with *N*-alkylmorpholinium based ionic liquids and recycling was not possible due to decomposition of the solvent.[13]

A wide selection of halogenation agents are available and as a reflection of their generally high reactivity, transition metals are seldom required to mediate such transformations. Even enantioselective halogenation can be performed in the absence of a catalyst. However, the chiral reagents required are fairly expensive and metal-catalysed asymmetric halogenation represents a relatively new methodology for the synthesis of enantiopure organohalides.[14]

Enantioselective fluorination is commonly conducted with chiral agents such as quinine-based $[N\text{-}F]^+$ compounds, and these have been successfully utilised in ionic liquids.[15,16] Very good yields and selectivities have been obtained in the enantioselective fluorination of β-ketoesters catalysed by the chiral palladium complex **57**, see Scheme 9.2.[17] Depending on the substrate employed, substantial acceleration of the reaction rate relative to that in ethanol was observed with yields and selectivities comparable to those obtained in water or ethanol. The reaction rate was found to depend on both the length of the alkyl substituent of the imidazolium cation as, well as on the type of anion present, whereas the selectivity was not affected by such variations. The products were extracted from the ionic liquid phase with diethyl ether, and in that manner catalytic activity was maintained for up to ten cycles.

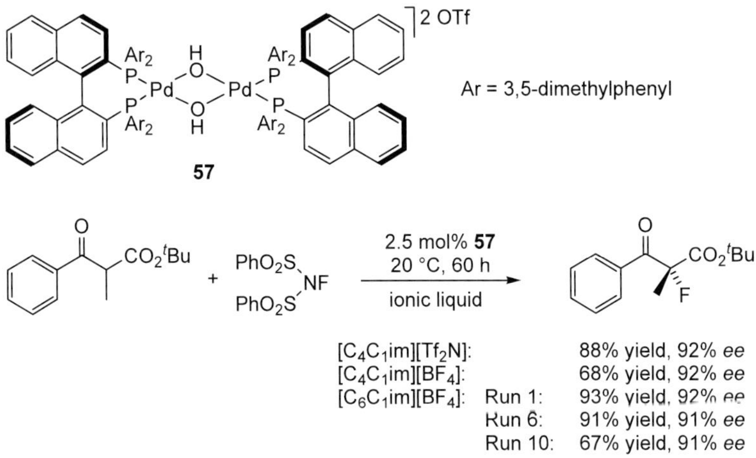

Scheme 9.2: Palladium-catalysed enantioselective fluorination of β-ketoesters

Dehalogenation reactions are particularly interesting as many halogenated aromatic compounds such as dioxins and polychlorinated biphenyls represent a serious environmental and toxicological hazard. The presence of the ionic liquid phase-transfer agent Aliquat 336 was found to have a marked influence on the selectivity of dehalogenation where Pd/C or Pt/C were employed as catalysts together with hydrogen gas as the reducing agent.[18] In the presence of Aliquat 336 the main products in the dechlorination of *p*-chlorobenzaldehyde with Pd/C were benzaldehyde and benzylalcohol, while in its absence, toluene was obtained almost exclusively.

Dehalogenation of polyhalogenated arenes and fullerenes was accomplished with palladium catalysts in an ionic liquid composed of a pyridinium cation and a carborane anion, *viz.* $[C_5py][closo\text{-}CB_{11}H_{12}]$.[19] For any substrate, the catalyst efficiency decreased in the order $PdCl_2(dppf) > PdCl_2(dppe) > PdCl_2(PPh_3)_2$. With an excess of TMEDA and $NaBH_4$ as reagents and 0.5 mol% of $[PdCl_2(dppf)]$, hexabromobenzene was converted quantitatively to benzene in less than three hours at room temperature. The beneficial effect of the ionic liquid is exemplified in the dechlorination of 1,2,4-trichlorobenzene with $PdCl_2(dppf)$. While a mixture of dichlorobenzene isomers was obtained when the reaction was conducted in organic solvents, 1,2-dichlorobenzene was formed with considerably higher selectivity in the ionic liquids, as shown in Scheme 9.3.

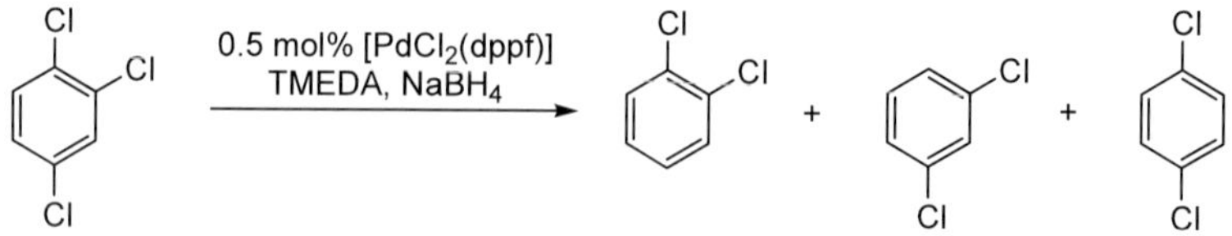

solvent	[%]	[%]	[%]
THF	40	20	40
CH_3CN	50	20	30
$[C_5py][CB_{11}H_{12}]$	95	4	1

Scheme 9.3: Effect of solvent on the dechlorination of 1,2,4-trichlorobenzene

The reaction products were isolated by distillation from the ionic liquid and all of the palladium catalysts could be reused several times before turnover frequencies noticeably deteriorated. However, once the ionic liquid catalyst solution was saturated with either NaBr or NaCl, it had to be regenerated by dissolving in chloroform or dichloromethane and filtering through celite. If this procedure was not followed then the precipitated sodium halide led to increased reaction times due to decreased diffusion of the substrate in the ionic liquid.

The dehalogenation of bromoketones in an organometallic ionic liquid is also worth mentioning here. The solvent is obtained from the reaction of $[C_4C_1im]Cl$ with $Na[Co(CO)_4]$ in acetone to afford a clear, blue liquid.[20] The debromination proceeded efficiently, but as no detailed mass balance studies were undertaken, it is not clear whether the ionic liquid really remains unchanged after the reaction or whether some of it is consumed in the course of the reaction.

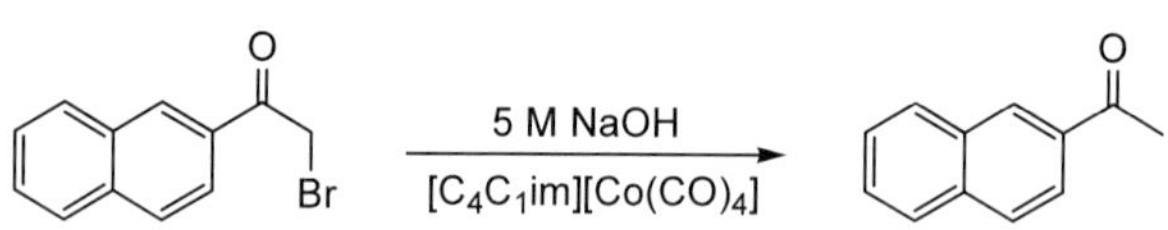

Scheme 9.4: Debromination with an organometallic ionic liquid as catalyst and reaction medium

9.2 Addition to Carbon-Hetero Multiple Bonds

9.2.1 Aldol-and Mannich-Type Reactions

Aldol-type reactions are among the most important methods for the formation of carbon-carbon bonds. A number of examples of aldol-reactions in ionic liquids have been reported,[21-24] of which those involving chiral organic catalysts are particularly promising.[25-27] While, for example, asymmetric cross-aldol reactions with a chiral organic catalyst have been performed in ionic liquid media with good yields and excellent stereoselectivity,[28] chiral metal catalysts have not yet been applied to the reaction under such biphasic conditions. The Mannich reaction provides a facile approach to the synthesis of natural products and pharmaceuticals and Brønsted-acidic ionic liquids have been used as both catalysts and reaction media for this reaction.[29]

The reaction between cycloalkanones and benzaldehydes affords *α,α*'-bis(substituted benzylindene)cycloalkanones in good to excellent yield in $[C_4C_1im][BF_4]$ with $FeCl_3$ as catalyst, see Scheme 9.5.[30,31] Despite the ease of the reaction, a very high catalyst loading is required, 50 mol%, together with the need for the presence of one equivalent of trimethylsilyl chloride as co-catalyst, making the system less attractive. Catalyst recycling was, however, possible for at least five runs with only a slight deterioration in activity.

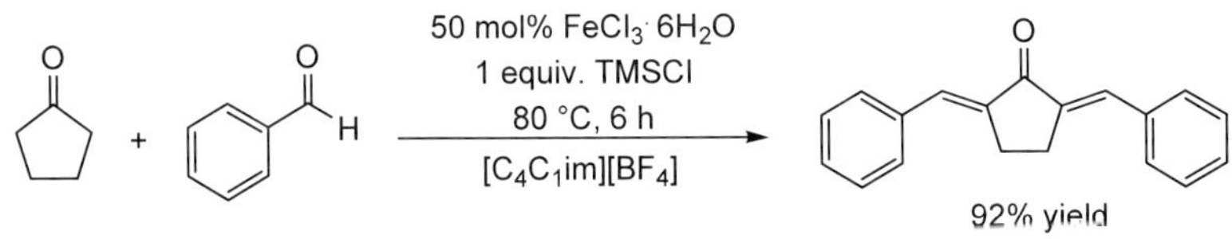

Scheme 9.5: Preparation of disubstituted cycloalkanones

The coupling of benzaldehydes with 3-buten-2-ol proceeds with moderate to good yield in $[C_4C_1im][PF_6]$ with $RuCl_2(PPh_3)_3$ as catalyst at elevated temperatures.[32,33] In contrast to the analogous reaction in conventional solvents, high yields were even obtained with electron-deficient aldehydes such as 2-naphthylaldehyde, but the reaction did not proceed with ketones as substrates. When *α*-vinylbenzyl alcohol is the substrate, addition of a Lewis acid co-catalyst is necessary in order to obtain good yields, and from screening a range of metal salts it was found that addition of 10 mol% $In(OAc)_3$ gives the best results. Compared with other solvents, the reaction between α-vinylbenzyl alcohol and the aldehyde progressed at higher rates and with better yields in the ionic liquid. Contrary to the case in aqueous or organic solvents, the *anti* products were favoured, although the selectivity was very low, as shown in Scheme 9.6.

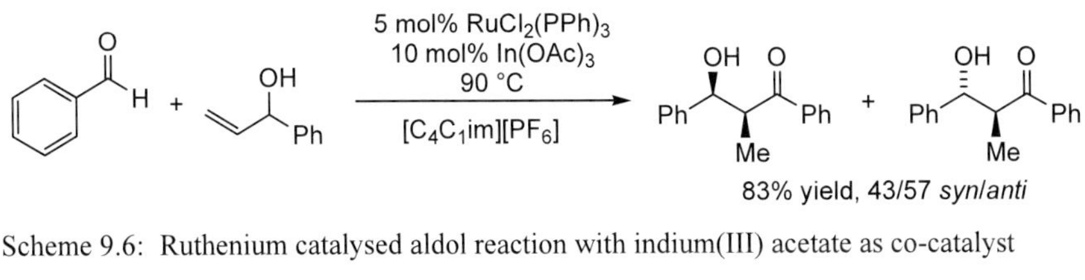

Scheme 9.6: Ruthenium catalysed aldol reaction with indium(III) acetate as co-catalyst

As in the above described aldol-reaction, $RuCl_2(PPh_3)_3$ also gave good results in the Mannich-type reaction between imines and allyl alcohols.[32,33] It was proposed that the ruthenium complex first isomerises the allyl alcohol to an enol which subsequently undergoes coupling with either an aldehyde or an imine to generate the aldol and Mannich-type products, respectively. An important advantage of performing the reaction in an ionic liquid is that, unlike under biphasic aqueous conditions, where aldol products are also formed as a result of decomposition of the imine, only Mannich products are observed. Catalyst recycling was tested with 2-naphthaldehyde, *p*-anisidine and 3-buten-2-ol as substrates, see Scheme 9.7. After extraction with diethyl ether, the ionic phase was dried and reloaded with substrate. When the catalyst was recycled in this manner, the activity was retained for at least five runs.

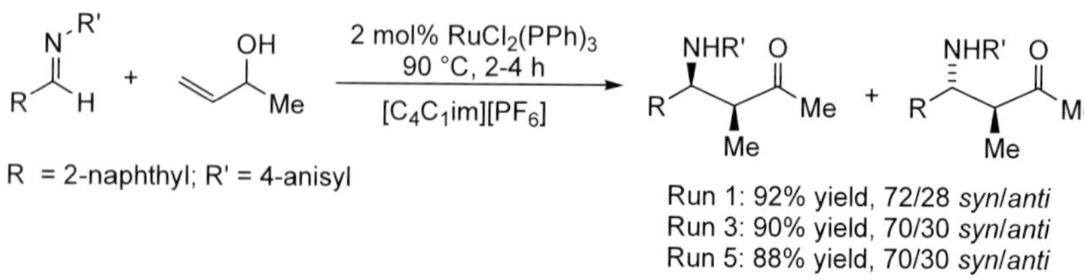

Scheme 9.7: Mannich-type reaction between 2-naphthylaldehyde, *p*-anisidine and 3-buten-2-ol.

Various aliphatic aldehydes, alkynes and dialkyl amines have been coupled using silver(I) iodide as catalyst in $[C_4C_1im][BF_4]$ with moderate to excellent yield, as illustrated in Scheme 9.8.[34] The nature of the aldehyde has a much greater influence on the outcome of the reaction in the ionic liquid than in water. With aliphatic aldehydes, the reaction proceeded cleanly with hardly any trimer formation whereas aromatic aldehydes afforded complex product mixtures.

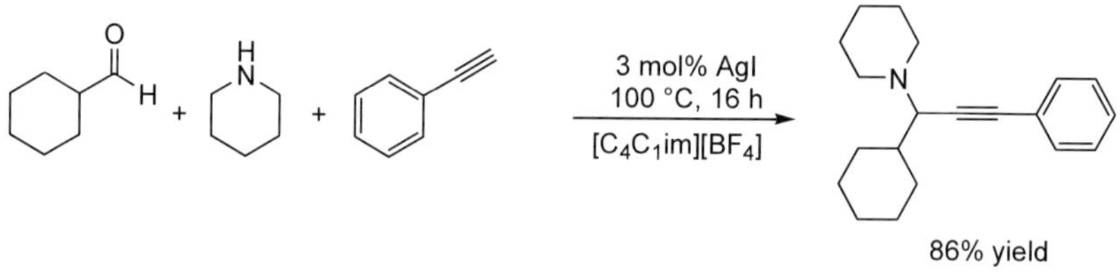

Scheme 9.8: Silver-catalysed formation of propargyl amines

Asymmetric Mannich-type reactions, using a chiral imine as substrate and high loadings of indium(III) chloride as catalyst, proceed with moderate yield and good diastereoselectivity in $[C_4C_1im][BF_4]$, see Scheme 9.9.[35] It is interesting to note that higher conversion with a reduced formation of the aldol product were observed in $[C_4C_1im][BF_4]$ relative to $[C_8C_1im][BF_4]$, and that the selectivity of the reaction can thus be tuned by altering the ionic liquid. An advantage of ionic liquids compared to water is that enolisable aldimines can be used as substrates. While efforts to recycle $InCl_3$ did not afford the desired Mannich products, $In(OTf)_3$ showed some recycling potential in $[C_6C_1im][BF_4]$, although by the third catalytic run very poor conversion was obtained.

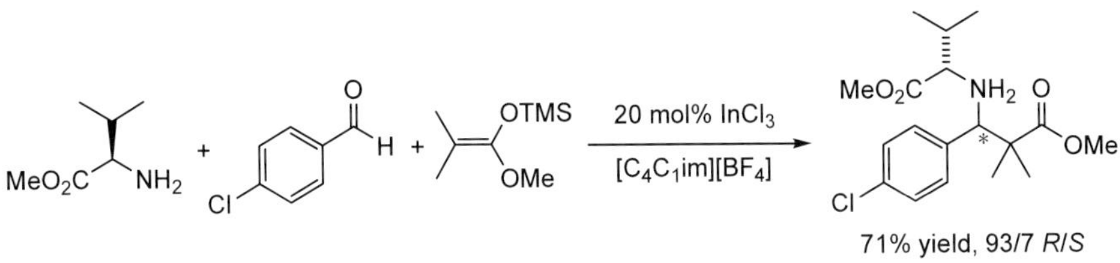

Scheme 9.9: Mannich-type reaction with a chiral substrate

Ytterbium(III) triflate has been found to be to be a highly efficient catalyst in the reaction between imines and trimethylsilyl enolates, affording good to excellent yields after 15 minutes at ambient temperature at low catalyst loadings (0.1-1 mol%), see Scheme 9.10.[36] Markedly higher reaction rates were observed in $[C_4C_1im][PF_6]$ compared to the corresponding $[BF_4]^-$ and $[SbF_6]^-$ salts, and it appears likely that fluoride generated from hydrolysis of the $[PF_6]^-$ anion acts as a co-catalyst in the reaction. Indeed, the reaction between benzaldehyde, aniline and acetophenone trimethylsilylenolate afforded the same product in 55% yield after 26 hours at room temperature even if no ytterbium salt was added to the ionic liquid.

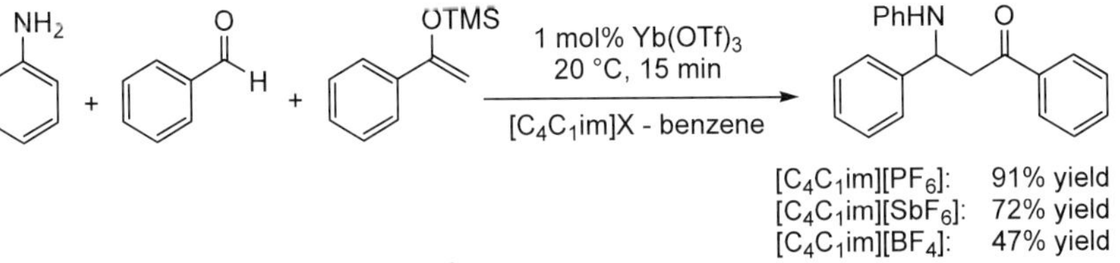

Scheme 9.10: Mannich-type reaction with trimethyl silyl enolate as substrate

9.2.2 Carbonylation Reactions

Imidazolium alkylselenite ionic liquids were found to be efficient catalysts/solvents for the oxidative carbonylation of amines,[37] but in general the synthesis of carbonyl compounds is carried out in the presence of a palladium catalyst. The first example of such a transformation stems from

2001, when arylhalides were carbonylated in an ionic liquid in the presence of either alcohols or amines.[38] The reaction between bromobenzene, methanol and CO in [C_4C_1im][BF_4], together with 0.5 mol% of $Pd(OAc)_2$ as catalyst and four equivalents of triphenylphosphine afforded methylbenzoate in much higher yield than in methanol, 82 *vs.* 30%. A higher ratio (20:1) of phosphine to palladium was required in order to accomplish recycling and reuse of the system, as shown in Scheme 9.11, while at a lower phosphine concentration the formation of colloidal palladium and rapid deterioration of catalytic activity was observed. In contrast to molecular solvents there was less double carbonylation under high CO-pressure in the ionic liquids. The mono-carbonylated product was formed almost exclusively in [C_4C_1im][PF_6] and this could be due to the generally low solubility of carbon monoxide in ionic liquids.

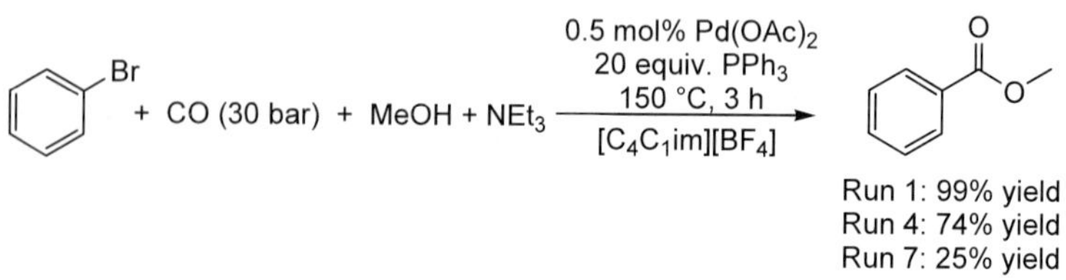

Scheme 9.11: Carbonylation of bromobenzene

Methylbenzoate was obtained from iodobenzene in moderate to good yield with polymer-stabilised palladium nanoparticles as the catalyst.[39] Of the ionic liquids screened in the reaction, those based on the pyridinium cation generally gave better results than imidazolium-type liquids. The latter probably deactivate the catalyst by formation of stable carbene complexes.

The same reaction was also investigated using the palladium-carbene complex, **58**, as catalyst (Scheme 9.12).[40] Of the different ionic liquids studied, tetrabutylammonium bromide gave by far the best results while imidazolium based solvents afforded only poor conversions. At atmospheric pressure, only iodobenzene was carbonylated. The conversion of less reactive arylhalides not only required higher CO-pressures, but also the addition of a phosphine ligand. Reuse of the catalyst after extraction of the product with diethyl ether was possible for at least 6 runs with only a moderate decrease in activity.

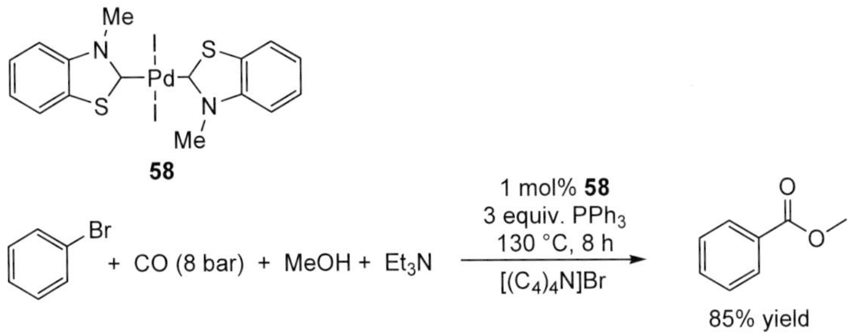

Scheme 9.12: Carbonylation of bromobenzene with a palladium-carbene catalyst

Palladium-catalysed carbonylation was also undertaken with amines as substrates.[41-43] Very good results were obtained in the aminocarbonylation of a steroid with morpholine using either $[C_4C_1im][BF_4]$ or $[C_4C_1im][PF_6]$ as solvent. Retention of catalytic activity was usually better with the tetrafluoroborate based ionic liquid. With 5 mol% $Pd(OAc)_2$ and six equivalents of triphenylphosphine, catalyst activity decreased from quantitative conversion in the first run to 94% in the fifth run at constant reaction time. Two equivalents of ligand had to be added in each run as some of the phosphine is lost during product extraction with toluene. This could probably easily be avoided if a charged phosphine ligand was to be used in place of the neutral phosphine.

The palladium(II) phenanthroline complex **59** (Scheme 9.13) was found to be a highly active catalyst for the oxidative carbonylation in ionic liquids with turnover frequencies above 4,000 $mol{\cdot}mol^{-1}{\cdot}h^{-1}$.[42] While aliphatic amines afforded ureas as products, carbamates were obtained with aromatic amines. With $[C_4C_1im]^+$ as cation, good conversions were obtained with both $[BF_4]^-$ and $[FeCl_4]^-$ whereas with $[PF_6]^-$ degradation of the ionic liquid occurred at the high reaction temperatures employed (175°C). The absence of an ionic liquid resulted in very poor conversion. The way the product was isolated is somewhat untypical in that addition of water causes the product to precipitate from the reaction mixture allowing for its facile isolation. Reuse of the ionic liquid after removal of water was possible, although the activity of the system decreases.

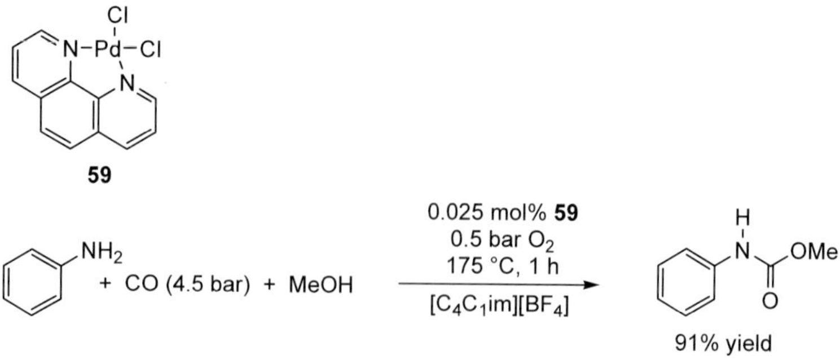

Scheme 9.13: Oxidative carboanylation of aniline

The aminocarbonylation of iodobenzene and iodo-alkenes with aminoacid methylesters proceeds well in $[C_4C_1im][BF_4]$ with $Pd(OAc)_2$ as catalyst.[43] Both PPh_3 as well as 4-(diphenylphosphino)benzoic acid as ligands gave good results. The hydroxycarbonylation of iodobenzene with $Pd(OAc)_2/PPh_3$ as catalyst system was compared in a range of different solvents including $[C_4C_1im][BF_4]$, $[C_4C_1im][PF_6]$ and Aliquat 336.[44] At 120°C and 30 bar CO pressure, 83% yield was obtained after 5 hours with $[C_4C_1im][PF_6]$. While higher reaction rates were observed with DMF and THF, catalyst-recycling was facile using the hydrophobic ionic liquid and total turnover numbers above 3,000 were achieved.

Carbonylation of 3-butyne-1-ol to yield the corresponding *exo*-methylene lactone proceeded well in both $[C_4C_1im][BF_4]$ and $[C_4C_1im][PF_6]$ catalysed by $Pd(OAc)_2/PPh_2Py$.[45] However, the catalytic activity decreases rapidly upon recycling and therefore there is no significant advantage compared to the reaction conducted in toluene. Carbonylation of methanol with $PdCl_2$ as the catalyst and $[C_4C_1im][PF_6]$ as the solvent to afford dimethyl carbonate was also attempted.[46] Relative to the reaction in the absence of the ionic liquid, a markedly higher selectivity towards the desired dimethyl carbonate was found. However, dimethyloxymethane remained the major product and methanol conversion was rather low.

There is only one example of a catalytic Pauson-Khand reaction in an ionic liquid[47] although the reaction has also been conducted in ionic liquids using stoichiometric amounts of $Co_2(CO)_8$.[48] In the catalysed reaction 10 mol% of $Co_2(CO)_8$ was used in $[C_4C_1im][PF_6]$ under 10 bar of CO. Under these conditions, diethylallyl malonates could be obtained in 90-99% yield within 90 minutes at 80°C, as shown in Scheme 9.14. However, with hetero-bridged enynes, as well as in the reaction between norbornene and phenylacetylene, only poor to moderate yields were achieved. A slight increase in activity was observed with the analogous tetrafluoroborate ionic liquid.

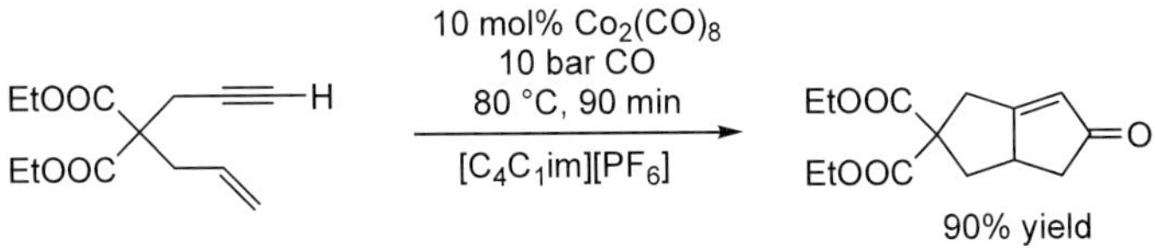

Scheme 9.14: Catalytic Pauson-Khand reaction

Recently, colloidal cobalt has been used for catalytic Pauson-Khand reactions.[49-53] Bearing in mind the encouraging results from other nanoparticle-catalysed reactions in ionic liquids, such catalysts may prove to be highly promising.

9.2.3 Reactions Employing CO_2 as Substrate

The remarkably high solubility of CO_2 in ionic liquids makes this class of solvent highly attractive for reactions involving this substrate. While there are plenty of examples where CO_2 is used as an extractant or a co-solvent in conjunction with ionic liquids, there are surprisingly few examples for reactions involving CO_2. One case, hydroformylation with CO_2 instead of CO as substrate has already been described in Chapter 4.[54]

The transformation of CO_2 to yield cyclic carbonates represents one of the best-known examples of successful chemical CO_2 fixation, yielding carbonates, which can be used, for example, in the synthesis of polymers. Examples include the cycloaddition of CO_2 to propylene oxide[55-58] and turnover frequencies above 400 mol·mol^{-1}·h^{-1} have been achieved in the absence of a catalyst in [C_4C_1im][BF_4] at 140°C. High turnover frequencies were also observed with [C_8C_1im][BF_4]-*sc*CO_2 at 100°C and 140 bar CO_2-pressure, leading to almost quantitative conversion to propylene oxide in five minutes.[59] With imidazolium based ionic liquids with zinc(II) halide anions, reaction rates approaching 5,000 mol·mol^{-1}·h^{-1} were observed.[57]

Chromium-salen complexes have been used for the reaction between styrene epoxide and *sc*CO_2 in [C_4C_1im][PF_6], as illustrated in Scheme 9.15.[60] At low catalyst concentrations, 1-phenyl-1,2-ethanediol was detected as a by-product while at a catalyst loading of ca. 0.35 mol%, 100% selectivity was obtained. Recycling of the catalyst was possible, but the ionic liquid phase needed to be exhaustively purified with volatile organic solvents prior to its reuse.

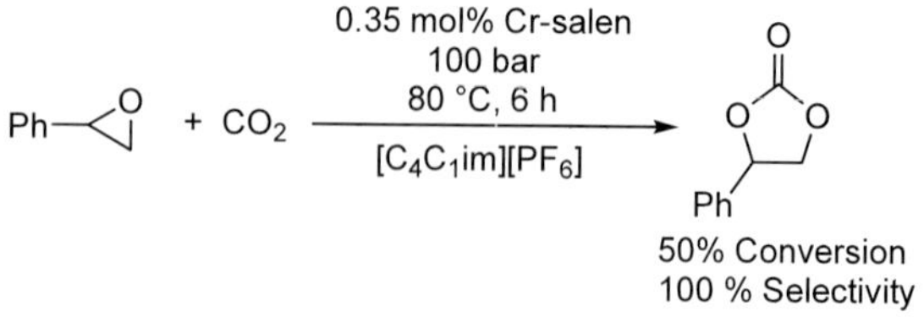

Scheme 9.15: CO_2-fixation *via* formation of cyclic carbonate

Numerous metal salts have been screened as potential catalysts for the reaction between propargyl alcohols and CO_2 to afford the corresponding α-methylene cyclic carbonates in a benzene sulfonate ionic liquid, *viz.* $[C_4C_1im][PhSO_3]$, see Scheme 9.16.[61] Highest activities were obtained with copper(I) halides while noble metal salts afforded poor results, since they catalysed the polymerisation of the alkyne-substrate. With 2-methyl-3-butyne-2-ol as substrate 97% yield was obtained in $[C_4C_1im][PhSO_3]$ after 8 hours at 120°C, which compares to 60% after 24 hours in DMF. When primary and secondary alcohols are used as substrates, the desired product is not formed. Recycling of the ionic liquid phase containing the catalyst was demonstrated for three runs without loss of activity.

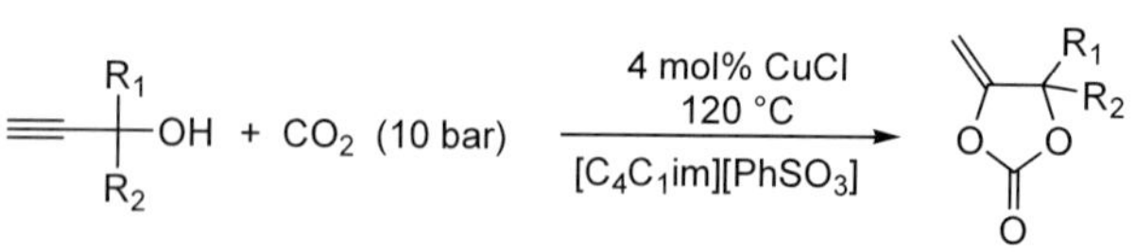

Scheme 9.16: CO_2-Fixation *via* reaction with propargyl alcohols

9.2.4 Synthesis of Bis(indoyl)methanes

The common route to bis(indoyl)methanes is via condensation of indoles with aldehydes or ketones in the presence of either protic or Lewis acids. The reaction has been evaluated in tetrafluoroborate and hexafluorophosphate ionic liquids and of the metal salts tested best results were obtained with $In(OTf)_3$ and $FeCl_3 \cdot 6H_2O$. Although $In(OTf)_3$ is somewhat more active, its higher price makes the use of iron(III)chloride more attractive. Furthermore, whereas the activity of $In(OTf)_3$ decreases quickly upon recycling, ionic liquid solutions of $FeCl_3$ remain reasonably active for at least four runs.[62] It was found that in hydrophilic ionic liquids the reaction did not proceed at all, whereas fast conversion was observed with the $[PF_6]^-$-anion, see Scheme 9.17. As water is produced in the course of the reaction it is possible that elimination of the water from the reaction medium helps to protect the catalyst, however, it cannot be excluded that at least some of the catalytic activity is due to the formation of HF. While the reaction proceeds well with a range of aldehydes, no coupling product was observed when benzophenone was used.

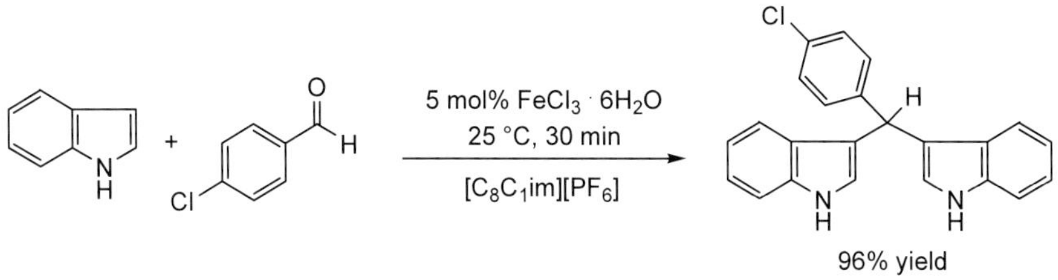

Scheme 9.17: Indole condensation with *p*-chlorobenzaldehyde

This limitation was overcome, when $Dy(OTf)_3$ was used in place of $In(OTf)_3$. This catalyst exhibited similar activity in both $[C_4C_1im][BF_4]$ and $[C_4C_1im][PF_6]$, and afforded better yields and reaction rates in these solvents relative to aqueous media.[63] With either hexanal or acetone as substrates, only a slight deterioration in activity occurred after six catalytic runs. The dysprosium salt also catalysed the reaction between the indole and various imines, but with relatively low selectivity towards the condensation product.

9.2.5 Prins-Cyclisation

The synthesis of tetrahydropyranols from the reaction between an aldehyde and a homoallylic alcohol with catalytic amounts of $Ce(OTf)_3$ immobilised in $[C_4C_1im][PF_6]$ affords the desired product in moderate yield after 12 hours at room temperature, see Scheme 9.18.[64] Addition of one equivalent of benzoic acid as co-catalyst was found to increase the isolated yield, but its role is not yet clear. A beneficial effect of performing the reaction in the ionic liquid is that, contrary to the reaction in chloroform, no ether derivatives were observed.

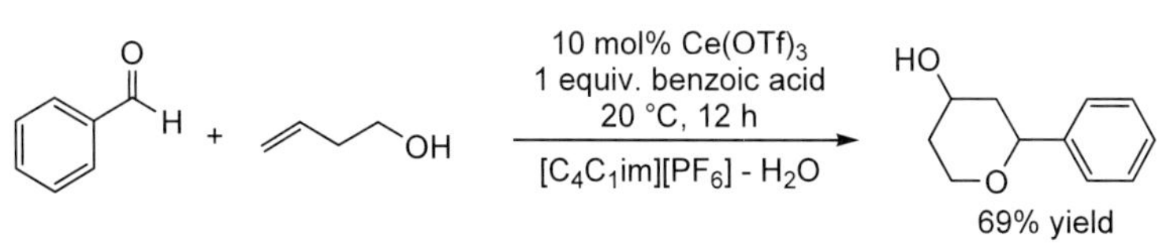

Scheme 9.18: Prins-cyclisation of benzaldehyde

9.2.6 Synthesis of Cyanohydrins

The synthesis of chiral cyanohydrins represents an important type of reaction with enantiopure mandelonitrile being produced on a scale of several hundred tons per year. With the chiral vanadium(IV) salen complex, **60a**, the reaction between benzaldehyde and trimethylsilyl cyanide proceeds with a comparable rate and selectivity in $[C_2C_1im][PF_6]$ relative to dichloromethane, see Scheme 9.19.[65] Inferior results were observed in $[C_4C_1im]Cl$ and $[C_4C_1im][BF_4]$. In the latter solvent, poor chiral induction (ca. 5% *ee*) was observed, thus the selectivity is highly dependent on the nature of the solvent. As $[C_4C_1im]Cl$ gave much better results, 35% *ee*, this observation cannot be attributed to halide impurities. After extraction with hexane, the yield and *ee* remained constant for at least five catalytic runs. Loss of vanadium catalyst was observed in $[C_4C_1im][PF_6]$ as evidenced by the colour of the organic phase, which was green after the extraction process.[66]

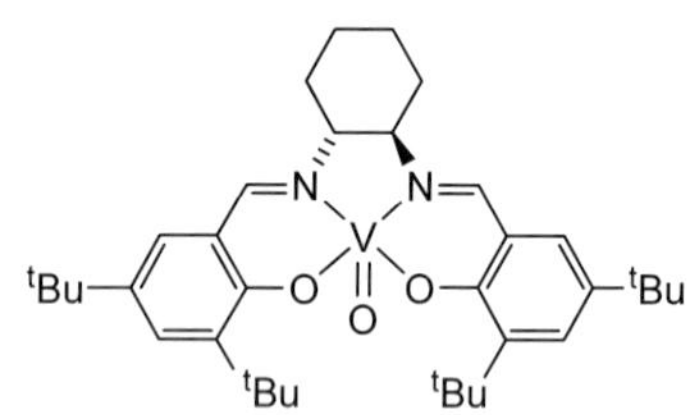

60a

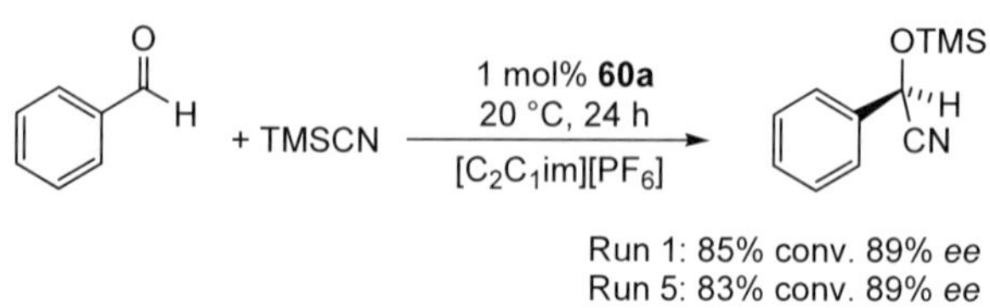

Scheme 9.19: Asymmetric cyano-silylation

Various strategies for catalyst recycling – *viz.* anchoring the catalyst onto a silica support, a single walled nanotube, activated charcoal and immobilisation in $[C_4C_1im][PF_6]$ – have been evaluated in the synthesis of cyanohydrines.[66] While the reaction rate in the ionic liquid was by far the highest, only moderate selectivity was achieved in this system. This was attributed to residual chloride present from the tagged catalyst **60b**, see Scheme 9.20. Therefore, immobilisation on silica was considered as the more promising solution. No catalyst leaching was observed when the catalyst was attached to a silica support while data for the imidazolium tagged complex were not given. It is conceivable that equally high enantioselectivity could be obtained with the ionic liquid immobilised catalyst if the chloride would be replaced for a non-coordinating anion via metathesis.

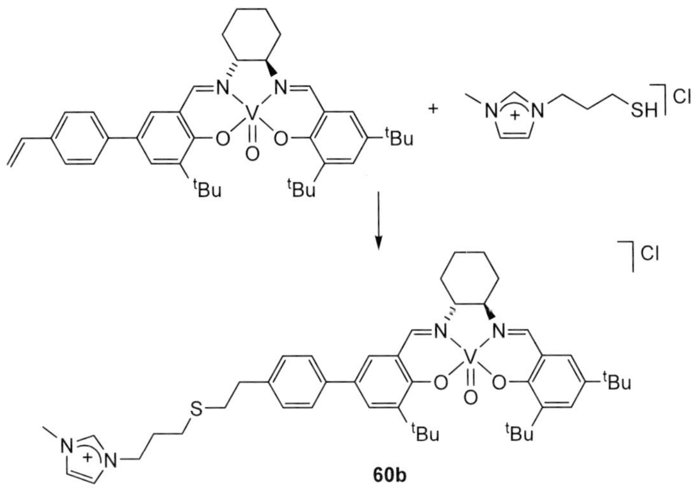

Scheme 9.20: Synthesis of a chiral vanadyl salen complex with an ionic tag

9.2.7 Acetylation and Thioacetylation

A number of Lewis acidic metal salts have been tested in the acetylation of alcohols with acetic anhydride and acetic acid.[67] Of these, copper(II) triflate showed by far the highest activity, but the recycling potential was low. The recyclability was found to be much better with either $Sc(OTf)_3$ or $Yb(OTf)_3$. The reaction between benzyl alcohol and acetic anhydride proceeded to completion within one hour, whereas with acetic acid two days were required under identical reaction conditions. Of the ionic liquids tested, the best activities were obtained in $[C_4C_1im][PF_6]$.

Carbonyl compounds as well as their *O,O*-acetals undergo thioacetalisation and *trans*-thioacetalisation in ionic liquids with scandium(III) triflate as catalyst under facile reaction conditions.[68] Both hydrophilic and hydrophobic ionic liquids give good results and the catalyst could be recycled at least three times without change in activity after extraction of the product with diethyl ether. Higher reaction rates were observed relative to the reaction in dichloromethane. The procedure is highly chemoselective in that in the presence of both an aldehyde and a ketone only the aldehyde reacts to form the corresponding thioacetal.

9.3 Aromatic Substitution Reactions

9.3.1 Friedel-Crafts-Type Reactions

The first example of a Friedel-Crafts type reaction in an ionic liquid medium dates back to 1976 when the electrochemical oxidation of hexamethylbenzene in $[C_2py]Br\text{-}AlCl_3$ afforded a mixture of alkylated polyphenyl compounds.[69] Other early examples include the alkylation of benzene in $[C_2C_1im]Cl\text{-}AlCl_3$[70] and the acylation of ferrocene in $[C_2C_1im]I\text{-}AlCl_3$.[71] There are now numerous examples of Lewis- or Brønsted acid-catalysed Friedel-Crafts type reactions in ionic liquids. These include alkylation,[72-76] acylation,[71,77-83] arylation,[77,84] sulfonylation,[85] sulfoamylation[86] and O-acetylation of alcohols.[87,88]

The mechanism of the $AlCl_3$ and $FeCl_3$ catalysed Friedel-Crafts acetylation of benzene by acetyl chloride has been studied *in situ* in $[C_4C_1im]Cl$ by means of IR and Mössbauer spectroscopy and was found to be identical to that in 1,2-dichloroethane, as shown in Scheme 9.21.[89,90] The species formed when $FeCl_3$ is mixed with an imidazolium chloride are similar to those in present in chloroaluminate liquids. Addition of $FeCl_3$ to $[C_4C_1im]Cl$ leads to an equilibrium mixture containing solid $FeCl_3$, $[C_4C_1im][Fe_2Cl_7]$, and Fe_2Cl_6 or $[C_4C_1im][FeCl_4]$, depending on the molar ratio of the two components.[90]

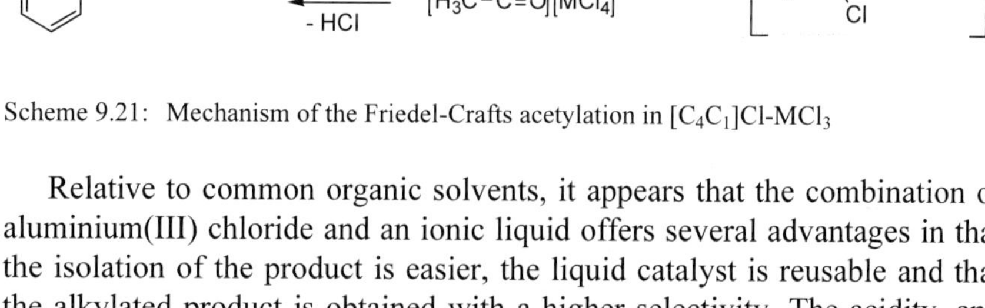

Scheme 9.21: Mechanism of the Friedel-Crafts acetylation in $[C_4C_1]Cl\text{-}MCl_3$

Relative to common organic solvents, it appears that the combination of aluminium(III) chloride and an ionic liquid offers several advantages in that the isolation of the product is easier, the liquid catalyst is reusable and that the alkylated product is obtained with a higher selectivity. The acidity, and hence the catalytic activity of the ionic liquid, depends on the ratio between [cation]Cl and $AlCl_3$, although the amount of $AlCl_3$ in chloroaluminate systems is limited by the liquid range of the ionic liquid to a maximum of 2 mol $AlCl_3$ per mol [cation]Cl. Much higher amounts, up to 5 mol $AlCl_3$, may be dissolved in $[cation][Tf_2N]$ ionic liquids, and recycling of this ionic reaction solution is considerably superior to the 'pure' ionic liquid.[91] One drawback to the use of chloroaluminate or chloroferrate ionic liquids lies in their susceptibility towards hydrolysis, forming HCl and aluminium oxides. Furthermore, not many functional groups are tolerant toward the extreme oxophilicity of these solvents, leading to adducts with C–O functionalities. Instead, Brønsted acidic ionic liquids consisting of *e.g.* hydrogensulfate and hydrogensulfato borate anions can be used,[76] and ionic liquids composed of perfluorinated anions, together with late transition-metal or lanthanide salts as catalysts represent another, easier to handle alternative.

One characteristic of Friedel-Crafts reactions in ionic liquids is their pronounced solvent dependence in that different anion-cation combinations can determine complete and fast conversion and total inactivity of a given catalyst. It is somewhat striking that the catalytic activity is often lowest in hydrophilic ionic liquids from which halide impurities are harder to remove. In any case, screening of a selection of cations and anions appears to be necessary in order to evaluate the suitability of any potential catalyst.

Several metal triflate salts have been tested in the Friedel-Crafts acylation of anisole with benzoyl chloride in $[C_4C_1im][BF_4]$. At a reaction temperature of 80°C and catalyst loadings of 10 mol%, all salts led to 100% conversion with reaction rates decreasing in the order $Cu(OTf)_2$ > $Zn(OTf)_2$ > $Sn(OTf)_2$ >> $Sc(OTf)_3$, see Scheme 9.22.[92]

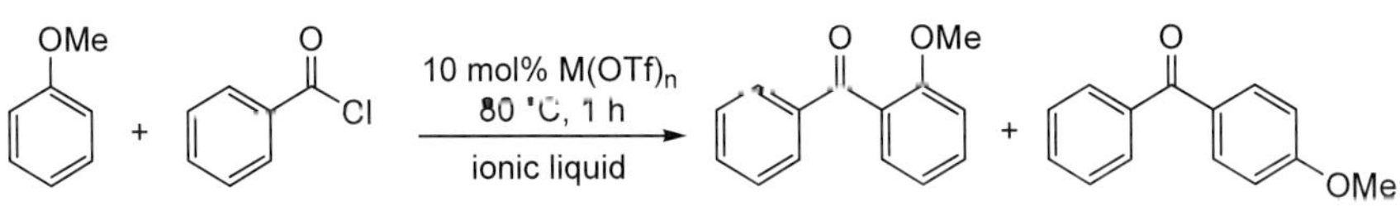

Catalyst	Solvent	Conversion [%]	*ortho*/*para*
$Cu(OTf)_2$	$[C_4C_1im][BF_4]$	100	4 / 96
$Cu(OTf)_2$	CH_3CN	64	7 / 93
$Zn(OTf)_2$	$[C_4C_1im][BF_4]$	87	4 / 96
$Zn(OTf)_2$	CH_3CN	82	8 / 92
$Sn(OTf)_2$	$[C_4C_1im][BF_4]$	74	25 / 75
$Sn(OTf)_2$	CH_3CN	80	7 / 93
$Sc(OTf)_2$	$[C_4C_1im][BF_4]$	10	16 / 84
$Sc(OTf)_2$	CH_3CN	53	9 / 91

Scheme 9.22: Solvent and catalyst effects in the benzoylation of anisole

The surprisingly low activity of $Sc(OTf)_3$ was markedly improved if triflic acid was added. Greater regioselectivities and enhanced reaction rates were achieved in the Friedel-Crafts acylation of activated aromatic compounds in the ionic liquid with $Cu(OTf)_2$, and to a lesser extent also with $Zn(OTf)_2$, relative to conventional solvents. However, deactivated aromatics gave less satisfactory results as was also the case when anhydrides, instead of acyl halides, were used for acylation. In $[C_4C_1im][BF_4]$ the copper(II) triflate reacts with the anhydride to form a precipitate of inactive $Cu(OAc)_2{\cdot}H_2O$, a process that is not observed in conventional solvents. Recycling of the catalyst led to a steady decrease of activity with each run whereas the selectivity remained unchanged.

Various bismuth(III) salts have also been found to act as catalysts for the acylation of benzoylchloride.[93] Complete conversion could be achieved within five minutes with 10 mol% Bi_2O_3 or $Bi(OTf)_3$, though at a relatively high reaction temperature of 150°C. The product was obtained quantitatively with $Bi(OTf)_3$ even at low catalyst loadings of 1 mol%, but reaction times of several hours were required under such conditions. As previously mentioned, catalytic activity is often highly sensitive to the nature of the ionic liquid. In this case, tetraalkylammonium and phosphonium cations led to much lower reaction rates while hydrophilic anions, explicitly $[BF_4]^-$ and $[OTf]^-$, led to poor levels of conversion. Recycling of the catalyst solution was possible after either extraction or distillation of the product *in vacuo*.

A series of imidazolium ionic liquids with various anions has been tested in the $Sc(OTf)_3$-catalysed Friedel-Crafts alkylation at a high catalyst loading of 20 mol%.[94] When hydrophobic liquids containing $[PF_6]^-$ or $[SbF_6]^-$ as anions were used, quantitative conversion was observed in the reaction between benzene and 1-hexene, although the scandium salt is only sparingly soluble in these liquids. The product could be isolated by simple decantation and the ionic phase reused at least two more times without decrease in catalytic activity. In contrast, no reaction took place with $[BF_4]^-$ and $[OTf]^-$ as anions as was the case when various common solvents were used. Rearrangement of the olefin took place prior to ring substitution, indicating that the carbonium ion is formed first.

Numerous transition and rare earth metal triflates were screened as catalysts for the alkenylation of arenes with alkynes, a transformation that can be performed only with difficulty in conventional solvents, see Scheme 9.23. Particularly with electron deficient alkynes such as *p*-trifluormethylphenylacetylene and *p*-chlorophenylacetylene the presence of an ionic liquid was found to be mandatory to observe any catalysis. Highest activity was obtained with $Hf(OTf)_3$, $In(OTf)_3$ and $Sc(OTf)_3$ and the reaction proceeded smoothly, even at relatively low catalyst loadings. It is proposed that the dramatic rate acceleration in the ionic liquid is due to stabilisation of the unstable cationic vinyl intermediate. Products were isolated via decantation and the catalyst could be reused, but the catalyst performance was markedly decreased in the second run. This protocol was also successfully extended to intra-molecular alkenylations of phenyl propiolates and aryl amides.

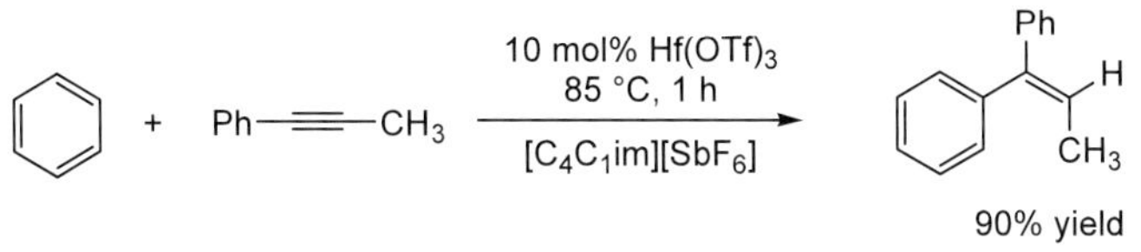

Scheme 9.23: Friedel-Crafts alkenylation of arenes

9.4 Cycloadditions

9.4.1 Diels-Alder-Reactions

The Diels-Alder reaction is one of the most important carbon-carbon bond forming reactions used to prepare cyclic structures. As an example, anthraquinone is synthesised via 1,4-cycloaddition of naphthoquinone and butadiene on an industrial scale. The reaction often affords a mixture of isomers and both selectivity and reaction rate show a strong solvent dependence.[95,96] A number of factors, including the polarity of the solvent,[97] hydrogen bonding interactions[98] and solvophobicity[99] are all important. In addition, the presence of Lewis acid catalysts can substantially affect the reaction and lead to very high selectivity.

The Diels-Alder reaction has been extensively studied in organic solvents and in water, and dramatic enhancements in rate and stereoselectivity are frequently observed in water compared to non-polar solvents. This has been attributed largely to enhanced hydrogen bonding between the solvent and the transition state, as well as to enforced hydrophobic association of reactants, essentially creating a hydrophobic pocket which facilitates alignment of the substrates.[100] The typical benchmark reaction for this transformation is the addition of cyclopentadiene to methyl acrylate and the main synthetic challenge is in achieving good *endo*:*exo* ratios, see Scheme 9.24.

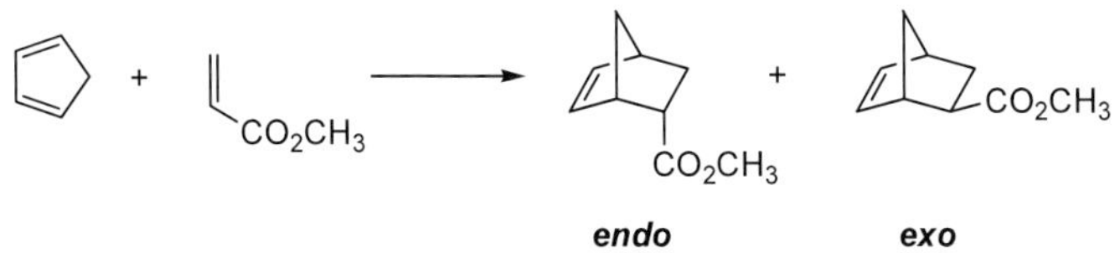

Scheme 9.24: The benchmark Diels-Alder reaction.

As results for the cycloaddition in water are often very good and catalysts used are frequently relatively cheap, performing the reaction in an ionic liquid has to offer significant advantages to be attractive from both an environmental and economical point of view.

The first example of a Diels-Alder reaction involving an ionic liquid, $[EtNH_3][NO_3]$, was published by Jaeger and Tucker in 1989.[101] In the cycloaddition of cyclopentadiene to methyl acrylate, better rates and selectivities were obtained in the ionic liquid relative to non-polar organic solvents, but reaction rates remained significantly below those observed in water at comparable selectivity. Subsequently, chloroaluminate ionic liquids were investigated as solvents for the reaction between cyclopentadiene and methyl acrylate and markedly higher rates and selectivities, *endo*:*exo* 19:1, relative to water were achieved when the mole fraction of $AlCl_3$ was sufficiently high (>0.5 relative to $[C_2C_1im]Cl$).[102]

The use of ionic liquids can in some cases lead to reverse selectivity compared to conventional molecular solvents. Commonly, the *exo*-product is favoured in the reaction between cyclopentadiene and methyl methacrylate, but in acidic $[C_4C_1im]^+$ or $[C_4py]^+$ chloroaluminate ionic liquids *endo:exo* ratios up to 4.55:1 were observed.[103] In another example, reaction between furan and methyl acrylate in either $[C_4C_1im][BF_4]$ or $[C_4C_1im][PF_6]$ also affords predominantly the *endo*-product, contrary to results in conventional solvents.[104]

Welton and co-workers tested a series of imidazolium-based ionic liquids with respect to their influence on the *endo*:*exo* ratio in the addition of methyl acrylate to cyclopentadiene.[105,106] With $[C_4C_1im]^+$, the selectivity decreased in the order $[CF_3CO_2]^- > [Tf_2N]^- > [OTf]^- > [BF_4]^- > [PF_6]^-$, which is in good agreement with the measured change of the charge-transfer absorption band of the nile red dye in these solvents, expressed in the E_T^N-value. More recently some 30 different ionic liquids have been screened with respect to solvent effects on the regioselectivity, see Table 9.1.[107] Although some results did not match previous findings, the most important observation is that even with functionalised imidazolium ionic liquids, the *endo*:*exo* ratio remained within a relatively narrow range, 3.8-5.6, which is clearly inferior to the results achieved in highly Lewis acidic media.

While it is clear that acidic chloroaluminate ionic liquids give very good results, their use is often not practicable and alternatives are desirable. Excellent stereoselectivity was obtained with phosphonium tosylate ionic liquids, $[R_3PR'][OTs]$, in the reaction between isoprene and methyl acrylate affording essentially exclusively the 1,4-isomer.[108] However, with cyclopentadiene as substrate, only poor to moderate selectivity was achieved.[109]

Table 9.1: Cation effect on *endo:exo* selectivities in the cycloaddition of cyclopentadiene to methyl acrylate in $[Tf_2N]^-$ based ionic liquids

R	R'	*endo:exo*
H	$-CH_3$	5.1
H	$-C_2H_5$	4.3
H	$-C_4H_9$	4.2
H	$-C_6H_{13}$	4.1
H	$-C_8H_{17}$	3.9
H	$-C_3H_6CO_2H$	5.4
H	$-C_3H_6CN$	4.4
H	$-CH_2C_6H_5$	5.1
CH_3	$-C_2H_5$	4.6
CH_3	$-C_4H_9$	4.4

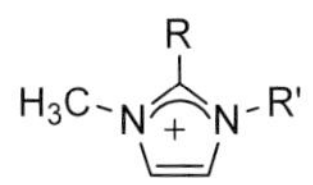

Good to excellent stereoselectivity has been obtained with $ZnCl_2$ as catalyst.[110,111] Relative to chloroaluminate ionic liquids, chlorozincate ionic liquids are insensitive towards water and thus far easier to handle. Choline-$ZnCl_2$ mixtures are fairly viscous (their viscosity typically exceeds 1,000 cP) and necessitate the use of either a co-solvent, heating or mechanical stirring,[110] whereas lower viscosities were observed when zinc(II) chloride was added to imidazolium chloride ionic liquids.[111] Here, best results were obtained with an imidazolium chloride to $ZnCl_2$ ratio of 1:2, in which case the ionic liquid anion is the Lewis acidic species $[Zn_2Cl_5]^-$ (*c.f.* $[Al_2Cl_7]^-$ and $[Fe_2Cl_5]^-$ which are observed in acidic chloroaluminate and chloroferrate ionic liquids, respectively).

Scandium triflate is a highly active catalyst for Diels-Alder[112] and hetero-Diels-Alder reactions.[113] With 0.2 mol% $Sc(OTf)_3$ the cycloaddition of a variety of substrates proceeds in $[C_4C_1im][OTf]$ with almost complete stereoselectivity and good rates at room temperature, see Scheme 9.25. After extraction with diethyl ether, the ionic liquid phase could be recycled with hardly any loss of activity even after the tenth run. Higher catalyst loadings of 3 mol% were used in the hetero-Diels-Alder reaction between aryl amines and cyclic enol ethers, affording the product in good yield and selectivity.

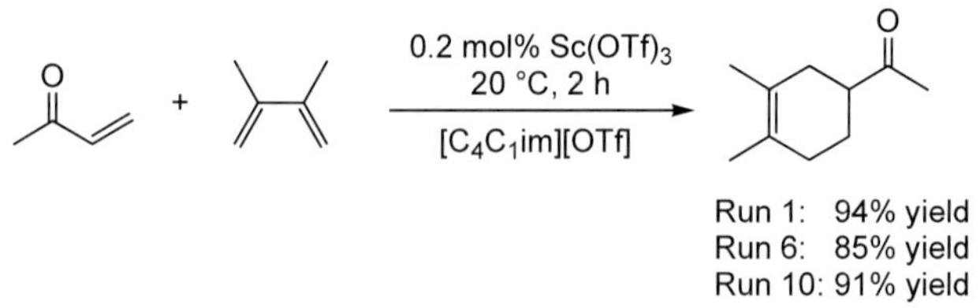

Scheme 9.25: Scandium catalysed Diels-Alder reaction

The cyclisation of a range of dienes and dienophiles has been evaluated in $[C_4C_1im][PF_6]$ using the tungsten(II) complex **61** as catalyst (Scheme 9.26).[114] A wide range of substrates was applied affording very good yield and selectivity in all cases. While reaction rates and selectivities in the ionic liquids were comparable to those in water, a wider scope of substrates could be used in $[C_4C_1im][PF_6]$. Furthermore, recyclability of the complex was found to be superior in the ionic liquid and even after ten runs only a moderate decrease in activity was observed. Extremely short reaction times were realised when the reaction mixture was conducted with microwave dielectric heating with the reaction complete in less than one minute.

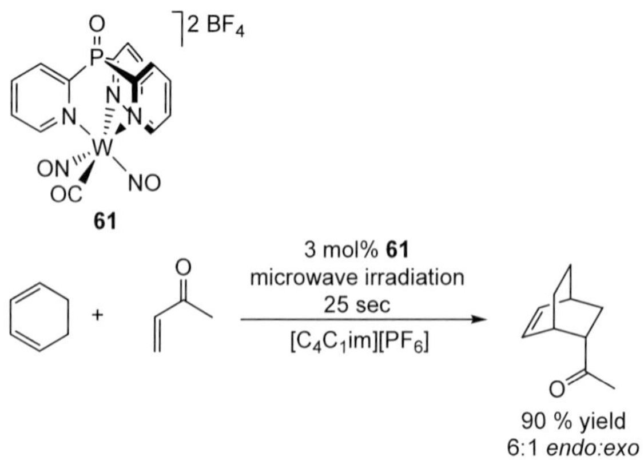

Scheme 9.26: Microwave-assisted Diels-Alder reaction

Very short reaction times were also observed in microwave-assisted hetero-Diels-Alder reactions carried out in solutions doped with an ionic liquid.[115]

Some attempts have been made towards asymmetric Diels-Alder reactions by using chiral ionic liquids. However, neither with a chiral cation, *viz.* *N*,*N*-di(2'*S*-2'-methylbutane)imidazolium,[116] nor a chiral anion, *viz.* *L*-lactate,[117] could any significant chiral induction be observed. However, substantial chiral induction was achieved with a chiral organic catalyst immobilised in either $[C_4C_1im][PF_6]$ or $[C_4C_1im][SbF_6]$, as shown in Scheme

9.27.[118] In the reaction between cyclohexadiene and acrylate, *ee*s as high as 93% could be observed. An ionic liquid composed of a structurally related chiral cation is thus likely to promote asymmetric Diels-Alder reactions.

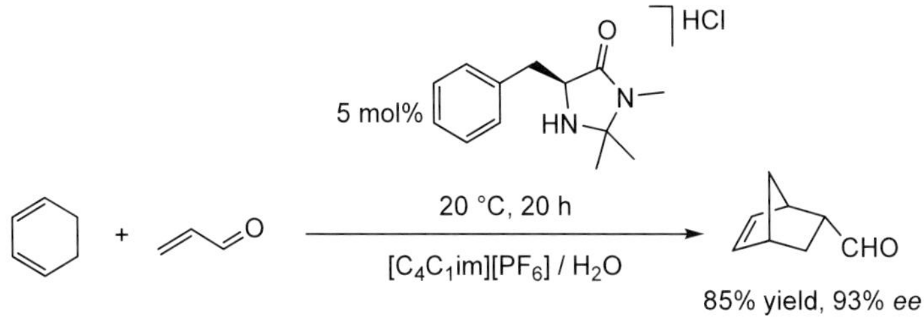

Scheme 9.27: Organo-catalysed asymmetric Diels-Alder reaction

Transition-metal catalysed asymmetric Diels-Alder reactions in ionic liquid media would certainly be an interesting area of study. Kündig and co-workers have recently shown that ruthenium complexes such as **62** afford enantiomeric excess greater than 90% in the reaction between cyclopentadiene and methyl acrylate with, at the same time, acceptable regioselectivity.[119] This system is particularly attractive as a strong anion-dependence on the rate and selectivity due to ion-pairing was observed in conventional solvents and it would be interesting to investigate the performance of such a compound in an ionic environment.

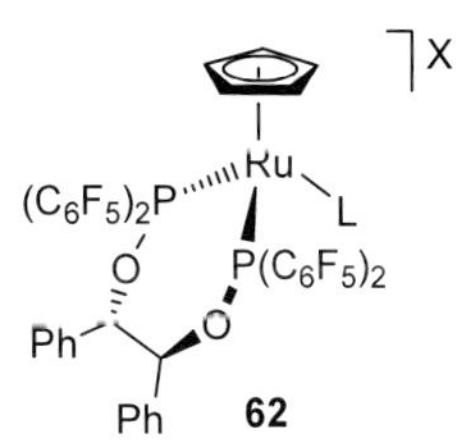

62

9.4.2 Cylcopropanation

Numerous synthetic methods have been developed for the synthesis of cyclopropanes, which represent an important core structure in a number of biologically active compounds. Of these techniques, metal-catalysed cyclopropanation of alkenes with ethyl diazoacetate constitutes a particularly simple and straightforward approach. The metal reacts with the azo compound to form a carbene complex which in turn reacts with the olefin, via formation of a metallabutacycle. Copper-complexes are most commonly employed, but other metals like rhodium and palladium are also used.

Chiral copper-bisoxalzoline complexes have successfully been used for the enantioselective cyclopropanation of styrene (Scheme 9.28) and other substrates with ethyl diazoacetate in an ionic liquid.[120-123] Results comparable in activity and selectivity to the same reaction performed in chloroform or dichloromethane were observed.

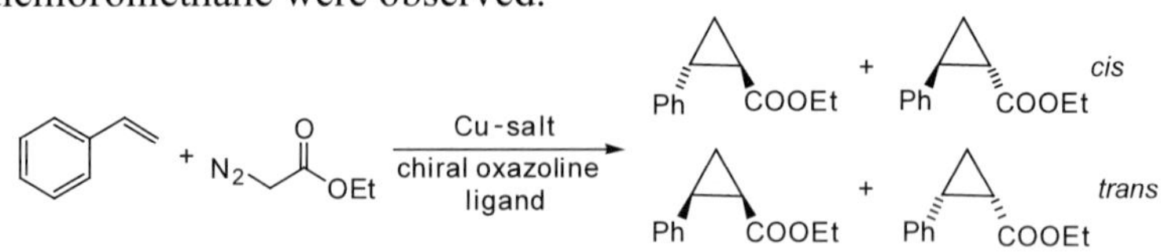

Scheme 9.28: Enantioselective cyclopropanation of styrene

However, high purity of the ionic liquid is crucial for good conversions as halide impurities were found to have a strong impact on the performance of the catalyst. Addition of 5% $[C_4C_1im]Cl$ or $[C_4C_1im]Br$ to the otherwise essentially halide-free ionic liquid solution led to almost complete inhibition of catalytic activity. An indication of the presence of halide impurities is given by the colour of the ionic liquid-catalyst solution, which is clear blue in colour. In the presence of Cl^-, the colour changes to green-yellow whereas upon addition of Br^-, a deep red-brown solution is obtained.[123] Catalyst recycling was superior with $[BF_4]^-$ and $[PF_6]^-$ as anions whereas some leaching was observed when $[OTf]^-$ or $[Tf_2N]^-$ were used. Activity and selectivity remain stable in $[C_4C_1im][BF_4]$ for four cycles after which some deterioration occurs.

The stability of the copper-oxazoline complex is the determining factor in terms of achieving good enantioselectivity.[121] In solution, an equilibrium between the free and complexed ligand exists, as shown in Scheme 9.29. If the catalyst solution is to be recycled, uncomplexed ligand is easily lost in a product extraction step. Of course, this problem can be overcome by adding further ligand after each catalytic cycle, but this is clearly not ideal.

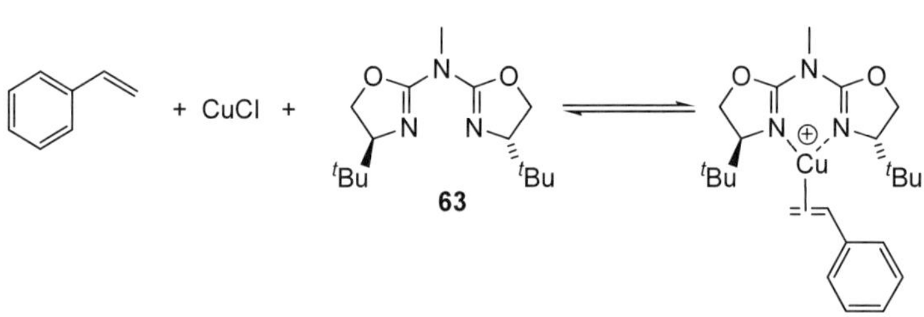

Scheme 9.29: Equilibrium between free and complexed azabis(oxazoline) ligand

Modification of the oxazoline ligand with a charged moiety is an obvious solution to the problem, but potentially not synthetically trivial. Increasing the binding constant of the ligand was found to be a suitable alternative. By using a more electron-rich azabis(oxazoline) ligand, **63**, the enantioselectivity remained stable at 90-92% towards the *trans* isomer for eight cycles. Only when the catalyst solution was exposed to a non-inert atmosphere for several days was a marked decrease in selectivity observed. This was attributed to an increasing water-content in the ionic liquid with the water partially replacing the chiral ligand. Indeed, when the wet ionic liquid was subsequently dried under vacuum, selectivities increased again.

At the present stage, the main advantage of performing asymmetric cyclopropantion with copper catalysts in ionic liquids is that the catalyst can be recycled a number of times. However, the high requirements on the purity of the solvent together with the additional cost and time during solvent recycling reduces this advantage to some degree.

Racemic cyclopropanation reactions have been performed with rhodium[124] and palladium[125] catalysts. The reaction between an olefin and ethyl diazoacetate in the presence of 1 mol% $Rh_2(OAc)_4$ in $[C_4C_1im][PF_6]$ proceeded with higher rates, improved yields and better *trans*-selectivity relative to conventional solvents for both electron-rich and electron-deficient styrene derivatives. The product was obtained by extraction with diethyl ether and in this manner only trace amounts of the catalyst were leached from the ionic liquid. Nevertheless, yields drop from 88% in the first run to 72% in the fifth run.

Simple palladium salts such as $PdCl_2$ and $Pd(OAc)_2$ are highly active catalysts for cyclopropanation. Whereas the use of palladium(II) chloride is not very efficient in common organic solvents due to its poor solubility, this salt shows considerable activity in $[C_8C_1im][BF_4]$. At low catalyst loadings, *ca.* 0.07 mol%, almost quantitative yields are obtained after 10 hours at ambient temperature and the product could be isolated by simple decantation of the organic layer. The recycling potential of $PdCl_2$ was found to be considerably higher than that of $Pd(OAc)_2$ as the latter is slowly reduced to colloidal palladium in the course of the reaction. Use of the cyclopalladated complexes **64** and **65** at low catalyst concentration of 0.03-0.07 mol% only led to a slight increase in reactivity at comparable selectivity, relative to the above palladium salts. As with $Pd(OAc)_2$, complex **64** is reduced to a Pd(0) species and thereby rapidly decreases in activity, whereas **65** shows a decrease in activity only after the sixth cycle.

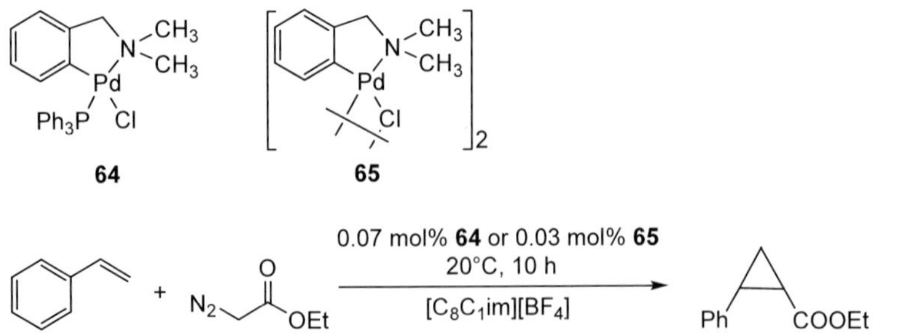

Catalyst:	64	65
Run 1:	98% yield	99% yield
Run 6:	8% yield	97% yield

Scheme 9.30: Palladium-catalysed cyclopropanation

9.4.3 Other [2+2]-Cycloadditions

In the iron-catalysed reaction between styrene derivatives and quinine, reaction rates increase dramatically on changing the solvent from acetonitrile to $[C_4C_1im][PF_6]$.[126] The desired cycloadduct in the reaction between quinone and 4-methoxystyrene was obtained within 10 minutes in the ionic liquid, while the reaction took 2.5 hours in acetonitrile under identical conditions, see Scheme 9.31. Highest reaction rates were observed with iron(II) tetrafluoroborate but $FeCl_3$ and $Fe(ClO_4)_3$-Al_2O_3 also gave good yields. Much lower rates were obtained in $[C_4C_1im][BF_4]$ while no reaction took place at all in $[C_4C_1im][OTf]$.

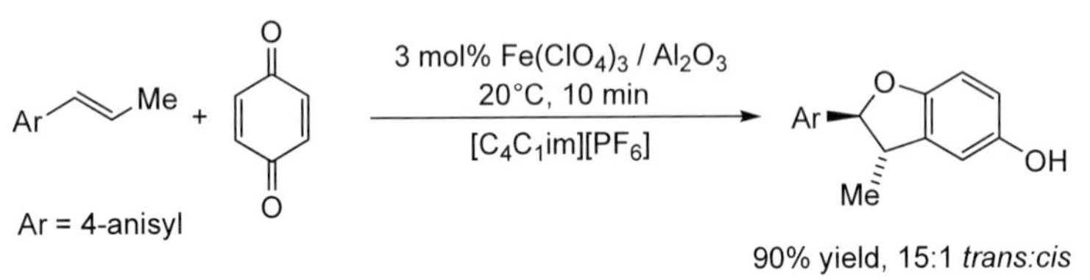

Scheme 9.31: Cycloadition of quinine with 4-methoxystyrene

Aziridines have been synthesised in ionic liquids from diazoacetate and imines.[127] The copper-mediated aziridination between [*N*-(*p*-tolylsufonyl)imino]phenyliodinane and an excess of olefin proceeded in good yield at room temperature (Scheme 9.32).[128] In most cases, the reaction progressed considerably faster in the ionic liquid than in acetonitrile. No marked differences were observed between the reaction rates in $[C_4C_1im][BF_4]$ and $[C_4C_1im][PF_6]$. After extraction with diethyl ether, the catalyst solution could be reused at least four more times with a consistent conservation of activity.

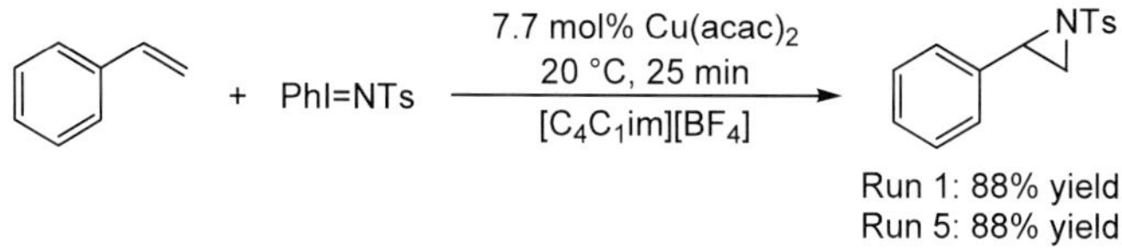

Scheme 9.32: Synthesis of aziridines with styrene as substrate

The reaction between an arylaldehyde, amine and ethyl diazoacetate proceeds in most cases rapidly in $[C_4C_1im][PF_6]$ at room temperature with $Bi(OTf)_3$ as catalyst, see Scheme 9.33.[129] The products were obtained with high stereoselectivity with the *cis*-aziridine carboxylates being formed predominantly. Both amines with electron-donating or withdrawing groups reacted readily and no carbene self-coupling products were detected under the reaction conditions. After extraction with diethyl ether, the catalyst solution could be reused, but with a gradual decrease in activity. Although the reaction proceeds smoothly in other solvents, better selectivity was obtained in the ionic liquid and the unwanted formation of enamines was reduced.

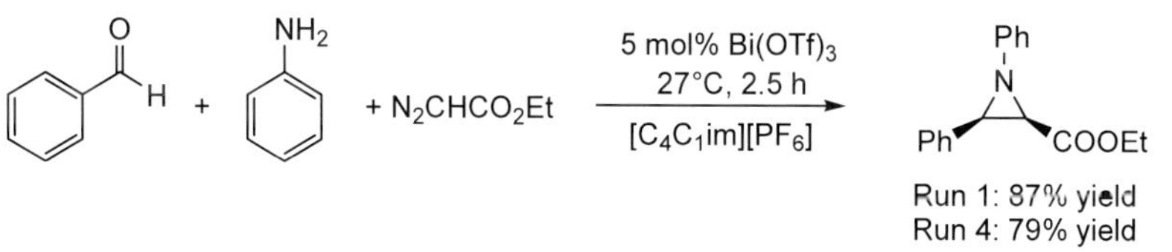

Scheme 9.33: Synthesis of *cis*-aziridine carboxylates

9.5 Isomerisation and Rearrangements

Not much attention has been directed to isomerisation and rearrangement reactions in ionic liquids. Beckmann rearrangements of ketoximes have been performed in neat ionic liquids catalysed by phosphorus(V) chlorides and oxides[130] and Fries rearrangements of phenylbenzoates have been carried out in chloroaluminate ionic liquids with moderate to good selectivity.[131]

The catalytic isomerisation of 2-methyl-3-butenenitrile into 3-pentenenitrile was studied with $Ni(cod)_2$ as catalyst, see Scheme 9.34.[132] After screening almost 20 different ionic liquids and an array of cationic and anionic phosphine ligands, $[C_4C_1C_1im][Tf_2N]$ as solvent and TPPMS as ligand were found to give the best results. Ionic liquids based on $[C_4C_1im]^+$ generally gave lower conversions than those where the acidic proton in the

2-position has been substituted by a methyl group. Formation of a nickel-carbene complex could be responsible for this observation although no such species were detected by *in situ* ^{13}C NMR spectroscopy. On the other hand, detection of such compounds via NMR is not trivial, especially at high ionic liquid/catalyst ratios. Although the catalyst could be successfully immobilised with charged phosphine ligands, recyclability remained poor.

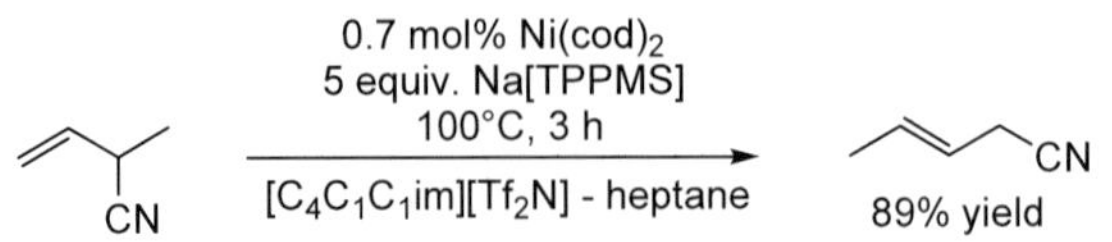

Scheme 9.34: Catalytic isomerisation of 2-methyl-3-butenenitrile

The skeletal reorganisation of enynes can be a useful transformation for the synthesis of natural products. Platinum(II) chloride represents one of the most widely used catalysts for this transformation, but due to its low solubility in organic solvents, usually toluene, high reaction temperatures are often necessary. While no 1-vinylcycloalkene was formed at 40°C in toluene with $PtCl_2$ as catalyst, the product was obtained with 92% yield at room temperature after 20 hours in $[C_4C_1im][PF_6]$, see Scheme 9.35.[133] In addition to $PtCl_2$, $[RuCl_2(CO)_3]_2$ and $AuCl_3$ also gave good conversions, of which the latter gradually decomposed in the course of the reaction. The product was extracted from the ionic liquid with diethyl ether and the slightly yellow colour of the organic phase indicates that some of the platinum catalyst is lost in the process. Accordingly, the catalytic activity slowly decreases upon repeated use of the catalyst solution.

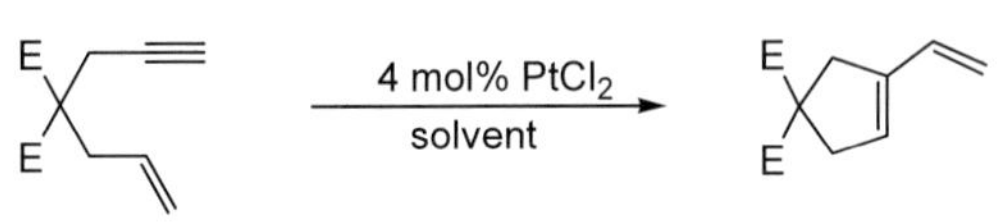

E = ethyl ester

toluene:	80°C	3 h	86% yield
$[C_4C_1im][PF_6]$:	80°C	15 min	79% yield
	40°C	5 h	81% yield
	23°C	20 h	92% yield

Scheme 9.35: Skeletal reorganisation of enynes

Another example describes the Claisen rearrangement of allylphenyl ethers in ionic liquids mainly based on the 1,8-diazabicyclo[5.4.0]-7-undecenium cation. Under the chosen conditions, with scandium(III) triflate as catalyst, these substrates were found the undergo a subsequent cyclisation reaction.[134]

Finally, (homo-)allylic alcohols were isomerised to their corresponding saturated ketones with good to excellent selectivity at 120°C in molten $[(C_4)_4N]Br$ with either $PdCl_2$, $Pd(OAc)_2$ or palladium black.[135]

9.6 Additions Across Carbon-Carbon Multiple Bonds

9.6.1 Hydroamination

Despite the industrial importance of amines and imines, hydroamination, i.e. the direct reaction of alkenes or alkynes with primary or secondary amines, is only used in one commercial process where isobutene and ammonia are converted in the presence of a zeolite catalyst to *t*-butylamine. Turnover frequencies are generally very low and consequently, high catalyst loadings are necessary, which in turn demands efficient recycling.

Many of the late transition metal catalysts that can be used for hydroamination reactions are cationic and thus well suited to the environment provided by ionic liquids. Inter- and intramolecular hydroamination in an ionic liquid was first reported in 2002.[136] The system, which could also be run continuously, comprised of the catalyst $Zn(OTf)_2$ immobilised in $[C_2C_1im][OTf]$ with heptane as the second phase, as depicted in Figure 9.1. Relative to the homogeneous reaction in toluene, markedly higher turnover rates were observed using the ionic liquid-toluene biphase, which were only somewhat lower than those achieved with the heterogeneous Zn-beta catalyst.[137] The concentration of zinc ions in the toluene phase was three orders of magnitude lower when the ionic liquid was present, implying that the reaction in the two-phase system occurs either in the ionic liquid or at the phase boundary. Retention of the zinc salt in the ionic liquid phase is very efficient and the amount of zinc found in the product phase is in the range of the ICP-OES detection limit.

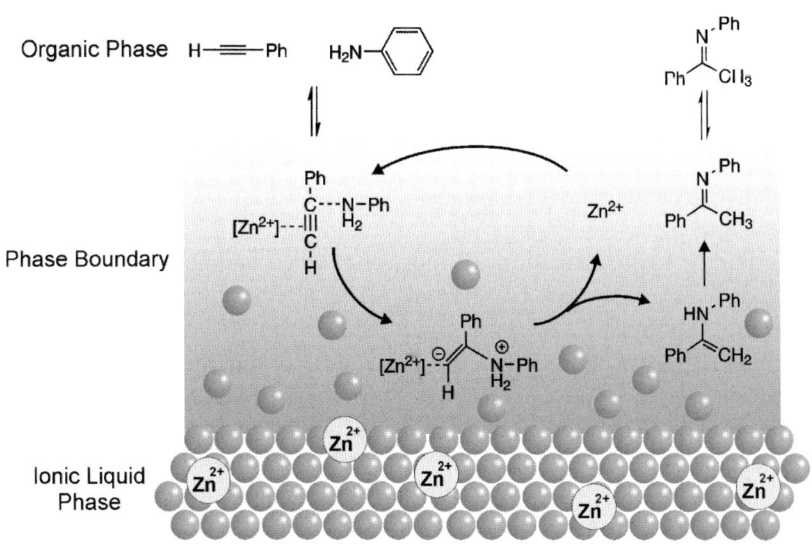

Figure 9.1: Proposed reaction mechanism for the Zn^{2+} catalysed hydroamination

The same catalytic system was also applied to the reaction of phenylacetylene or cyclohexadiene with primary amines.[138] Whereas the substrates dissolve well in the ionic liquid, the formed products are only poorly soluble. With phenylacetylene, electron rich anilines react more slowly than electron poor anilines such as 4-fluoro-phenylamine. With a two-fold excess of amine the reaction proceeds much faster than under stoichiometric conditions. With *n*-octylamine, the reaction proceeds smoothly at low to medium conversion, but after longer reaction times an increasing amount of by-products are formed.

In an extension of this work, either zinc(II), palladium(II), rhodium(I) or copper(I) salts were immobilised in an ionic liquid film (SILP, *vide supra*) onto diatomic earth and the catalysts tested for activity in the reaction between phenylacetylene and 4-isopropyl-phenylamine.[139,140] The supported rhodium, ruthenium and zinc complexes afford higher rates and selectivities relative to their use under homogenous reaction conditions. Lower rates are, however, observed with the copper salt, which is rationalised by strong complexation of the ionic liquid to the Cu(I) centre.

A series of rhodium and platinum compounds have been tested in the hydroamination of norbornene with aniline, as shown in Scheme 9.36.[141] Selectivity and activity were highly dependent on the nature of the ionic liquid, but were always superior to those observed in THF. Solvents with chloride anions led to essentially no catalytic activity, whereas $[PF_6]^-$ or Br^- afforded some catalysis. Nonetheless, even with the best solvent/catalyst combination, less than 40 turnovers are achieved after 6 days at 140°C.

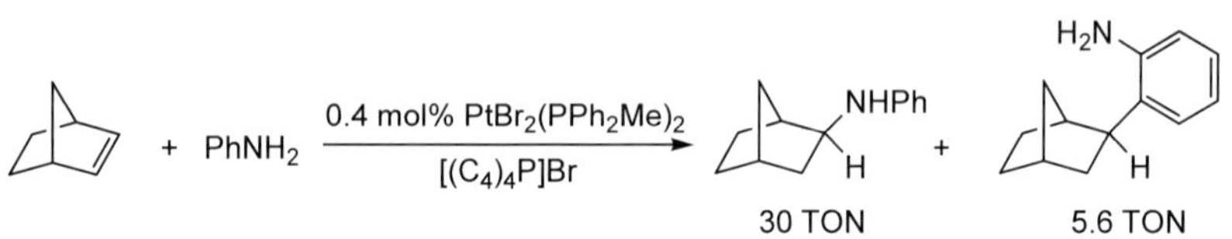

Scheme 9.36: Hydroamination of norbornene with aniline

Similarly, hydroamination of ethylene with aniline and $PtBr_2$ was achieved.[142] The influence of a number of parameters such as reaction time, temperature, ethylene pressure and aniline/catalyst ratio have been investigated. It was found that the catalyst is gradually poisoned by the formed *N*-alkylated aniline. Accordingly, a biphasic system with $[(C_4)_4P]Br$ and decane as the organic phase afforded higher turnover numbers, see Scheme 9.37.

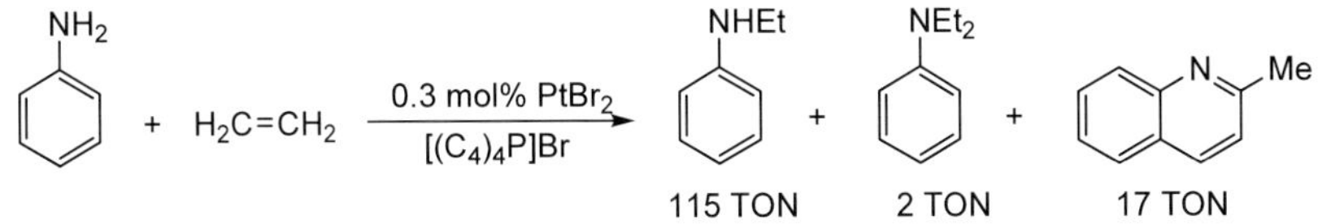

Scheme 9.37: Hydroamination of ethylene with aniline

Asymmetric hydroamination using a chiral PIGIPHOS-Ni(II) complex has also been achieved in ionic liquids, as shown in Scheme 9.38.[143] A number of different imidazolium and picolinium ionic liquids were tested and relative to THF, much higher turnover numbers (300 *vs.* 20) were observed in the reaction between methacrylonitrile and morpholine at comparable selectivity (64% ee *vs.* 69%). Moreover, the catalyst in the ionic liquid solution is less sensitive to air and moisture so that non-distilled reagents can be used.

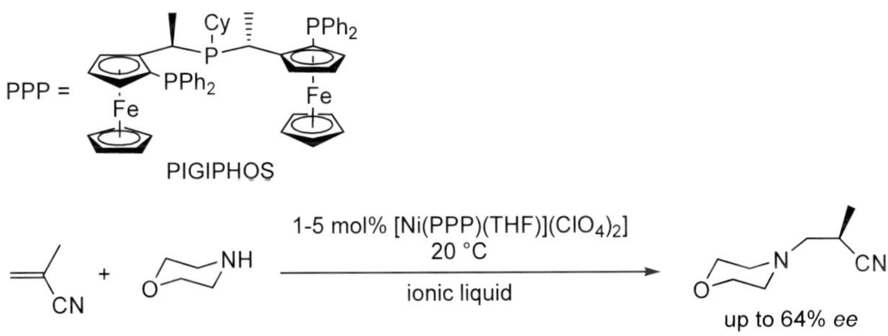

Scheme 9.38: Hydroamination of methacrylonitrile with morpholine

Finally, though not strictly a hydroamination reaction, the asymmetric addition of alkynes to imines with a copper-bis(oxazoline) complex is worth briefly mentioning.[144] The nature of the ionic liquid cation has a strong effect on the enantioselectivity of the reaction and it appears that a good balance between hydrophobicity and acidity play an important role with best results obtained with $[C_4C_1im][Tf_2N]$.

9.6.2 Hydroboration and Hydrosilylation

Very little has been reported so far on catalytic hydroboration and hydrosilylation in ionic liquids, which makes it difficult to judge whether they could offer significant improvements over existing methods. Platinum-catalysed hydrosilylation, for example, is a highly efficient process with catalyst loadings as low as 1 ppm relative to the substrate. Due to the low catalyst loading and the fact that hydrosilylation products are not for human consumption, efforts to remove the catalyst are from the final product are seldom made. However, the biphasic process could be very attractive if it was found to be economically and environmentally advantageous.

The hydroborylation of terminal alkynes with pinacolborane (PinBH) and $RhCl(PPh_3)_3$ or $NiCl_2$(dppe) as catalyst has been studied in a variety of ionic liquids including chloroaluminates and chlorozincates, see Scheme 9.39.[145] In the absence of a catalyst, the reaction between 1-hexyne and PinBH is 100% selective whereas in the presence of a catalyst, faster rates are achieved at the cost of selectivity.

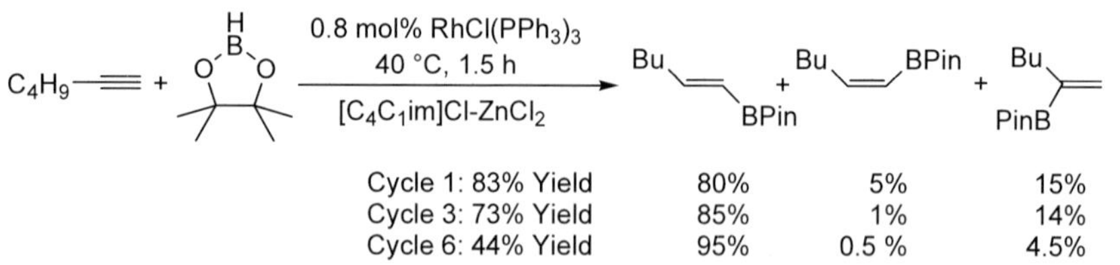

Scheme 9.39: Hydroborylation of 1-hexyne

A slight increase in temperature led to a significant reduction of the reaction time (1.5 h at 40°C *vs.* 48 h at 20°C), which might to some extent be due to the different viscosities of the ionic liquid at these temperatures. After product isolation via distillation, reuse of the catalyst solution was feasible, but with a steady decrease in activity.

While the rhodium catalyst $RhCl(PPh_3)_3$ gives good results in the hydrosilylation of 1-octene in benzene, its recycling potential, when immobilised in $[C_4C_1im][BF_4]$, is low because the catalyst dissolves preferentially in the product phase. By synthesising an analogue of Wilkinson's catalyst, bearing fluorinated phosphine ligands, the stability of the system could be markedly improved, see Scheme 9.40.[146] Crucial to the retention of the catalyst was the use of an imidazolium-based ionic liquid with a highly fluorinated anion, $[B\{C_6H_4(SiMe_2CH_2CH_2C_6F_{13})\text{-}p\}_4]^-$. In this manner a total turnover number of 4,000 after 15 cycles was achieved at *ca.* 4% loss of rhodium per cycle. In contrast, rhodium leaching with the same catalyst in biphasic fluorous catalysis was 12% per cycle.

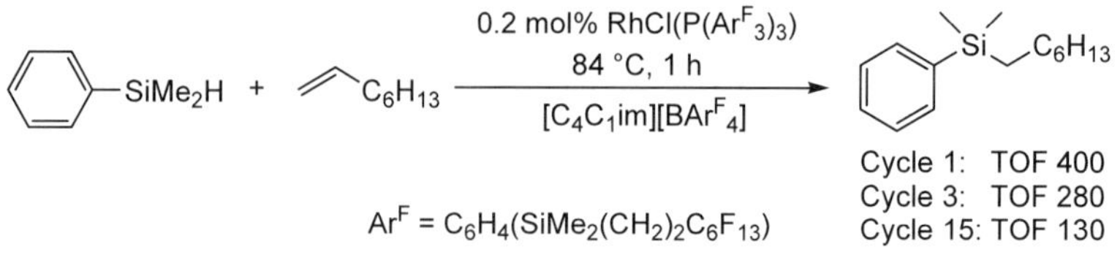

Scheme 9.40: Hydrosilylation of 1-octene

A series of ionic liquids have been tested in the hydrosilylation of monosubstituted alkynes with H_2PtCl_6 (Speier's catalyst), as shown in Scheme 9.41.[145] After completion of the reaction, the products were isolated by vacuum distillation. The catalyst solution was reused two times and average yields for the isolated vinylsilanes are in a range of 80-96%. Reaction rates are highest in $[(C_8)_3C_1N][Tf_2N]$ which is a reflection of the homogeneity of the reaction mixtures with this particular ionic liquid. No reaction was observed with $[(C_8)_3C_1N]Cl$ (Aliquat 336), probably due to catalyst deactivation from the nucleophilic chloride.

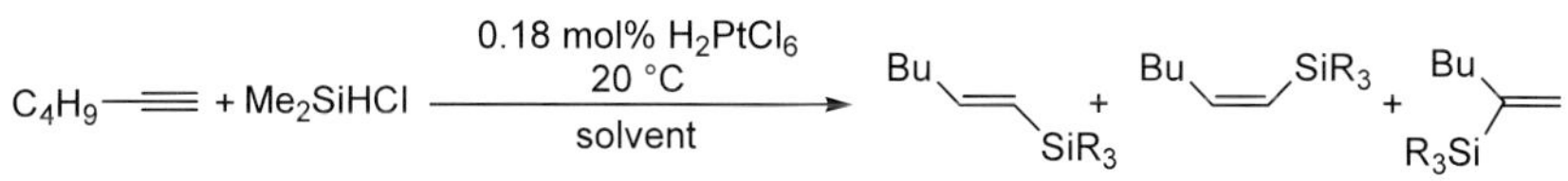

CH_2Cl_2:	0.5 h	92% yield	98%	0%	2%
$[C_2C_1im][Tf_2N]$:	48 h	95% yield	94%	0%	6%
$[C_4C_1im][Tf_2N]$:	24 h	87% yield	85%	0%	14%
$[(C_8)_3C_1N][Tf_2N]$:	1 h	95% yield	95%	0 %	5%

Scheme 9.41: Hydrosilylation of 1-hexyne

Additionally, $AlCl_3$ was also tested as catalyst in $[C_4C_1im][BF_4]$ for the hydrosilylation of 1-hexyne, but the reaction was rather difficult to control, and formation of polymers was frequently observed.

9.6.3 Hydrovinylation and Hydroarylation

Hydrovinylation, or co-dimerisation of alkenes with ethylene to yield 3-substituted 1-butenes, is a reaction that is catalysed with high enantioselectivity by Wilkes' catalyst, **66**, shown in Scheme 9.42. Usually, a chloride-abstracting agent, such as an alkyl aluminium chloride, is required to activate the catalyst. Ionic liquids were found to activate the complex, but the effectiveness depends highly on the nature of the anion of the ionic liquid.[147] Highest activities were obtained with $[BARF]^-$ and $[Al\{OC(CF_3)_2Ph\}_4]^-$ while $[BF_4]^-$ gave only poor results. This is expressed in different enantioselectivities, which are highest in $[C_2C_1im][BARF]$, and this specific ionic liquid also led to the highest degree of consecutive reactions *i.e.* isomerisation and oligomerisation. A cation effect on the selectivity was also observed in that pyridinium-type ionic liquids afforded better results than those based on the imidazolium cation. If the reaction was run under batch-conditions, catalyst degradation occurred after three to four catalytic runs. It was speculated that the catalytically active species decomposed in the absence of substrate and accordingly a continuous-flow process was developed in which the product is extracted with *sc*CO_2. In this manner, the catalyst remains active for at least 61 hours with only a small decrease in selectivity during that period.

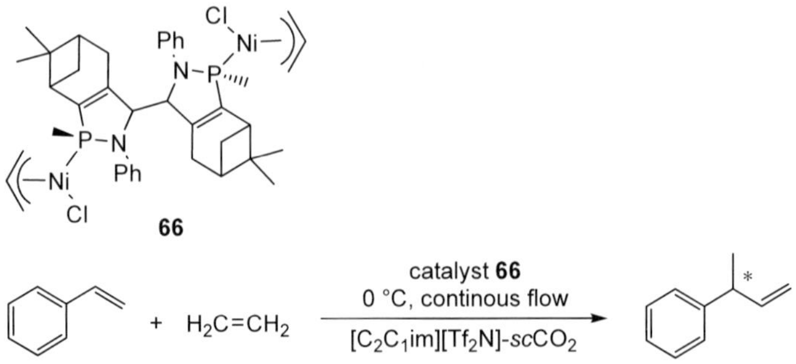

Scheme 9.42: Asymmetric hydrovinylation

The hydroarylation of diphenylacetylene proceeds well with a palladium(0) catalyst bearing a triazacyclopentadecatriene ligand in $[C_4C_1im][BF_4]$ (Scheme 9.43).[148] Inferior conversions were obtained with either $Pd(PPh_3)_4$ or $Pd_2(dba)_3$ as catalyst or with $[C_4C_1im][PF_6]$ as solvent. After extraction of the product with diethyl ether the catalyst remained relatively stable for several runs. Likewise, propargylic alcohols were used as substrate and the reaction with $Pd(OAc)_2$ in $[C_4C_1im][BF_4]$ proceeded under mild conditions (40°C) at higher rates and with better regioselectivity than in common molecular solvents.[149]

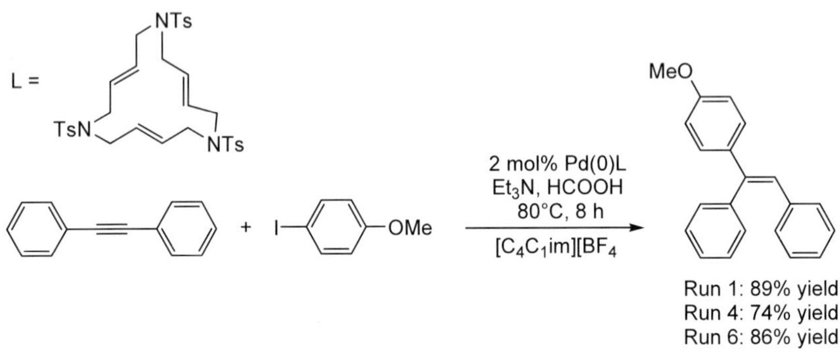

Scheme 9.43: Hydroarylation of diphenylacetylene with *p*-iodoanisole

9.6.4 Silylstannation

Tributylsilyltin readily reacts with terminal alkynes in $[C_4C_1im][PF_6]$ to afford the corresponding 1-trialkylsilyl-2-tributylstannyl-1Z-alkenes regio- and stereoselectively in near quantitative yield, as shown in Scheme

9.44.[150,151] Depending on the substrate, fairly long reaction times, frequently longer than 24 hours, were required at 70°C in either $[C_4C_1im][PF_6]$ or $[C_4C_1im][BF_4]$ to obtain good conversion. While reactions proceeded considerably faster in THF, catalyst stability appeared to be superior in the ionic liquid and catalyst recycling was possible for up to 10 runs without loss of activity. Diethyl ether as co-solvent serves the purpose to "wash" substrate condensing at the walls of the flask during the reaction back into the ionic liquid. Palladium nanoparticles as well as Pd-imidazolylidene carbene complexes have been suggested to be the active species in the reaction, but their exact nature remains unclear.

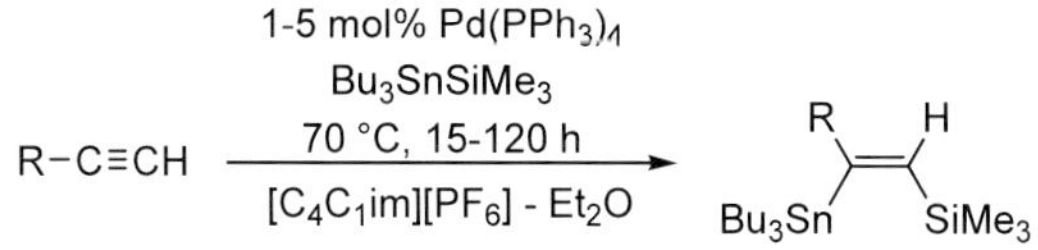

Scheme 9.44: Silylstannation of terminal alkynes

9.6.5 Sakurai-Reaction

The reaction of several *α*,*β*-unsaturated ketones with allyltrimethyl silane in either $[C_4C_1im][BF_4]$ or $[C_4C_1im][PF_6]$ has been compared with several classical solvents, see Scheme 9.45.[152] With indium(III) chloride as catalyst, only in some cases was a higher activity observed in the ionic liquids, but the differences relative to the commonly used solvent, dichloromethane, were small and no obvious advantage in performing the reaction in the ionic media was evident. Very high catalyst loadings, ca. 20-50 mol%, were used but no attempts were made to recycle the catalyst phase.

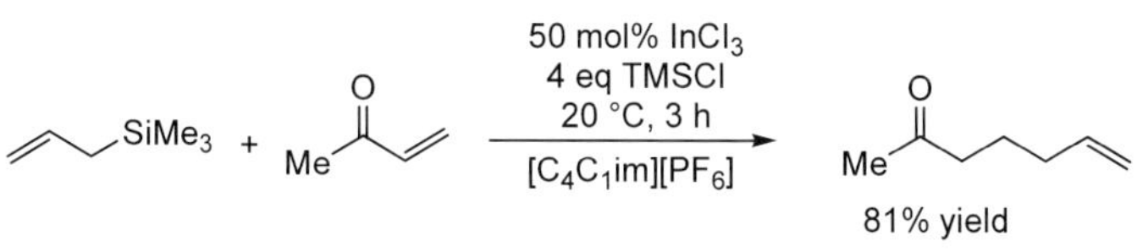

Scheme 9.45: Sakurai reaction to afford 6-hepten-2-one

9.6.6 Ferrier-Rearrangement

Glycals react with nucleophiles such as allylsilanes and alcohols under Lewis-acid catalysed conditions. The reaction is usually highly stereoselective in favour of the *α*-enantiomer. Several lanthanide salts, $Dy(OTf)_3$, $Sc(OTf)_3$ and $Yb(OTf)_3$ have been successfully used as catalysts in the reaction between 2,4,6-tri-*O*-acetyl-D-glucal and different nucleophiles in $[C_4C_1im][BF_4]$ and $[C_4C_1im][PF_6]$, see Scheme 9.46.[153,154]

With alcohols as nucleophiles, superior results were obtained in hydrophobic ionic liquid although the catalyst $Dy(OTf)_3$ formed a suspension in the liquid.[153] However, at elevated temperatures, the reaction proceeds in $[C_4C_1im][BF_4]$ in the absence of a catalyst. After extraction with diethyl ether reuse of the catalyst was possible without deterioration of activity. With allyl and propargyl silanes as nucleophiles and $Yb(OTf)_3$ as catalyst in $[C_4C_1im][BF_4]$, re-addition of catalyst was necessary after each run to drive the reaction to completion.[154]

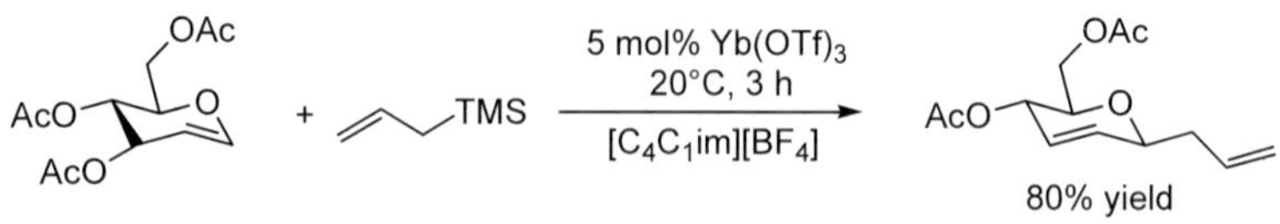

Scheme 9.46: Carbon-Ferrier rearrangement with allyl silane as nucleophile.

9.7 Concluding Remarks

There seems to be no limit to the type of reactions that can be conducted in ionic liquids, although it is not always clear what advantage is offered by the ionic environment. However, the overwhelming majority of reactions have been evaluated in a very narrow range of ionic liquids, notably $[C_4C_1im][BF_4]$ and $[C_4C_1im][PF_6]$. With more than one trillion ionic liquids potentially available,[155] it is highly likely that considerable improvements in the reactions described in this chapter, as well as in many other reactions not yet evaluated in ionic liquids, will emerge over the next few years.

9.8 References

[1] Yoshino, H.; Matsubara, S.; Oshima, K.; Matsumoto, K.; Hagiwara, R.; Ito, Y. *J. Fluorine Chem.* **2004**, *125*, 455-458.

[2] Yoshino, H.; Nomura, K.; Matsubara, S.; Oshima, K.; Matsumoto, K.; Hagiwara, R.; Ito, Y. *J. Fluorine Chem.* **2004**, *125*, 1127-1129.

[3] Yadav, J. S.; Reddy, B. V. S.; Reddy, R. S. R.; Basak, A. K.; Narsaiah, A. V. *Adv. Synth. Catal.* **2004**, *346*, 77-82.

[4] Wu, H. H.; Sun, J.; Yang, F.; Tang, J.; He, M. Y. *Chin. J. Chem.* **2004**, *22*, 619-621.

[5] Salaheldin, A. M.; Yi, Z.; Kitazume, T. *J. Fluorine Chem.* **2004**, *125*, 1105-1110.

[6] Pascali, G.; Kiesewetter, D. O.; Salvadori, P. A.; Eckelman, W. C. *J. Labelled Compd. Radiopharm.* **2004**, *47*, 373-383.

[7] Nguyen, H. P.; Kirilov, P.; Matondo, H.; Baboulene, M. *J. Mol. Catal. A* **2004**, *218*, 41-45.

[8] Togo, H.; Hirai, T. *Synlett* **2003**, 702-704.

[9] Rajagopal, R.; Jarikote, D. V.; Lahoti, R. J.; Daniel, T.; Srinivasan, K. V. *Tetrahedron Lett.* **2003**, *44*, 1815-1817.

[10] Salazar, J.; Dorta, R. *Synlett* **2004**, 1318-1320.

[11] Bortolini, O.; Bottai, M.; Chiappe, C.; Conte, V.; Pieraccini, D. *Green Chem.* **2002**, *4*, 621-627.

[12] Murray, C. B.; Sandford, G.; Korn, S. R. *J. Fluorine Chem.* **2003**, *123*, 81-84.

[13] Kim, J.; Shreeve, J. M. *Org. Biomol. Chem.* **2004**, *2*, 2728-2734.

[14] Ibrahim, H.; Togni, A. *Chem. Commun.* **2004**, 1147-1155.

[15] Baudequin, C.; Plaquevent, J. C.; Audouard, C.; Cahard, D. *Green Chem.* **2002**, *4*, 584-586.

[16] Baudequin, C.; Loubassou, J. F.; Plaquevent, J. C.; Cahard, D. *J. Fluorine Chem.* **2003**, *122*, 189-193.

[17] Hamashima, Y.; Takano, H.; Hotta, D.; Sodeoka, M. *Org. Lett.* **2003**, *5*, 3225-3228.

[18] Evdokimova, G.; Zinovyev, S.; Perosa, A.; Tundo, P. *Appl. Catal. A* **2004**, *271*, 129-136.

[19] Zhu, Y. H.; Ching, C. B.; Carpenter, K.; Xu, R.; Selvaratnam, S.; Hosmane, N. S.; Maguire, J. A. *Appl. Organomet. Chem.* **2003**, *17*, 346-350.

[20] Brown, R. J. C.; Dyson, P. J.; Ellis, D. J.; Welton, T. *Chem. Commun.* **2001**, 1862-1863.

[21] Mehnert, C. P.; Dispenziere, N. C.; Cook, R. A. *Chem. Commun.* **2002**, 1610-1611.

[22] Hamaya, J.; Suzuki, T.; Hoshi, T.; Shimizu, K.; Kitayama, Y.; Hagiwara, H. *Synlett* **2003**, 873-875.

[23] Chen, S. L.; Ji, S. J.; Loh, T. P. *Tetrahedron Lett.* **2004**, *45*, 375-377.

[24] Hagiwara, H.; Sayuri, T.; Okabe, T.; Hoshi, T.; Suzuki, T.; Suzuki, H.; Shimizu, K.; Kitayama, Y. *Green Chem.* **2002**, *4*, 461-463.

[25] Kotrusz, P.; Kmentova, I.; Gotov, B.; Toma, S.; Solcaniova, E. *Chem. Commun.* **2002**, 2510-2511.

[26] Loh, T. P.; Feng, L. C.; Yang, H. Y.; Yang, J. Y. *Tetrahedron Lett.* **2002**, *43*, 8741-8743.

[27] Gruttadauria, M.; Riela, S.; Lo Meo, P.; D'Anna, F.; Noto, R. *Tetrahedron Lett.* **2004**, *45*, 6113-6116.

[28] Cordova, A. *Tetrahedron Lett.* **2004**, *45*, 3949-3952.

[29] Zhao, G. Y.; Jiang, T.; Gao, H. X.; Han, B. X.; Huang, J.; Sun, D. H. *Green Chem.* **2004**, *6*, 75-77.

[30] Zhang, J.; Bond, A. M. *Anal. Chem.* **2003**, *75*, 6938-6948.

[31] Zhang, X. Y.; Fan, X. S.; Niu, H. Y.; Wang, J. J. *Green Chem.* **2003**, *5*, 267-269.

[32] Yang, X. F.; Wang, M. W.; Varma, R. S.; Li, C. J. *Org. Lett.* **2003**, *5*, 657-660.

[33] Yang, X. F.; Wang, M. W.; Varma, R. S.; Li, C. J. *J. Mol. Catal. A* **2004**, *214*, 147-154.

[34] Li, Z.; Wei, C.; Chen, L.; Varma, R. S.; Li, C. J. *Tetrahedron Lett.* **2004**, *45*, 2443-2446.

[35] Chen, S. L.; Ji, S. J.; Loh, T. P. *Tetrahedron Lett.* **2003**, *44*, 2405-2408.

[36] Lee, S. G.; Park, J. H. *Bull. Korean Chem. Soc.* **2002**, *23*, 1367-1368.

[37] Kim, H. S.; Kim, Y. J.; Bae, J. Y.; Kim, S. J.; Lah, M. S.; Chin, C. S. *Organometallics* **2003**, *22*, 2498-2504.

[38] Mizushima, E.; Hayashi, T.; Tanaka, M. *Green Chem.* **2001**, *3*, 76-79.

[39] Woitków, W.; Trzeciak, A. M.; Choukroun, R.; Pellegatta, J. L. *J. Mol. Catal. A* **2004**, *224*, 81-86.
[40] Calò, V.; Giannoccaro, P.; Nacci, A.; Monopoli, A. *J. Organomet. Chem.* **2002**, *645*, 152-157.
[41] Skoda-Földes, R.; Takács, E.; Horváth, J.; Tuba, Z.; Kollár, L. *Green Chem.* **2003**, *5*, 643-645.
[42] Shi, F.; Peng, J. J.; Deng, Y. Q. *J. Catal.* **2003**, *219*, 372-375.
[43] Müller, E.; Peczely, G.; Skoda-Foldes, R.; Takacs, E.; Kokotos, G.; Bellis, E.; Kollar, L. *Tetrahedron* **2005**, *61*, 797-802.
[44] Mizushima, E.; Hayashi, T.; Tanaka, M. *Top. Catal.* **2004**, *29*, 163-166.
[45] Consorti, C. S.; Ebeling, G.; Dupont, J. *Tetrahedron Lett.* **2002**, *43*, 753-755.
[46] Jiang, T.; Han, B. X.; Zhao, G. Y.; Chang, Y. H.; Gao, L.; Zhang, J.; Yang, G. Y. *J. Chem. Res.* **2003**, 549-551.
[47] Mastrorilli, P.; Nobile, C. F.; Paolillo, R.; Suranna, G. P. *J. Mol. Catal. A* **2004**, *214*, 103-106.
[48] Becheanu, A.; Laschat, S. *Synlett.* **2002**, 1865-1867.
[49] Kim, S. W.; Son, S. U.; Lee, S. S.; Hyeon, T.; Chung, Y. K. *Chem. Commun.* **2001**, 2212-2213.
[50] Son, S. U.; Lee, S. I.; Chung, Y. K.; Kim, S. W.; Hyeon, T. *Org. Lett.* **2002**, *4*, 277-279.
[51] Son, S. U.; Park, K. H.; Chung, Y. K. *Org. Lett.* **2002**, *4*, 3983-3986.
[52] Park, K. H.; Son, S. U.; Chung, Y. K. *Chem. Commun.* **2003**, 1898-1899.
[53] Park, K. H.; Jung, I. G.; Chung, Y. K. *Org. Lett.* **2004**, *6*, 1183-1186.
[54] Tominaga, K.; Sasaki, Y. *Chem. Lett.* **2004**, *33*, 14-15.
[55] Peng, J. J.; Deng, Y. Q. *Chin. J. Catal.* **2001**, *22*, 598-600.
[56] Peng, J. J.; Deng, Y. Q. *New J. Chem.* **2001**, *25*, 639-641.
[57] Kim, H. S.; Kim, J. J.; Kim, H.; Jang, H. G. *J. Catal.* **2003**, *220*, 44-46.
[58] Kim, Y. J.; Cheong, M. *Bull. Korean Chem. Soc.* **2002**, *23*, 1027-1028.
[59] Kawanami, H.; Sasaki, A.; Matsui, K.; Ikushima, Y. *Chem. Commun.* **2003**, 896-897.
[60] Alvaro, M.; Baleizao, C.; Das, D.; Carbonell, E.; Garcia, H. *J. Catal.* **2004**, *228*, 254-258.
[61] Gu, Y. L.; Shi, F.; Deng, Y. Q. *J. Org. Chem.* **2004**, *69*, 391-394.
[62] Ji, S.-J.; Zhou, M.-F.; Gu, D.-G.; Jiang, Z.-Q.; Loh, T.-P. *Eur. J. Org. Chem.* **2004**, 1584-1587.
[63] Mi, X.; Luo, S.; He, J.; Cheng, J.-P. *Tetrahedron Lett.* **2004**, 4567-4570.
[64] Keh, C. C. K.; Namboodiri, V. V.; Varma, R. S.; Li, C. J. *Tetrahedron Lett.* **2002**, *43*, 4993-4996.
[65] Baleizao, C.; Gigante, B.; Garcia, H.; Corma, A. *Green Chem.* **2002**, *4*, 272-274.
[66] Baleizao, C.; Gigante, B.; Garcia, H.; Corma, A. *Tetrahedron* **2004**, *60*, 10461-10468.
[67] Lee, S. G.; Park, J. H. *J. Mol. Catal. A* **2003**, *194*, 49-52.
[68] Kamal, A.; Chouhan, G. *Tetrahedron Lett.* **2003**, *44*, 3337-3340.
[69] Koch, V. R.; Miller, L. L.; Osteryoung, R. A. *J. Am. Chem. Soc.* **1976**, *98*, 5277-5284.
[70] Boon, J. A.; Levisky, J. A.; Pflug, J. L.; Wilkes, J. S. *J. Org. Chem.* **1986**, *51*, 480-483.
[71] Surette, J. K. D.; Green, L.; Singer, R. D. *J. Chem. Soc., Chem. Commun.* **1996**, 2753-2754.
[72] Formentin, P.; Garcia, H. *Catal. Lett.* **2002**, *78*, 115-118.
[73] DeCastro, C.; Sauvage, E.; Valkenberg, M. H.; Holderich, W. F. *J. Catal.* **2000**, *196*, 86-94.
[74] Piao, L.; Fu, X.; Yang, Y.; Tao, G. H.; Kou, Y. *Catal. Today* **2004**, *93-95*, 301-305.

[75] Qiao, K.; Deng, Y. Q. *J. Mol. Catal. A* **2001**, *171*, 81-84.
[76] Wasserscheid, P.; Sesing, M.; Korth, W. *Green Chem.* **2002**, *4*, 134-138.
[77] Mohile, S. S.; Potdar, M. K.; Salunkhe, M. M. *J. Chem. Res. S* **2003**, 650-651.
[78] Sarca, V. D.; Laali, K. K. *Green Chem.* **2004**, *6*, 245-248.
[79] Stark, A.; MacLean, B. L.; Singer, R. D. *J. Chem. Soc., Dalton Trans.* **1999**, 63-66.
[80] Valkenberg, M. H.; deCastro, C.; Holderich, W. F. *App. Catal. A* **2001**, *215*, 185-190.
[81] Wu, J.; Zhang, J.; Zhang, H.; He, J. S.; Ren, Q.; Guo, M. *Biomacromol.* **2004**, *5*, 266-268.
[82] Yeung, K. S.; Farkas, M. E.; Qiu, Z. L.; Yang, Z. *Tetrahedron Lett.* **2002**, *43*, 5793-5795.
[83] Adams, C. J.; Earle, M. J.; Roberts, G.; Seddon, K. R. *J. Chem. Soc., Chem. Commun.* **1998**, 2097-2098.
[84] Hardacre, C.; Katdare, S. P.; Milroy, D.; Nancarrow, P.; Rooney, D. W.; Thompson, J. M. *J. Catal.* **2004**, *227*, 44-52.
[85] Nara, S. J.; Harjani, J. R.; Salunkhe, M. M. *J. Org. Chem.* **2001**, *66*, 8616-8620.
[86] Naik, P. U.; Harjani, J. R.; Nara, S. J.; Salunkhe, M. M. *Tetrahedron Lett.* **2004**, *45*, 1933-1936.
[87] Forsyth, S. A.; MacFarlane, D. R.; Thomson, R. J.; von Itzstein, M. *Chem. Commun.* **2002**, 714-715.
[88] Gholap, A. R.; Venkatesan, K.; Daniel, T.; Lahoti, R. J.; Srinivasan, K. V. *Green Chem.* **2003**, *5*, 693-696.
[89] Csihony, S.; Mehdi, H.; Horvath, I. T. *Green Chem.* **2001**, *3*, 307-309.
[90] Csihony, S.; Mehdi, H.; Homonnay, Z.; Vertes, A.; Farkas, O.; Horvath, I. T.*Dalton Trans.* **2002**, 680 685.
[91] Brausch, N.; Metlen, A.; Wasserscheid, P. *Chem. Commun.* **2004**, 1552-1553.
[92] Ross, J.; Xiao, J. L. *Green Chem.* **2002**, *4*, 129-133.
[93] Gmouth, S.; Yang, H. L.; Vaultier, M. *Org. Lett.* **2003**, *5*, 2219-2222.
[94] Song, C. E.; Shim, W. H.; Roh, E. J.; Choi, J. H. *Chem. Commun.* **2000**, 1695-1696.
[95] Cativiela, C.; Garcia, J. I.; Mayoral, J. A.; Royo, A. J.; Salvatella, L.; Assfeld, X.; Ruiz-Lopez, M. F. *J. Phys. Org. Chem.* **1992**, *5*, 230-238.
[96] Blokzijl, W.; Blandammer, M.; Engberts, J. *J. Am. Chem. Soc.* **1991**, *113*, 4241-4246.
[97] Berson, J. A.; Hamlet, Z.; Müller, W. *J. Am. Chem. Soc.* **1962**, *84*, 297-304.
[98] Otto, S.; Blokzijl, W.; Engberts, J. B. F. N. *J. Org. Chem.* **1994**, *62*, 5372-5376.
[99] Schneider, H.-J.; Sangwan, N. K. *J. Chem. Soc., Chem. Commun.* **1986**, 1787-1788.
[100] Breslow, R. *Acc. Chem. Res.* **1991**, *24*, 159-164.
[101] Jaeger, D. A.; Tucker, C. E. *Tetrahedron Lett.* **1989**, *30*, 1785-1788.
[102] Lee, C. W. *Tetrahedron Lett.* **1999**, *40*, 2461-2464.
[103] Kumar, A.; Pawar, S. S. *J. Org. Chem.* **2004**, *69*, 1419-1420.
[104] Hemeon, I.; DeAmicis, C.; Jenkins, H.; Scammells, P.; Singer, R. D. *Synlett* **2002**, 1815-1818.
[105] Fischer, T.; Sethi, A.; Welton, T.; Woolf, J. *Tetrahedron Lett.* **1999**, *40*, 793-796.
[106] Aggarwal, A.; Lancaster, N. L.; Sethi, A. R.; Welton, T. *Green Chem.* **2002**, *4*, 517-520.
[107] Vidis, A.; Ohlin, C. A.; Laurenczy, G.; Küsters, E.; Sedelmeier, G.; Dyson, P. J. *Adv Synth Catal* **2005**, *347*, 266-274.
[108] Ludley, P.; Karodia, N. *Tetrahedron Lett.* **2001**, *42*, 2011-2014.
[109] Ludley, P.; Karodia, N. *Arkivoc* **2002**, 172-175.
[110] Abbott, A. P.; Capper, G.; Davies, D. L.; Rasheed, R. H.; Tambyrajah, V. *Green Chem.* **2002**, *4*, 24-26.

[111] Sun, I. W.; Wu, S. Y.; Su, C. H.; Shu, Y. L.; Wu, P. L. *J. Chin. Chem. Soc.* **2004**, *51*, 367-370.

[112] Song, C. E.; Shim, W. H.; Roh, E. J.; Lee, S. G.; Choi, J. H. *Chem. Commun.* **2001**, 1122-1123.

[113] Yadav, J. S.; Reddy, B. V. S.; Gayathri, K. U.; Prasad, A. R. *Synthesis* **2002**, 2537-2541.

[114] Chen, I.-H.; Young, J.-N.; Yu, S. J. *Tetrahedron* **2004**, *60*, 11903-11909.

[115] Van der Eycken, E.; Appukkuttan, P.; De Borggraeve, W.; Dehaen, W.; Dallinger, D.; Kappe, C. O. *J. Org. Chem.* **2002**, *67*, 7904-7907.

[116] Howarth, J.; Hanlon, K.; Fayne, D.; McCormac, P. *Tetrahedron Lett.* **1997**, *38*, 3097-3100.

[117] Earle, M. J.; McCormac, P. B.; Seddon, K. R. *Green Chem.* **1999**, *1*, 23-25.

[118] Park, J. K.; Sreekanth, P.; Kim, B. M. *Adv. Synth. Catal.* **2004**, *346*, 49-52.

[119] Kündig, E. P.; Saudan, C. M.; Bernardinelli, G. *Angew. Chem. Int. Ed.* **1999**, *38*, 1220-1223.

[120] Fraile, J. M.; Garcia, J. I.; Herrerias, C. I.; Mayoral, J. A.; Carrié, D.; Vaultier, M. *Tetrahedron: Asymmetry* **2001**, *12*, 1891-1894.

[121] Fraile, J. M.; Garcia, J. I.; Herrerias, C. I.; Mayoral, J. A.; Reiser, O.; Vaultier, M. *Tetrahedron Lett.* **2004**, *45*, 6765-6768.

[122] Fraile, J. M.; Garcia, J. I.; Herrerias, C. I.; Mayoral, J. A.; Gmough, S.; Vaultier, M. *Green Chem.* **2004**, *6*, 93-98.

[123] Davies, D. L.; Kandola, S. K.; Patel, R. K. *Tetrahedron: Asymmetry* **2004**, *15*, 77-80.

[124] Yadav, J. S.; Reddy, B. V. S.; Reddy, P. N. *Adv. Synth. Catal.* **2004**, *346*, 53-56.

[125] Yang, F.; Zhang, Y. M.; Qiu, W. W.; Tang, J.; He, M. Y. *Chin. J. Chem.* **2002**, *20*, 114-116.

[126] Ohara, H.; Kiyokane, H.; Itoh, T. *Tetrahedron Lett.* **2002**, *43*, 3041-3044.

[127] Sun, W.; Xia, C. G.; Wang, H. W. *Tetrahedron Lett.* **2003**, *44*, 2409-2411.

[128] Kantam, M. L.; Neeraja, V.; Kavita, B.; Haritha, Y. *Synlett* **2004**, 525-527.

[129] Yadav, J. S.; Reddy, B. V. S.; Reddy, P. N.; Rao, M. S. *Synthesis* **2003**, 1387-1390.

[130] Peng, J. J.; Deng, Y. Q. *Tetrahedron Lett.* **2001**, *42*, 403-405.

[131] Harjani, J. R.; Nara, S. J.; Salunkhe, M. M. *Tetrahedron Lett.* **2001**, *42*, 1979-1981.

[132] Vallee, C.; Valerio, C.; Chauvin, Y.; Niccolai, G. P.; Basset, J. M.; Santini, C. C.; Galland, J. C.; Didillon, B. *J. Mol. Catal. A* **2004**, *214*, 71-81.

[133] Miyanohana, Y.; Inoue, H.; Chatani, N. *J. Org. Chem.* **2004**, *69*, 8541-8543.

[134] Zulfiqar, F.; Kitazume, T. *Green Chem.* **2000**, *2*, 296-297.

[135] Ganchegui, B.; Bouquillon, S.; Henin, F.; Muzart, J. *J. Mol. Catal. A* **2004**, *214*, 65-69.

[136] Neff, V.; Müller, T. E.; Lercher, J. A. *Chem. Commun.* **2002**, 906-907.

[137] Penzien, J.; Müller, T. E.; Lercher, J. A. *Microporous Mesoporous Mater.* **2001**, *48*, 285-291.

[138] Bodis, J.; Müller, T. E.; Lercher, J. A. *Green Chem.* **2003**, *5*, 227-231.

[139] Breitenlechner, S.; Fleck, M.; Müller, T. E.; Suppan, A. *J. Mol. Catal. A* **2004**, *214*, 175-179.

[140] Müller, T. E. *Abstr. Pap. Am. Chem. Soc.* **2003**, *226*, U621-U621.

[141] Brunet, J. J.; Chu, N. C.; Diallo, O.; Mothes, E. *J. Mol. Catal. A* **2003**, *198*, 107-110.

[142] Brunet, J. J.; Cadena, M.; Chu, N. C.; Diallo, O.; Jacob, K.; Mothes, E. *Organometallics* **2004**, *23*, 1264-1268.

[143] Fadini, L.; Togni, A. *Chimia* **2004**, *58*, 208-211.

[144] Rosa, J. N.; Santos, A. G.; Afonso, C. A. M. *J. Mol. Catal. A* **2004**, *214*, 161-165.

[145] Aubin, S.; Le Floch, F.; Carrie, D.; Guegan, J. P.; Vaultier, M. In: *Ionic Liquids, Industrial Applications for Green Chemistry*, Rogers, R. D., Seddon, K. R. Eds.; ACS Symposium Series No. 818, **2002**, 334-346.
[146] van den Broeke, J.; Winter, F.; Deelman, B. J.; van Koten, G. *Org. Lett.* **2002**, *4*, 3851-3854.
[147] Bösmann, A.; Franciò, G.; Janssen, E.; Solinas, M.; Leitner, W.; Wasserscheid, P. *Angew. Chem. Int. Ed.* **2001**, *40*, 2697-2699.
[148] Cacchi, S.; Fabrizi, G.; Goggiamani, A.; Moreno-Manas, M.; Vallribera, A. *Tetrahedron Lett.* **2002**, *43*, 5537-5540.
[149] Cacchi, S.; Fabrizi, G.; Goggiamani, A. *J. Mol. Catal. A* **2004**, *214*, 57-64.
[150] Hemeon, I.; Singer, R. D. *J. Mol. Catal. A* **2004**, *214*, 33-44.
[151] Hemeon, I.; Singer, R. D. *Chem. Commun.* **2002**, 1884-1885.
[152] Howarth, J.; James, P.; Dai, J. *J. Mol. Catal. A* **2004**, *214*, 143-146.
[153] Yadav, I. S.; Reddy, B. V. S.; Reddy, J. S. S. *Perkin Trans. 1* **2002**, 2390-2394.
[154] Anjaiah, S.; Chandrasekhar, S.; Grée, R. *J. Mol. Catal. A* **2004**, *214*, 133-136.
[155] Holbrey, J. D.; Seddon, K. R. *Clean Prod. Process.* **1999**, *1*, 223-236.

INDEX

1

B

C

D

E

I

J

K

O

P

Q

R

S

U

V

W

X

Y

Z

ABBREVIATIONS

$[(C_2)_4N]$	tetraethylammonium
$[(C_4)_3C_1P]$	tributyl(methyl)phosphonium
$[(C_4)_4N]$	tetrabutylammonium
$[(C_4)_4P]$	tetrabutylphoshonium
$[(C_5O_2)C_1im]$	1-(1-ethoxy-3-methylimidazolium)-2-methoxyethane
$[(C_6)_3C_{14}P]$	trihexyl(tetradecyl)phosphonium
$[(C_6)_3C_{16}P]$	trihexyl(hexadecyl)phosphonium
$[(C_8)_3C_1N]$	trioctylmethylammonium
$[(CC{=}C)C_1im]$	1-propene-3-methylimidazolium
$[(Mes)_2im]$	1,3-dimesityl-imidazolium
$[C_{10}C_1im]$	1-decyl-3-methylimidazolium
$[C_{12}C_1im]$	1-dodecyl-3-methylimidazolium
$[C_1C_1C_1im]$	1,2,3-trimethylimidazolium
$[C_1C_1im]$	1-methyl-3-methylimidazolium
$[C_2C_1C_1im]$	1-ethyl-2,3-dimethylimidazolium
$[C_2C_1im]$	1-ethyl-3-methylimidazolium
$[C_2py]$	*N*-ethylpyridinium
$[C_3C_1C_1im]$	1-propyl-2,3-dimethylimidazolium
$[C_3CNpy]$	*N*-butyronitrile-pyridinium
$[C_4C_1C_1im]$	1-butyl-2,3-dimethylimidazolium
$[C_4C_1C_1im]$	1-butyl-2,3-dimethylimidazolium
$[C_4C_1im]$	1-butyl-3-methylimidazolium
$[C_4C_1py]$	*N*-butyl-4-methylpyridinium
$[C_4C_4pyrr]$	*N*,*N*-dibutylpyrrolidinium
$[C_4py]$	*N*-butylpyridinium
$[C_6C_1im]$	1-hexyl-3-methylimidazolium
$[C_6py]$	*N*-hexylpyridinium
$[C_8C_1im]$	1-octyl-3-methylimidazolium

[HC_4im]	1-butylimidazolium
[*S*-C_4C_1im]	1-(*S*-2-methylpropane)-3-methylimidazolium
4-PPNO	4-phenylpyridine *N*-oxide
acac	acetylacetonate
AD	asymmetric dihydroxylation
AIBN	2,2'-azabisisobutyronitrile
Aliquat 336	tricaprylmethylammonium chloride
Ar	aryl
ATRP	atom transfer radical polymerisation
BARF	tetrakis-(3,5-bis(trifluoromethyl)phenyl)borate
BI	branch index
BINAP	2,2-bis(diphenylphosphino)-1,1'-binaphthyl
BINAS	sulfonated 2,2-bis(diphenylphosphino)-1,1'-binaphthyl
bipy	bispyridine
Bu	butyl
C_4C_1imy	1-butyl-3-methyl-imidazolylidene
C_4im	1-butylimidazole
cod	cyclooctadiene
cot	cyclooctatriene
cP	centi Poise
Cp	cyclopentadienyl
Cp*	1,2,3,4,5-pentamethylcyclopentadienyl
Cy	cyclohexyl
dba	di(benzylidene)acetone
DBP	dibutylphosphate
DHM	dihydromuconate
DHQ	dihydroquinine
DHQD	dihydrquinidine
DIOP	*O*-isopropylidene-2,3-dihydroxy-1,4-bis(diphenylphosphino)butane
DMA	*N,N*-dimethylacetamide
DMAP	dimethylaminopyridine
DMF	*N,N*-dimethylformamide
DMPA	4-(dimethylamino)-pyridine
DMSO	dimethylsulfoxide
DPEN	1,2-diphenylethylenediamine
DPP-[C_1C_1py]	[PPh_2(2-(1,6-dimethyl-pyridinyl))]$^+$
DPP-[C_2C_1py]	diphenylphosphine-1-ethyl-3-methylimidazolium
DPP-[C_4C_1im]	[PPh_2(2-(1-butyl-3-methyl-imidazolyl))]$^+$
dppe	1,2-bis(diphenylphosphino)ethane
dppf	1,1-bis(diphenylphosphino)ferrocene
dppp	1,3-bis(diphenylphosphino)propane

DuPHOS	1,2-bisphospholano-benzene
ee	enantiomeric excess
ETM	electron transfer mediator
E_T^N	transition energy of nile red dye
h	hour(s)
hfacac	hexafluoroacetylacetonate
ICP-OES	inductive coupled plasma optical emission spectroscopy
IFP	Institut Francais de Petrol
IL	ionic liquid
IMes	*N,N'*-dimesityl-imidazolinylidene
*i*Pr	isopropyl
kDa	kilo Dalton
l:b	linear to branched ratio
M3P	methyl-3-pentenoate
MA	methacrylate
Me	methyl
MeDuPHOS	1,2-bis(2,5-dimethylphospholano)benzene
mM	milli molar
MMA	methyl methacrylate
MTO	methyltrioxorhenium(VII)
M_W	molecular weight
ndb	norbornadiene
nm	nanometer
NMM	*N*-methyl(morpholine)
NMO	*N*-morpholine-*N'*-oxide
NMR	nuclear magnetic resonance
NORBOS	sulfonated trimethyl-dimethyl-phosphanorbonadiene
OAc	acetate
OTf	trifluoromethylsulfonate
OTs	tosylate
PEG	polyethylene glycole
Ph	phenyl
PHAL	phthalazine
ppm	parts per million
Pr	propyl
PTA	1,3,5-triaza-7-phosphaadamantane
PVP	poly(*N*-vinyl-2-pyrrolidone)
Py	pyridine
PYR	pyrimidine
QN	dihydroquininyl
RCM	ring-closing metathesis

ROMP	ring-opening polymerisation
scCO_2	supercritical carbon dioxide
SHOP	Shell Higher Olefin Process
SILP	supported ionic liquid phase
TBHP	*tert*-butyl hydrogenperoxide
*t*Bu	*tert*-butyl
TEAA	tetraethylammonium acetate
TEM	Transmission Electron Microscopy
TEMPO	2,2,6,6-tetramethylpiperidinyl-1-oxy
Tf_2N	bis(trifluoromethylsulfonyl)imide
TMEDA	tetramethylethylene diamine
TMGL	1,1,3,3-tetramethylguanidinium lactate
$TMSN_3$	trimethylsilylazide
TOF	turnover frequency
tol	tolyl
TON	turnover number
TPPD-Guanidine	$[PPh_2(m\text{-}C_6H_4\text{-guanidinium})_2]^{2+}$
TPPDS	$[PPh(o\text{-}C_6H_4\text{-}SO_3)_2]^{2-}$
TPPM-$[C_1C_1im]$	$[PPh_2(p\text{-}C_6H_4\text{-2-(1,3-dimethyl-imidazolyl)})]^+$
TPPM-Guanidine	$[PPh_2(m\text{-}C_6H_4\text{-guanidinium})]^+$
TPPMS	$[PPh_2(o\text{-}C_6H_4\text{-}SO_3)_3]^-$
TPPOTS	$[P(O\text{-}(o\text{-}C_6H_4\text{-}SO_3))_3]^{3-}$
TPPTS	$[P(o\text{-}C_6H_4\text{-}SO_3)_3]^{3-}$
TsDPEN	*N*-toluenesulfonyl-1,2-diphenylethylenediamine
UHP	urea hydrogen peroxide

0—0—60—30